Francesco Bottacin

Linear Algebra and Geometry

Società Editrice
ESCULAPIO

ISBN 978-88-9385-384-2

Società Editrice Esculapio s.r.l.
Via Terracini, 30 - 40131 Bologna
www.editrice-esculapio.com - info@editrice-esculapio.it

Cover layout: Carlotta Lenzi

Printed in Italy by Digital Team - Fano (PU)

Società Editrice
ESCULAPIO

Foreword

Linear algebra is a fundamental branch of mathematics that provides a powerful toolkit for solving problems in a wide range of fields, including physics, engineering, computer science, economics, and more. Its concepts and methods form the foundation of much of modern mathematics and its applications.

This book aims to provide a comprehensive introduction to linear algebra at the university level, suitable for students majoring in mathematics, science, or engineering. It covers the essential concepts of linear algebra, including vectors, matrices, linear transformations, eigenvalues and eigenvectors, determinants, and more.

The book is organized into chapters that build on each other, with each chapter introducing new concepts and building on the ones before it. Each chapter also includes numerous examples and exercises to reinforce the material covered and help students develop their problem-solving skills.

Throughout the book, we strive to balance rigor and accessibility, presenting the material in a way that is both mathematically precise and easy to understand. We hope that this book will be a valuable resource for students studying linear algebra, as well as for instructors looking for a comprehensive textbook that covers the essential concepts of the subject.

Note to the reader: the above lines were not written by the author, but by ChatGPT, an artificial intelligence chatbot!

I chose to ask ChatGPT to write a foreword for this book to give an example of the practical importance of linear algebra. In fact, linear algebra is a crucial tool for artificial intelligence (AI) and machine learning (ML) algorithms. Linear algebra provides the mathematical foundation for many key operations in AI and ML, such as data preprocessing, feature extraction, dimensionality reduction, and model training. For example, matrix operations such as matrix multiplication, matrix inversion, and eigenvalue decomposition are used extensively in AI and ML algorithms.

In addition, linear algebra provides a way to represent and manipulate high-dimensional data, which is common in many AI and ML applications. The use of vector and matrix notation allows us to compactly represent large amounts of data, making it easier to analyze and process.

Furthermore, linear algebra is also used in the development of deep learning models, which have revolutionized the field of AI in recent years. Deep learning models, such as artificial neural networks, rely heavily on linear algebra for their design and training.

Overall, the importance of linear algebra in the field of AI cannot be overstated, and any student or practitioner of AI and ML should have a strong foundation in this area.

(By the way, this explanation was also written by ChatGPT ;-)

Francesco Bottacin

Contents

1

Vector Spaces

In this chapter we introduce the notion of vector space. As a motivation we will start by discussing the case of "geometric vectors." After that we will give the general definition of a vector space over a field, and study its main properties. We will introduce the fundamental notions of linear independence, systems of generators, and basis of a vector space. Finally we will carry out a detailed study of finite dimensional vector spaces.

Before going into the study of vector spaces it is useful to briefly recall some definitions and properties of systems of linear equations.

1.1 Systems of linear equations

In this section we will study systems of linear equations with coefficients in a field,[1] and describe an elementary method of solution. The in-depth treatment of linear systems will be developed in a subsequent chapter, when we will have the necessary linear algebra tools at our disposal.

Definition 1.1.1. A *system of linear equations* with coefficients in a field K is a set S of equations of the form

$$S : \begin{cases} a_{11}\, x_1 + a_{12}\, x_2 + \cdots + a_{1n}\, x_n = b_1 \\ a_{21}\, x_1 + a_{22}\, x_2 + \cdots + a_{2n}\, x_n = b_2 \\ \cdot \quad \cdot \quad \cdot \quad \cdot \quad \cdot \quad \cdot \quad \cdot \quad \cdot \quad \cdot \quad \cdot \\ a_{m1}\, x_1 + a_{m2}\, x_2 + \cdots + a_{mn}\, x_n = b_m \end{cases} \tag{1.1.1}$$

with $a_{ij}, b_h \in K$. The elements a_{ij} are the *coefficients* of the system, $b_1, \ldots, b_m$ are the *constant terms*, and $x_1, \ldots, x_n$ are the *unknowns* of the

[1]For the definition of *field* see the Appendix "Basic notions of Algebra."

system. To solve the system S means to determine all the values of the unknowns $x_1, \dots, x_n$ such that all the equations are simultaneously satisfied. If all the constant terms are zero the system is *homogeneous.*

We will now describe a simple method for solving a generic system of m linear equations with n unknowns. This method, known as the "substitution method," can be described as follows:

Step 1. Given the linear system S, we choose any of its equations, for example the i-th,

$$a_{i1}\, x_1 + a_{i2}\, x_2 + \cdots + a_{ij}\, x_j + \cdots + a_{in}\, x_n = b_i$$

and choose one of the unknowns x_j that appear in this equation. We express x_j in terms of the remaining unknowns:

$$x_j = \frac{b_i - a_{i1}\, x_1 - \cdots - a_{in}\, x_n}{a_{ij}}.$$

Step 2. We substitute the expression for x_j in the remaining $m-1$ equations, thus obtaining a new system S' consisting of $m-1$ equations with $n-1$ unknowns.
Step 3. If the system S' still contains some unknowns we go back to Step 1 with S replaced by S'. Otherwise the solutions of the system S, if they exist, can be found with a simple back substitution starting from the last equation.

Let's now try to clarify the algorithm just described by analyzing in detail three concrete examples, which represent the three typical situations that may arise.

Example 1. (*Linear system with no solutions*) Let's consider the following linear system, with coefficients in $\mathbb{Q}$:

$$S : \begin{cases} 2x_1 - x_2 = 1 \\ x_1 + 4x_2 = -2 \\ 3x_1 - 2x_2 = 3 \end{cases}$$

From the second equation we obtain $x_1 = -2 - 4x_2$. We substitute this expression in the first and third equations, obtaining the system

$$S' : \begin{cases} -9x_2 = 5 \\ -14x_2 = 9 \end{cases}$$

Now we consider the system S' and from the first equation we obtain $x_2 = -5/9$. By substituting this value in the third equation we obtain the equality $-14\,(-5/9) = 9$, which is obviously false! It follows that the linear system S has no solutions.

Example 2. (*Linear system with a single solution*) Let's consider now the following linear system, with coefficients in $\mathbb{Q}$:

$$S : \begin{cases} 3x_1 - x_2 = 2 \\ 2x_1 + 5x_2 = -3 \end{cases}$$

From the first equation we get $x_2 = 3x_1 - 2$. By substituting this expression in the second equation we obtain the "system" (which consists of a single equation)

$$S' : 17x_1 = 7.$$

From this equation we get $x_1 = 7/17$. Now we can "replace backwards" the value of x_1 just found in the previous expression for x_2 thus obtaining $x_2 = -13/17$. This means that the linear system S admits a single solution given by

$$x_1 = 7/17, \quad x_2 = -13/17.$$

Example 3. (*Linear system with infinite solutions*) Let's consider the following linear system, with coefficients in $\mathbb{Q}$:

$$S : \begin{cases} 3x_1 + 2x_2 - x_3 + x_4 = 2 \\ 2x_1 + 2x_3 - 3x_4 = 0 \end{cases}$$

From the first equation we get $x_3 = 3x_1 + 2x_2 + x_4 - 2$. By substituting this expression in the second equation we obtain the "system"

$$S' : 8x_1 + 4x_2 - x_4 = 4.$$

From this equation we obtain, for instance, $x_4 = 8x_1 + 4x_2 - 4$. At this point we must stop, as there are no other equations that can be used. We have therefore determined the value of x_4 as a function of the unknowns x_1 and x_2, which can have any value. Substituting back the expression of x_4 into the previous expression for the unknown x_3, we get $x_3 = 11x_1 + 6x_2 - 6$.

It follows that the system S admits infinite solutions, which depend on two parameters:[2]

$$\begin{cases} \forall x_1 \\ \forall x_2 \\ x_3 = 11x_1 + 6x_2 - 6 \\ x_4 = 8x_1 + 4x_2 - 4 \end{cases}$$

Of course we could have followed a different path expressing, for example, the unknowns x_1 and x_2 as a function of x_3 and x_4, thus obtaining a different expression for the solutions of the system S. Needless to say, all possible expressions for solutions of the same system are equivalent to one another.

Exercises

Exercise 1.1.1. Solve the following system with coefficients in $\mathbb{Q}$:

$$\begin{cases} 3x - 2y = 1 \\ 2x + 5y = -1. \end{cases}$$

Exercise 1.1.2. Solve the following system with coefficients in $\mathbb{Q}$:

$$\begin{cases} 2x + 3y + 4z = 3 \\ 4x - y - 2z = 1 \\ 5x - 3y + z = -2. \end{cases}$$

Exercise 1.1.3. Solve the following system with coefficients in $\mathbb{Q}$:

$$\begin{cases} 5x - 2y + z = 1 \\ 3x - 4y - 2z = 3. \end{cases}$$

Exercise 1.1.4. Solve the following system, where $m \in \mathbb{Q}$ is a parameter, and analyze how the solutions depend on m.

$$\begin{cases} x + (m+1)y = m + 2 \\ mx + (m+4)y = 3. \end{cases}$$

Exercise 1.1.5. Solve the following system, where $m \in \mathbb{R}$ is a parameter, and analyze how the solutions depend on m.

$$\begin{cases} mx + (m-1)y = m + 2 \\ (m+1)x - my = 5m + 3. \end{cases}$$

[2] This is sometimes expressed by saying that the system admits ∞^2 solutions.

Exercise 1.1.6. Solve the following system, where $a \in \mathbb{R}$ is a parameter, and analyze how the solutions depend on a.

$$\begin{cases} x + (a-1)y + (2-a)z = a+5 \\ x + ay + 2z = 4 \\ x + (a-2)y + (2-2a^2)z = 6. \end{cases}$$

Exercise 1.1.7. Solve the following system, where $\lambda \in \mathbb{Q}$ is a parameter, and analyze how the solutions depend on λ.

$$\begin{cases} (\lambda-1)x_1 + 2x_2 + 3x_4 = 0 \\ \lambda x_2 + (\lambda+1)x_4 = 1 \\ x_1 + \lambda x_3 + x_4 = 0 \\ (\lambda-1)x_1 + x_4 = 0. \end{cases}$$

1.2 Geometric vectors

The notion of vector is often introduced using motivations coming from physics. In fact, in physics, alongside quantities that can be adequately expressed with a single number, such as temperature or time, there are others whose description requires more information. To describe, for example, the displacement of a point, the only numerical information about the "measure" of such displacement is not enough; it is also necessary to specify the line along which the movement takes place and, finally, the direction of travel along this line. Similarly, to specify a force it is necessary to provide the numerical value of the magnitude of the force (in some unit of measurement) together with the direction of the force (in many cases this is not yet sufficient, and it is necessary to also specify the point of application of the force).

To justify the considerations that we will make in the following, we will refer to the geometric notion of "movement" or, more precisely, to the notion of *translation* in a plane.

As we have said before, to describe a translation it is necessary to specify a straight line in which the movement occurs, a *direction of movement* along this straight line and, finally, a *number* that, in some way, measures the magnitude of this translation.

A particularly convenient way to express all this information graphically is to use an *oriented segment* v.

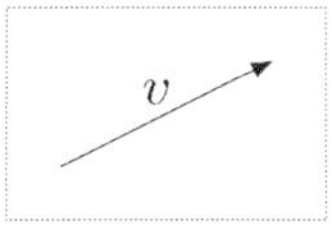

The line on which this segment lies, together with the arrow placed in one of the two ends, specify the direction of travel, and the length of the segment itself (expressed in some units of measurement) determines the extent of the displacement.

Such an object is called a *vector.* Given a vector v, represented by an oriented segment as above, the length of the segment is called the *norm* of v, and it is denoted by $\|v\|$.

At this point a clarification is perhaps necessary. A vector v does not describe the displacement of some point A towards some other point B; it describes a translation of the whole plane (or of the whole space), hence it does not matter "where" the segment that graphically represents the vector is drawn. In other words, two different oriented segments that have the same direction and orientation and the same length are two different graphical representations of the *same* vector.

This idea can be made mathematically precise in the following way.

Definition 1.2.1. Two oriented segments are said to be *equivalent* if they have the same direction (i.e., they lie on parallel lines), the same orientation and the same length.

It is easy to verify that this is actually an equivalence relation[3] in the set of all oriented segments. We can therefore give the following definition of vector:

Definition 1.2.2. A (*geometric*) *vector* is an equivalence class of oriented segments.

Remark 1.2.3. As we have already pointed out, it is sometimes important to also specify the point at which a vector is "applied" (as in the case of a force). This leads to the definition of an *applied vector*, which is a pair (P, v) consisting of a point P (the application point) and a vector v.

In the set of geometric vectors, two operations are defined in a natural way. The first is the multiplication of a vector v by a (real) number λ. If v represents a certain translation and if $\lambda > 0$, the vector λv represents a translation that occurs in the same direction as that represented by v, but determines a displacement equal to λ times that performed by the translation represented by v. The vector λv is therefore represented by an oriented segment that has the same direction and orientation as v, but a length equal

[3]For the definition of *equivalence relation* see the Appendix "Basic notions of Algebra."

to the length of v multiplied by λ.

If $\lambda = 0$ we get a vector of null length, which corresponds to a "null translation." In this case the notions of direction and orientation lose their meaning because the oriented segment is reduced to a point.

If $\lambda < 0$ the vector λv represents a translation that occurs in the opposite direction to that represented by v, with a displacement equal to the absolute value of λ times the displacement represented by v.

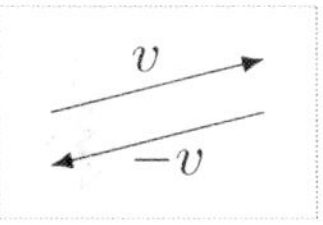

It follows that the composition of the translations represented by the vectors λv and $-\lambda v$ is the null translation: $\lambda v + (-\lambda v) = 0$.

The second operation is the *sum* of two vectors; it corresponds to the composition of two translations. If u and v are two vectors, their sum $w = u+v$ is, by definition, the vector representing the translation obtained by first carrying out the translation represented by u and then that represented by v. The effect of this composition of translations is shown in the figure on the right.

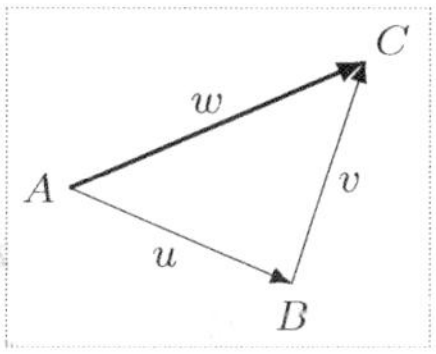

If the translation represented by u brings the point A to the point B and the translation represented by v brings the point B to the point C, then the composition of the two translations, represented by $w = u + v$, moves the point A to the point C.

It is immediately verified that the sum of vectors is commutative, that is $u + v = v + u$, as can be seen in the figure on the right. This figure illustrates the so-called *parallelogram rule*: the vector $w = u + v$ is the diagonal of the parallelogram whose sides are the vectors u and v.

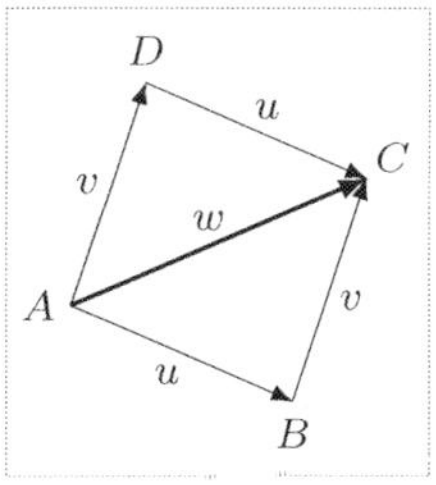

If we fix an orthogonal Cartesian coordinate system OXY in the plane, each vector v can be represented by a pair of real numbers (v_x, v_y), which identify the projections of v on the coordinate axes (see figure 1.1).

We can therefore identify the vector v with the pair $(v_x, v_y) \in \mathbb{R}^2$. With this identification, the sum of the two vectors $u = (u_x, u_y)$ and $v = (v_x, v_y)$ is given by

$$u + v = (u_x + v_x, u_y + v_y),$$

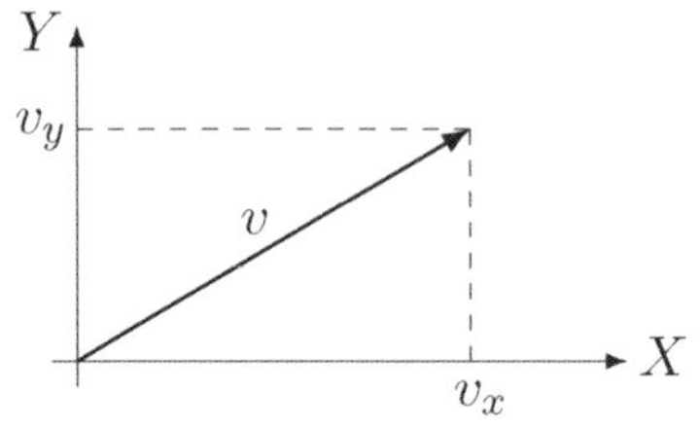

Figure 1.1: Decomposition of a vector into its components

while the product of the real number λ and the vector $v = (v_x, v_y)$ is

$$\lambda v = (\lambda v_x, \lambda v_y).$$

Using these formulas it is now immediate to verify that the sum of vectors is associative and commutative. There exists a neutral element for the sum, the *null vector*, whose components are all zero, and which we will denote by $\mathbf{0} = (0,0)$. Moreover, for each vector $v = (v_x, v_y)$ there exists its opposite $-v = (-v_x, -v_y)$, such that $v + (-v) = \mathbf{0}$.

All this can be summarized by saying that the set of vectors, with the addition operation, forms an *abelian group*.[4]

Let us now consider the multiplication of a vector by a real number. It is easy to verify that this operation satisfies the following properties:

(*i*) $(\lambda\mu)v = \lambda(\mu v)$,

(*ii*) $\lambda(u + v) = \lambda u + \lambda v$,

(*iii*) $(\lambda + \mu)v = \lambda v + \mu v$,

(*iv*) $1v = v$,

for any $\lambda, \mu \in \mathbb{R}$ and any vectors u and v.

The set of vectors therefore has a richer structure than that of an abelian group. This structure is called a *vector space*.

Before concluding we observe that quite similar considerations can be made for vectors in the usual three-dimensional space. To each such vector v it is possible to associate three numbers $(v_x, v_y, v_z) \in \mathbb{R}^3$, which represent the projections of v on the three coordinate axes of a suitable coordinate system $OXYZ$, as shown in figure 1.2.

[4]For the definition of *abelian group* see the Appendix "Basic notions of Algebra."

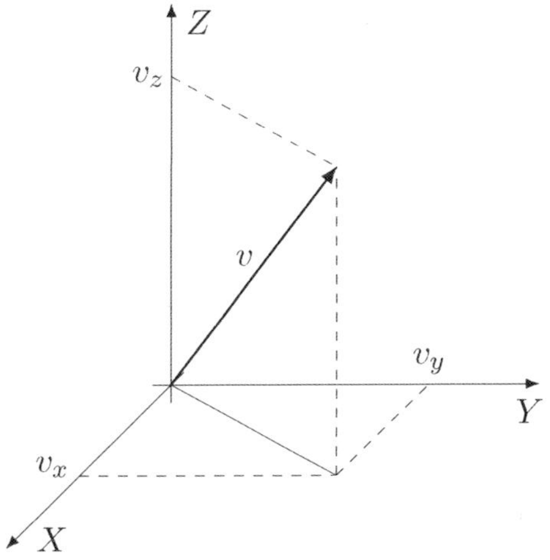

Figure 1.2: Decomposition of a vector into its components

Thus we obtain an identification between vectors in three-dimensional space and triples of real numbers, in terms of which the sum of two vectors $u = (u_x, u_y, u_z)$ and $v = (v_x, v_y, v_z)$ is given by

$$u + v = (u_x + v_x, u_y + v_y, u_z + v_z),$$

and the product of a real number λ and a vector $v = (v_x, v_y, v_z)$ is

$$\lambda v = (\lambda v_x, \lambda v_y, \lambda v_z).$$

Remark 1.2.4. Taking inspiration from the previous considerations we can *define* vectors with n components simply by identifying them with n-tuples of real numbers, $v = (a_1, a_2, \dots, a_n) \in \mathbb{R}^n$. The operations of adding two vectors and multiplying a real number and a vector will be defined in a similar way to what we have already seen for $\mathbb{R}^2$ and $\mathbb{R}^3$:

$$(a_1, a_2, \dots, a_n) + (b_1, b_2, \dots, b_n) = (a_1 + b_1, a_2 + b_2, \dots, a_n + b_n)$$

and

$$\lambda (a_1, a_2, \dots, a_n) = (\lambda a_1, \lambda a_2, \dots, \lambda a_n).$$

This construction will be extensively studied (and suitably generalized) in the next sections.

1.3 Vector spaces

In the previous section we have seen that vectors are "objects" that can be added together and can also be multiplied by numbers, so that the operations thus defined satisfy some properties, essentially analogous to the usual properties which are valid for the sum and the product of numbers.

If we now focus our attention not so much on the nature of such "objects" (previously defined as equivalence classes of oriented segments), but rather on the existence of certain operations between them and on the properties that, reasonably, should be satisfied by these operations, we can provide an abstract definition of a "set of vectors" as some set in which a sum of elements and a product between an element of that set and a number are defined, so that properties similar to those listed in the previous section are satisfied.

Let us now try to clarify the ideas expressed so far. Let K be a field[5] (for simplicity, we can suppose that K is the field $\mathbb{Q}$ of rational numbers, or the field $\mathbb{R}$ of real numbers, or the field $\mathbb{C}$ of complex numbers).

Definition 1.3.1. A *vector space* over K is a non-empty set V endowed with an operation $+_V$, called *sum*,

$$+_V : V \times V \to V, \quad (v_1, v_2) \mapsto v_1 +_V v_2,$$

and an operation $\cdot_V$

$$\cdot_V : K \times V \to V, \quad (\lambda, v) \mapsto \lambda \cdot_V v,$$

called *scalar multiplication*, that satisfy the following properties. For every $\lambda, \lambda_1, \lambda_2 \in K$ and every $v, v_1, v_2 \in V$, we have:

(1) $(v_1 +_V v_2) +_V v_3 = v_1 +_V (v_2 +_V v_3)$;

(2) $v_1 +_V v_2 = v_2 +_V v_1$;

(3) there exists an element $\mathbf{0}_V \in V$ such that $v +_V \mathbf{0}_V = \mathbf{0}_V +_V v = v$;

(4) for every $v \in V$ there exists an element $v' \in V$ such that $v +_V v' = v' +_V v = \mathbf{0}_V$. Such v' is denoted by $-v$ and is called the opposite of v;

(5) $\lambda \cdot_V (v_1 +_V v_2) = (\lambda \cdot_V v_1) +_V (\lambda \cdot_V v_2)$;

[5] A field is a set with two operations, which we will indicate with $+$ and $\cdot$, which satisfy properties quite similar to those of the sum and product of rational numbers. More precisely, a field is a commutative ring with unit in which every element other than 0 admits a multiplicative inverse.

(6) $(\lambda_1 + \lambda_2) \cdot_V v = (\lambda_1 \cdot_V v) +_V (\lambda_2 \cdot_V v)$;

(7) $(\lambda_1\lambda_2) \cdot_V v = \lambda_1 \cdot_V (\lambda_2 \cdot_V v)$;

(8) $1 \cdot_V v = v$.

The elements of a vector space V are called *vectors*. The elements of the field K are called *scalars*.

Remark 1.3.2. From the properties listed above it follows that, in any vector space V, we have $0 \cdot_V v = \mathbf{0}_V$, for any $v \in V$. In fact we have:

$$v + 0 \cdot_V v = 1 \cdot_V v + 0 \cdot_V v = (1+0) \cdot_V v = v.$$

By adding to both sides of this equality the opposite of v, we obtain

$$-v + v + 0 \cdot_V v = -v + v = \mathbf{0}_V,$$

from which it follows $0 \cdot_V v = \mathbf{0}_V$. From this we can now deduce that $(-1) \cdot_V v = -v$. In fact we have:

$$\mathbf{0}_V = 0 \cdot_V v = (1-1) \cdot_V v = 1 \cdot_V v + (-1) \cdot_V v = v + (-1) \cdot_V v,$$

from which it follows that the vector $(-1) \cdot_V v$ is the opposite of v.

From now on, the addition operation in a vector space V will be denoted simply with $+$ while the symbol of the multiplication by a scalar will be omitted: we will therefore write $v_1 + v_2$ instead of $v_1 +_V v_2$ and λv instead of $\lambda \cdot_V v$.

Example 1.3.3. Let $V = K^n$ and let us define the sum of two elements of V by setting

$$(a_1, a_2, \ldots, a_n) + (b_1, b_2, \ldots, b_n) = (a_1 + b_1, a_2 + b_2, \ldots, a_n + b_n).$$

Let us define the product of $\lambda \in K$ and $(a_1, \ldots, a_n) \in V$ as follows:

$$\lambda (a_1, a_2, \ldots, a_n) = (\lambda a_1, \lambda a_2, \ldots, \lambda a_n).$$

It is easy to verify that V is a vector space over K.

Example 1.3.4. Let K be a field and let us denote by $K[X]$ the set of polynomials with coefficients in K in the indeterminate X. A generic element of $K[X]$ is written in the form

$$p(X) = a_0 + a_1 X + a_2 X^2 + \cdots + a_n X^n,$$

for some $n \geq 0$, where all the coefficients a_i are elements of K.

With respect to the operations of sum of polynomials and product of a polynomial and an element of K, the set $K[X]$ is a vector space.

Example 1.3.5. Let K be a field and S a non-empty set. Let us denote by K^S the set of all functions $f\colon S \to K$.

Given two functions $f, g \in K^S$ we can define their sum by setting

$$(f+g)(s) = f(s) + g(s).$$

We can define the product of a function f and a scalar $\lambda \in K$ by setting

$$(\lambda f)(s) = \lambda(f(s)),$$

for all $s \in S$. It is easy to verify that the set K^S is a vector space over K.

Example 1.3.6. Let $K = \mathbb{Q}$ be the field of rational numbers and let $V = \mathbb{R}$. With respect to the usual operations of addition and multiplication of numbers, V turns out to be a vector space over K.

More generally, for every field K and every extension of fields $K \subset L$, L is a vector space over K.

Remark 1.3.7. We want to point out that, in the definition of a vector space, the property (8) is needed. In fact, the equality $1 \cdot_V v = v$ does not follow from the first seven properties, as can be seen from the following example.

Let $V = K^n$. Let us define the sum of vectors componentwise (as in the Example 1.3.3) and the product of a vector and an element of K as follows:

$$\lambda\,(a_1, a_2, \ldots, a_n) = (0, 0, \ldots, 0),$$

for every $\lambda \in K$ and every $(a_1, a_2, \ldots, a_n) \in V$.

It is easy to check that these operations verify all the properties listed in the definition of a vector space, with the exception of (8).

We end this section with the following definition:

Definition 1.3.8. Let V be a vector space over K. A *linear combination* of elements of V is a finite sum

$$\lambda_1 v_1 + \lambda_2 v_2 + \cdots + \lambda_n v_n,$$

with $\lambda_1, \ldots, \lambda_n \in K$ and $v_1, \ldots, v_n \in V$.

1.3.1 Vector subspaces

Let V be a vector space over a field K.

Definition 1.3.9. A *vector subspace* W of V is a non-empty subset $W \subseteq V$ such that the restriction to W of the operations of sum and product defined in V turns W into a vector space over K.

From the definition we deduce that, for a non-empty subset W of V to be a vector subspace, it is necessary and sufficient that the following properties hold:

(1) for every $w_1, w_2 \in W$, we have $w_1 + w_2 \in W$;

(2) for every $w \in W$, we have $-w \in W$;

(3) $\mathbf{0}_V \in W$;

(4) for every $\lambda \in K$ and every $w \in W$, we have $\lambda w \in W$.

Actually, it is enough to require that W be *closed* for the addition and multiplication by a scalar, i.e., that we have

$$w_1 + w_2 \in W, \qquad \forall w_1, w_2 \in W$$

and

$$\lambda w \in W, \qquad \forall \lambda \in K, \forall w \in W.$$

These two conditions can be grouped into one:

Proposition 1.3.10. *A non-empty subset W of a vector space V over the field K is a vector subspace of V if and only if*

$$\lambda_1 w_1 + \lambda_2 w_2 \in W,$$

for every $\lambda_1, \lambda_2 \in K$ and every $w_1, w_2 \in W$.

Proof. It is immediate to verify that, under this hypothesis, the operations of sum of vectors and product of a vector by a scalar defined in V make W a vector subspace. □

Remark 1.3.11. Every vector space V is obviously a vector subspace of itself. The set $\{\mathbf{0}_V\}$ is trivially a vector subspace of V, called the *null subspace*. Sometimes we will just write 0 to denote the subspace $\{\mathbf{0}_V\}$ (the meaning will be clear from the context).

Example 1.3.12. Let $V = K^n$ and W be the set of vectors $w = (x_1, x_2, \dots, x_n)$ that are solutions of a linear equation of the form

$$a_1 x_1 + a_2 x_2 + \cdots + a_n x_n = 0,$$

where $a_1, \dots, a_n \in K$.

It is easy to verify that a linear combination of two elements of W still provides a solution of the previous equation, so it belongs to W. This means that W is a vector subspace of V.

Note that, if $k \neq 0$, the set of solutions of an equation of the form

$$a_1 x_1 + a_2 x_2 + \cdots + a_n x_n = k,$$

is not a vector subspace of V since it does not contain the null vector $\mathbf{0}$.

Proposition 1.3.13. *If $\{W_i\}_{i \in I}$ is a collection of vector subspaces of a vector space V, their intersection*

$$W = \bigcap_{i \in I} W_i$$

is a vector subspace of V.

Proof. Let $\lambda_1, \lambda_2 \in K$ and $w_1, w_2 \in W$. Then $w_1, w_2 \in W_i$, for every $i \in I$, hence we also have $\lambda_1 w_1 + \lambda_2 w_2 \in W_i$, because W_i is a vector subspace of V. It follows that $\lambda_1 w_1 + \lambda_2 w_2 \in W$. □

Remark 1.3.14. Note that a similar property does not hold for the union: if W_1 and W_2 are two vector subspaces of V, in general their union $W_1 \cup W_2$ is not a vector subspace of V.

As an example we can consider the vector space $V = K^2$. Let $W_1 = \{(a, 0) \mid a \in K\}$ and $W_2 = \{(0, b) \mid b \in K\}$. It is immediate to verify that W_1 and W_2 are two vector subspaces of V. We have $(1, 0) \in W_1$ and $(0, 1) \in W_2$, however their sum $(1, 1)$ belongs neither to W_1 nor to W_2. This shows that the set $W_1 \cup W_2$ is not closed for the addition operation, hence it is not a vector subspace of V.

Definition 1.3.15. Let S be a subset of a vector space V. The *vector subspace generated by* S, that will be denoted by $L(S)$, is the smallest[6] vector subspace of V containing S (if S is the empty set we have $L(S) = \{\mathbf{0}\}$).

[6] Smallest with respect to the order relation given by the inclusion.

Since the intersection of a family of vector subspaces of V is a vector subspace of V, it is easy to verify that $L(S)$ coincides with the intersection of all the vector subspaces of V that contain S.

A more explicit description of $L(S)$ is given by the following result.

Proposition 1.3.16. *The vector subspace $L(S)$ generated by S is the set of all finite linear combinations of elements of S, i.e.,*

$$L(S) = \Big\{ \sum_{i=1}^{n} \lambda_i v_i \,|\, n \in \mathbb{N}, \lambda_i \in K, v_i \in S \Big\},$$

where the linear combination of zero *elements of S is meant to represent the zero vector $\mathbf{0} \in V$.*

Proof. Let us set

$$\Lambda(S) = \Big\{ \sum_{i=1}^{n} \lambda_i v_i \,|\, n \in \mathbb{N}, \lambda_i \in K, v_i \in S \Big\}.$$

Obviously $S \subseteq \Lambda(S)$. Note that every vector subspace of V containing S also contains all finite linear combinations of elements of S, hence it contains $\Lambda(S)$. It follows that $\Lambda(S)$ is contained in the intersection of all the vector subspaces of V that contain S, hence $\Lambda(S) \subseteq L(S)$.

On the other hand it is evident that $\Lambda(S)$ is a vector subspace of V: in fact the linear combination of two finite linear combinations of elements of S is just another finite linear combination of elements of S. Since $L(S)$ is the smallest vector subspace of V containing S, we have $L(S) \subseteq \Lambda(S)$, hence we conclude that $L(S) = \Lambda(S)$. □

Remark 1.3.17. If $S = \{v_1, v_2, \ldots, v_n\}$, the vector subspace $L(S)$ is also denoted by $\langle v_1, v_2, \ldots, v_n \rangle$.

As we have already observed, in the context of vector spaces the operation of union of two subspaces does not have good properties: in fact, the union of two vector subspaces is not, in general, a vector subspace (cf. Remark 1.3.14). This operation is then replaced by the sum:

Definition 1.3.18. If W_1 and W_2 are vector subspaces of V, their *sum* $W_1 + W_2$ is the vector subspace $L(W_1 \cup W_2)$ generated by $W_1 \cup W_2$. This definition can be generalized, in an obvious way, to the sum of an arbitrary (possibly infinite) family of subspaces of V.

An explicit description of the sum of two vector subspaces is provided by the following proposition:

Proposition 1.3.19. *We have*

$$W_1 + W_2 = \{w_1 + w_2 \,|\, w_1 \in W_1, w_2 \in W_2\}.$$

Proof. It is easy to verify that the set

$$\{w_1 + w_2 \,|\, w_1 \in W_1, w_2 \in W_2\}$$

is a vector subspace of V that contains W_1 and W_2, hence it contains also their union. Since, by definition, $W_1 + W_2$ is the smallest vector subspace of V that contains $W_1 \cup W_2$, we have the inclusion

$$W_1 + W_2 \subseteq \{w_1 + w_2 \,|\, w_1 \in W_1, w_2 \in W_2\}.$$

On the other hand, every vector of the form $w_1 + w_2$ belongs to $W_1 + W_2$. This proves the opposite inclusion, hence the equality. □

Remark 1.3.20. A completely analogous result also holds for the sum of a finite family of vector subspaces of V. We have

$$W_1 + \cdots + W_n = \{w_1 + \cdots + w_n \,|\, w_i \in W_i, \text{ for } i = 1, \ldots, n\}.$$

For an infinite family $\{W_i\}_{i \in I}$ of vector subspaces it is easy to verify that the sum

$$\sum_{i \in I} W_i$$

is equal to the set of all *finite* sums of vectors $w_i \in W_i$.

Definition 1.3.21. The sum of two vector subspaces W_1 and W_2 of V is called *direct*, and is denoted by $W_1 \oplus W_2$, if $W_1 \cap W_2 = \{\mathbf{0}\}$.

An important characterization of the direct sum of two vector subspaces is given by the following result:

Proposition 1.3.22. *Every vector $v \in W_1 \oplus W_2$ can be written in a unique way as $v = w_1 + w_2$, for some $w_1 \in W_1$ and $w_2 \in W_2$.*

Proof. In the previous proposition we saw that each $v \in W_1 \oplus W_2$ can be written as $v = w_1 + w_2$, for some $w_1 \in W_1$ and $w_2 \in W_2$. We just have to prove that this decomposition is unique.

Suppose we have

$$v = w_1 + w_2 = w_1' + w_2',$$

with $w_1, w_1' \in W_1$ and $w_2, w_2' \in W_2$. Then

$$w_1 - w_1' = w_2' - w_2 \in W_1 \cap W_2.$$

Since the sum of W_1 and W_2 is direct, we have $W_1 \cap W_2 = \{\mathbf{0}\}$, hence $w_1 - w_1' = w_2' - w_2 = \mathbf{0}$, from which it follows that $w_1 = w_1'$ e $w_2 = w_2'$. □

This result gives us the correct idea to extend the definition of direct sum to an arbitrary family of vector subspaces of a vector space V.

Definition 1.3.23. Let $\{W_i\}_{i \in I}$ be a family of vector subspaces of V. The subspace $W \subseteq V$ is the *direct sum* of the subspaces W_i if every vector $w \in W$ can be written in a unique way as the sum of a finite number of vectors $w_i \in W_i$.

To denote that W is the direct sum of the family of subspaces $\{W_i\}_{i \in I}$ we will write

$$W = \bigoplus_{i \in I} W_i.$$

It is easy to prove that the sum of a family $\{W_i\}_{i \in I}$ of subspaces of V is direct if and only if we have

$$W_i \cap \sum_{j \neq i} W_j = \{\mathbf{0}\}, \qquad \text{for every } i \in I.$$

Remark 1.3.24. The direct sum of two subspaces of a vector space V is also called the *internal direct sum.* Now we will see how it is also possible to define the direct sum of any two vector spaces V and W, so that V and W can be identified with two vector subspaces of $V \oplus W$. Such a sum is called the *external direct sum.*

Let V and W be two vector spaces over the field K. On the Cartesian product $V \times W$ we define the addition operation by setting

$$(v_1, w_1) + (v_2, w_2) = (v_1 + v_2, w_1 + w_2),$$

and the operation of multiplication by an element of K by setting

$$\lambda(v_1, w_1) = (\lambda v_1, \lambda w_1),$$

for every $(v_1, w_1), (v_2, w_2) \in V \times W$ and every $\lambda \in K$. It is easy to verify that these operations define a structure of vector space on $V \times W$. Let us denote by $V \oplus W$ this vector space.

The two functions $i_V : V \to V \oplus W$, $v \mapsto (v, \mathbf{0}_W)$ and $i_W : W \to V \oplus W$, $w \mapsto (\mathbf{0}_V, w)$ are injective and allow us to identify the two vector spaces V and W with the vector subspaces $i_V(V) = V \times \{\mathbf{0}_W\}$ and $i_W(W) = \{\mathbf{0}_V\} \times W$ of $V \oplus W$. It is easy to verify that the vector space $V \oplus W$ constructed in this way is equal to the (internal) direct sum of its two subspaces $i_V(V)$ and $i_W(W)$.

More generally, if $V_1, V_2, \ldots, V_n$ are a finite family of vector spaces over the field K, it is possible to define, in a natural way, a vector space structure on the Cartesian product $V_1 \times V_2 \times \cdots \times V_n$ by setting

$$(v_1, v_2, \ldots, v_n) + (w_1, w_2, \ldots, w_n) = (v_1 + w_1, v_2 + w_2, \ldots, v_n + w_n)$$

and

$$\lambda(v_1, v_2, \ldots, v_n) = (\lambda v_1, \lambda v_2, \ldots, \lambda v_n),$$

for every $\lambda \in K$ and every $(v_1, \ldots, v_n), (w_1, \ldots, w_n) \in V_1 \times \cdots \times V_n$. Each V_i can be identified in a natural way with the subspace V_i' of the Cartesian product $V_1 \times \cdots \times V_n$ formed by all elements of the form $(0, \ldots, 0, v, 0, \ldots, 0)$, for $v \in V_i$ (the vector v is in the i-th position). Now it is easy to verify that the direct sum of all these subspaces V_i' is equal to the Cartesian product $V_1 \times \cdots \times V_n$. We therefore define the external direct sum of the family of vector spaces $V_1, V_2, \ldots, V_n$ by setting

$$\bigoplus_{i=1}^{n} V_i = \prod_{i=1}^{n} V_i.$$

We can summarize what we have seen so far saying that, in the case of a *finite* family of vector spaces, the direct sum coincides with the Cartesian product.

1.3.2 Sets of generators and bases

Let V be a vector space over a field K.

Definition 1.3.25. A subset $S \subseteq V$ is called a *set of generators* of V if $L(S) = V$. In this case we also say that S *generates*, or *spans*, V.

We note that every vector space has sets of generators: the entire space V is trivially a set of generators of V.

From Proposition 1.3.16 it follows that, if S is a set of generators of V, every vector $v \in V$ can be written as a finite linear combination of elements of S:

$$v = \lambda_1 v_1 + \lambda_2 v_2 + \cdots + \lambda_n v_n,$$

for some $v_1, \ldots, v_n \in S$ and $\lambda_1, \ldots, \lambda_n \in K$. However, such an expression is not unique, in general.

Definition 1.3.26. A subset $S \subseteq V$ is called a *free set* of vectors if it has the following property: a finite linear combination of elements of S is the zero vector if and only if all the coefficients λ_i are zero, i.e., if

$$\lambda_1 v_1 + \lambda_2 v_2 + \cdots + \lambda_n v_n = \mathbf{0}$$

with $v_1, \ldots, v_n \in S$, implies $\lambda_1 = \lambda_2 = \cdots = \lambda_n = 0$.

If $S = \{v_1, v_2, \ldots, v_n\}$ is a free set, we say that the vectors $v_1, v_2, \ldots, v_n$ are *linearly independent*.

Hence the vectors $v_1, v_2, \ldots, v_n \in V$ are linearly independent if and only if the equation

$$\lambda_1 v_1 + \lambda_2 v_2 + \cdots + \lambda_n v_n = \mathbf{0}$$

has only the trivial solution $\lambda_1 = \lambda_2 = \cdots = \lambda_n = 0$.

Remark 1.3.27. If S contains only one vector v, to say that S is free means that $v \neq \mathbf{0}$. Similarly, if the vectors $v_1, v_2, \ldots, v_n$ are linearly independent, they must all be non-zero.

Definition 1.3.28. The vectors $v_1, v_2, \ldots, v_n \in V$ are *linearly dependent* if they are not linearly independent, i.e., if there exists $\lambda_1, \lambda_2, \ldots, \lambda_n \in K$, not all zero, such that

$$\lambda_1 v_1 + \lambda_2 v_2 + \cdots + \lambda_n v_n = \mathbf{0}.$$

Proposition 1.3.29. *The vectors $v_1, v_2, \ldots, v_n \in V$ are linearly dependent if and only if one of them can be written as a linear combination of the remaining ones, i.e., if and only if there exists an index i such that*

$$v_i = \sum_{j=1,\, j \neq i}^{n} \alpha_j v_j,$$

with $\alpha_j \in K$.

Proof. If the vectors $v_1, v_2, \ldots, v_n$ are linearly dependent, there exists a linear combination

$$\lambda_1 v_1 + \lambda_2 v_2 + \cdots + \lambda_n v_n = \mathbf{0}$$

whose coefficients λ_j are not all zero. Let i be an index such that $\lambda_i \neq 0$. Then we can write

$$\lambda_i v_i = -\lambda_1 v_1 - \cdots - \lambda_{i-1} v_{i-1} - \lambda_{i+1} v_{i+1} - \cdots - \lambda_n v_n,$$

hence

$$v_i = -\frac{\lambda_1}{\lambda_i} v_1 - \cdots - \frac{\lambda_{i-1}}{\lambda_i} v_{i-1} - \frac{\lambda_{i+1}}{\lambda_i} v_{i+1} - \cdots - \frac{\lambda_n}{\lambda_i} v_n.$$

Conversely, suppose that one of the vectors v_i is a linear combination of the remaining ones

$$v_i = \alpha_1 v_1 + \cdots + \alpha_{i-1} v_{i-1} + \alpha_{i+1} v_{i+1} + \cdots + \alpha_n v_n.$$

Then we have

$$\alpha_1 v_1 + \cdots + \alpha_{i-1} v_{i-1} - v_i + \alpha_{i+1} v_{i+1} + \cdots + \alpha_n v_n = \mathbf{0},$$

which proves that the vectors $v_1, \ldots, v_n$ are linearly dependent. □

We now show that if a vector can be written as a linear combination of a set of linearly independent vectors, this expression is unique.

Proposition 1.3.30. *Let $v_1, v_2, \ldots, v_n \in V$ be linearly independent vectors. If $v \in V$ can be written as a linear combination of the vectors $v_1, v_2, \ldots, v_n$,*

$$v = \lambda_1 v_1 + \lambda_2 v_2 + \cdots + \lambda_n v_n,$$

then the scalars $\lambda_1, \lambda_2, \ldots, \lambda_n$ are uniquely determined.

Proof. Suppose that we can write v in two ways, as

$$v = \lambda_1 v_1 + \lambda_2 v_2 + \cdots + \lambda_n v_n,$$

and also as

$$v = \mu_1 v_1 + \mu_2 v_2 + \cdots + \mu_n v_n.$$

Then we have

$$\lambda_1 v_1 + \lambda_2 v_2 + \cdots + \lambda_n v_n = \mu_1 v_1 + \mu_2 v_2 + \cdots + \mu_n v_n,$$

hence

$$(\lambda_1 - \mu_1)v_1 + (\lambda_2 - \mu_2)v_2 + \cdots + (\lambda_n - \mu_n)v_n = \mathbf{0}.$$

Since the vectors $v_1, v_2, \ldots, v_n$ are linearly independent, it follows that

$$\lambda_1 - \mu_1 = 0, \lambda_2 - \mu_2 = 0, \ldots, \lambda_n - \mu_n = 0,$$

which proves that $\lambda_i = \mu_i$, for every $i = 1, \ldots, n$. □

From this result it follows that, if we consider a set S of generators of V with the additional property that the vectors of S are linearly independent, then every vector of V can be written as a finite linear combination of elements of S, and this expression is *unique*.

Definition 1.3.31. A free set of generators of a vector space V is called a *basis* of V. In other words, a basis of V is a set of linearly independent vectors which generate the entire space V.

Remark 1.3.32. Usually a basis of a vector space V is considered as an *ordered* set of linearly independent generators of V. This means that if the set of the two vectors v_1 and v_2 is a basis of a vector space V, then $\mathbf{v} = \{v_1, v_2\}$ and $\mathbf{v}' = \{v_2, v_1\}$ are two *different* bases of V.

From what we have seen it follows that:

Corollary 1.3.33. *Let S be a basis of V. Every vector $v \subset V$ can be written as a finite linear combination of elements of S, and this expression in unique.*

Remark 1.3.34. Suppose that $S = \{v_1, v_2, \ldots, v_n\}$ is a basis of a vector space V. Then, for every $v \in V$, we have

$$v = \lambda_1 v_1 + \lambda_2 v_2 + \cdots + \lambda_n v_n,$$

and the scalars $\lambda_i \in K$ are uniquely determined by v. These scalars are called the *coordinates* of v with respect to the basis $v_1, v_2, \ldots, v_n$.

1.3.3 Finitely generated vector spaces

In the previous section we did not make any assumptions about the number of generators of a vector space. Now we will deal in detail with the case in which that number is finite.

Definition 1.3.35. A vector space V is *finitely generated* if there exists a finite set of generatos of V.

We begin by proving that every finitely generated vector space V admits a basis. More precisely, we will prove that from every finite[7] set of generators of V it is possible to extract a basis.

Proposition 1.3.36. *Let $S = \{v_1, v_2, \ldots, v_n\}$ be a set of generators of V. Then S contains some vectors $v_{i_1}, v_{i_2}, \ldots, v_{i_r}$, for some $r \leq n$, that form a basis of V.*

Proof. If the vectors $v_1, v_2, \ldots, v_n$ are linearly independent, they are a basis of V and we are done. If, on the other hand, they are linearly dependent, one of them can be expressed as a linear combination of the others. Up to a reordering of the vectors, we can assume that this vector is v_n, hence we can write

$$v_n = \lambda_1 v_1 + \lambda_2 v_2 + \cdots + \lambda_{n-1} v_{n-1}.$$

It follows that the vectors $v_1, v_2, \ldots, v_{n-1}$ generate the vector space V; in fact, every vector that can be written as a linear combination of $v_1, v_2, \ldots, v_n$ can also be written as a linear combination of the vectors $v_1, v_2, \ldots, v_{n-1}$. Now, if the vectors $v_1, v_2, \ldots, v_{n-1}$ are linearly independent, they are a basis of V and we are done. Otherwise one of them can be written as a linear combination of the remaining ones. Once more, up to reordering the vectors, we can assume that v_{n-1} can be written as a linear combination of $v_1, v_2, \ldots, v_{n-2}$, and this implies that the vectors $v_1, v_2, \ldots, v_{n-2}$ are a set of generators of V.

By repeating the reasoning described above, sooner or later we will arrive at a set of vectors $v_1, v_2, \ldots, v_r$, for some $r \leq n$, that generate the vector space V and are linearly independent, hence they are a basis of V. □

The following result, also known as *Steinitz Lemma* or *Exchange Theorem*, clarifies the relationships that exist between sets of linearly independent vectors, bases and sets of generators.

Proposition 1.3.37 (Exchange Theorem). *Let V be a finitely generated vector space and $\{v_1, v_2, \ldots, v_n\}$ a set of generators of V. If $w_1, w_2, \ldots, w_r$ are linearly independent vectors, then $r \leq n$.*

Proof. Since the vectors $v_1, v_2, \ldots, v_n$ generate V, the vector w_1 can be written as a linear combination of them,

$$w_1 = \lambda_1 v_1 + \lambda_2 v_2 + \cdots + \lambda_n v_n.$$

[7] A similar result also holds for infinite sets of generators and for non-finitely generated vector spaces. The proof, however, is different.

Since $w_1 \neq \mathbf{0}$, the scalars $\lambda_1, \ldots, \lambda_n$ are not all zero, hence there exists an index i such that $\lambda_i \neq 0$. It follows that the vector v_i can be written as a linear combination of the vectors $w_1, v_1, \ldots, v_{i-1}, v_{i+1}, \ldots, v_n$.

Up to reordering the vectors v_j, we can assume that $i = n$, i.e., that v_n can be written as a linear combination of the vectors $w_1, v_1, \ldots, v_{n-1}$; but this implies that $\{w_1, v_1, \ldots, v_{n-1}\}$ is a set of generators of V. Hence the vector w_2 can be written as a linear combination of the vectors $w_1, v_1, \ldots, v_{n-1}$:

$$w_2 = \alpha_1 w_1 + \lambda_1 v_1 + \cdots + \lambda_{n-1} v_{n-1},$$

and the scalars $\lambda_1, \ldots, \lambda_{n-1}$ cannot be all zero, because otherwise the vectors w_1 and w_2 would be linearly dependent.

It follows that there exists an index i such that $\lambda_i \neq 0$ and, once again, we can assume that it is $i = n-1$ (up to reordering the vectors v_j). It follows that v_{n-1} can be written as a linear combination of $w_1, w_2, v_1, \ldots, v_{n-2}$, hence also $\{w_1, w_2, v_1, \ldots, v_{n-2}\}$ is a set of generators of V. Continuing in this way, we prove that all sets

$$\{w_1, w_2, \ldots, w_h, v_1, \ldots, v_{n-h}\}$$

are sets of generators of V.

If $n < r$, by setting $h = n$ we would find that the vectors $w_1, w_2, \ldots, w_n$ generate the vector space V, hence the vector w_{n+1} could be written as a linear combination of the vectors $w_1, w_2, \ldots, w_n$, which contradicts the hypothesis that the vectors $w_1, w_2, \ldots, w_r$ are linearly independent. Hence we must have $r \leq n$. □

Corollary 1.3.38. *Let V be a finitely generated vector space and let us consider a basis $\{v_1, v_2, \ldots, v_n\}$ of V. Then, for every set of linearly independent vectors $\{w_1, w_2, \ldots, w_r\}$, we have $r \leq n$, and for every set $\{u_1, u_2, \ldots, u_s\}$ of generators of V, we have $s \geq n$.*

Proof. This result is an immediate consequence of the previous proposition; just remember that the vectors $v_1, v_2, \ldots, v_n$ are linearly independent and are also a set of generators of V. □

Corollary 1.3.39. *Any two bases of a (finitely generated) vector space V have the same number of elements.*

Proof. Let $\{v_1, v_2, \ldots, v_r\}$ and $\{w_1, w_2, \ldots, w_s\}$ be two bases of V. Since the vectors $v_1, v_2, \ldots, v_r$ are linearly independent and the vectors $w_1, w_2, \ldots, w_s$ are generators of V, we have $r \leq s$. By swapping the role of the two bases, we also get $s \leq r$, hence the equality $r = s$. □

The number of vectors that form a basis of a finitely generated vector space V is independent of the chosen basis, hence it depends only on the space V. We can therefore give the following definition:

Definition 1.3.40. The *dimension* of a (finitely generated) vector space V, denoted by $\dim V$, is the number of elements of a basis of V.

Example 1.3.41. Let's consider the vector space $V = K^n$, introduced in Example 1.3.3. For every $i = 1, \dots, n$, we denote by e_i the n-tuple of elements of K whose components are all zero except the i-th, that is 1:

$$\begin{aligned} e_1 &= (1, 0, 0, 0, \dots, 0, 0), \\ e_2 &= (0, 1, 0, 0, \dots, 0, 0), \\ e_3 &= (0, 0, 1, 0, \dots, 0, 0), \\ &\cdot \quad \cdot \quad \cdot \quad \cdot \quad \cdot \quad \cdot \\ e_n &= (0, 0, 0, 0, \dots, 0, 1). \end{aligned}$$

Note that, for any $\lambda_1, \dots, \lambda_n \in K$, we have

$$\lambda_1 e_1 + \lambda_2 e_2 + \dots + \lambda_n e_n = (\lambda_1, \lambda_2, \dots, \lambda_n).$$

It follows that the vectors $e_1, e_2, \dots, e_n$ are linearly independent and they generate the vector space V, hence they are a basis of $V = K^n$. This basis is called the *canonical basis* of K^n. It follows that $\dim K^n = n$.

Example 1.3.42. The null vector space, $V = \{\mathbf{0}\}$, has dimension zero. In fact it contains only one vector $v = \mathbf{0}$, but this vector is not a basis of V because it is not linearly independent, since a set containing just one vector v is a free set (i.e., v is linearly independent) if and only if $v \neq \mathbf{0}$.

From Proposition 1.3.37 we also derive the next two results.

Corollary 1.3.43. *Let V be a finitely generated vector space. Then every vector subspace W of V is finitely generated and* $\dim W \leq \dim V$.

Proof. Let $n = \dim V$. If $\{w_1, \dots, w_r\}$ is a set of linearly independent vectors of W, they are also linearly independent vectors of V; hence $r \leq n$. If these vectors do not generate W, this means that there exists a vector $w_{r+1} \in W$ that cannot be written as a linear combination of $w_1, \dots, w_r$. It follows that the vectors $w_1, \dots, w_r, w_{r+1}$ are linearly independent. If they still do not generate W, there exists a vector $w_{r+2} \in W$ that cannot be written as a linear combination of $w_1, \dots, w_r, w_{r+1}$. This implies that the vectors $w_1, \dots, w_r, w_{r+1}, w_{r+2}$ are linearly independent.

Since the number of linearly independent vectors cannot be bigger than n, repeating the previous reasoning we construct a set of linearly independent vectors $\{w_1, \ldots, w_s\}$, with $s \leq n$, that generate the subspace W, hence they are a basis of W. This proves that $\dim W \leq \dim V$. □

Corollary 1.3.44. *Let V be a finitely generated vector space. Every set of linearly independent vectors $v_1, \ldots, v_r$ can be completed to a basis of V. In other words, there exist $v_{r+1}, \ldots, v_n$ such that $\{v_1, \ldots, v_r, v_{r+1}, \ldots, v_n\}$ is a basis of V. Moreover, the vectors $v_{r+1}, \ldots, v_n$ can always be chosen from the vectors of any fixed basis of V.*

Proof. Let us fix a basis $\{u_1, \ldots, u_n\}$ of V and suppose that $v_1, \ldots, v_r \in V$ are linearly independent vectors. If $v_1, \ldots, v_r$ generate V then they are a basis of V (and in this case we have $r = n$).

Let us suppose that $v_1, \ldots, v_r$ do not generate V. This means that there exists a vector u_i of the basis of V that cannot be written as a linear combination of $v_1, \ldots, v_r$. In fact if all the vectors $u_1, u_2, \ldots, u_n$ could be written as a linear combination of $v_1, \ldots, v_r$ then, since every vector $v \in V$ can be written as a linear combination of the basis vectors $u_1, \ldots, u_n$, it would follow that every vector of V could be written as a linear combination of $v_1, \ldots, v_r$, hence these vectors would generate V, thus contradicting the hypothesis.

Let's set $v_{r+1} = u_i$. From the above it follows that $v_1, \ldots, v_r, v_{r+1}$ are linearly independent. If they generate V then they are a basis and we are done, otherwise we can repeat the reasoning just made. In this case there must exist a vector u_j of the basis of V that cannot be written as a linear combination of $v_1, \ldots, v_r, v_{r+1}$. Hence we set $v_{r+2} = u_j$. As before, the vectors $v_1, \ldots, v_r, v_{r+1}, v_{r+2}$ are linearly independent. Continuing in this way, we must necessarily obtain a set of linearly independent vectors $\{v_1, \ldots, v_r, v_{r+1}, \ldots, v_n\}$ that are also a set of generatos of V, otherwise an infinite set of linearly independent vectors would be obtained, against the hypothesis that V is finitely generated. □

If the dimension of a vector space V is known, the verification that a given set of vectors of V forms a basis is simplified.

Proposition 1.3.45. *Let V be a vector space of dimension n and $v_1, \ldots, v_n$ be vectors of V.*

- (*i*) *If $v_1, \ldots, v_n$ are linearly independent, then they generate V, hence they are a basis of V.*
- (*ii*) *If $v_1, \ldots, v_n$ generate V, then they are also linearly independent, hence they are a basis V.*

Proof. (*i*) Suppose that the vectors $v_1, \dots, v_n$ are linearly independent. By the previous corollary, they are contained in a basis

$$\{v_1, \dots, v_n, v_{n+1}, \dots, v_{n+r}\}$$

of V. But, since V has dimension n, every basis of V must have n elements. It follows that $r = 0$, hence the vectors $v_1, \dots, v_n$ are a basis of V.

(*ii*) Suppose that the vectors $v_1, \dots, v_n$ generate V. If they are linearly dependent, one of them is a linear combination of the remaining ones. Up to reordering the vectors, it is not restrictive to assume that v_n is a linear combination of $v_1, \dots, v_{n-1}$. But in this case the vectors $v_1, \dots, v_{n-1}$ would generate V. However, this is false; in fact the number of generators of V must be $\geq \dim V$ (cf. Corollary 1.3.38). It follows that $v_1, \dots, v_n$ are linearly independent, hence they are a basis of V. □

Corollary 1.3.46. *Let V be a finitely generated vector space and $W \subset V$ be a proper subspace. Then* $\dim W < \dim V$.

Proof. In Corollary 1.3.43 we proved that $\dim W \leq \dim V$. Let us suppose that $\dim W = \dim V = n$, and let $w_1, \dots, w_n$ be a basis of W. Since these are n linearly independent vectors of V, and since $\dim V = n$, by (*i*) of the previous proposition they are a basis of V. But this implies that $W = V$, which contradicts the hypothesis that W is a proper subspace of V. □

Remark 1.3.47. If $K \subset L$ is a field extension,[8] every vector space V over the field L can also be considered as a vector space over the field K. If $\{v_1, \dots, v_n\}$ is a basis of V as a K-vector space, these vectors also generate V as a vector space over L. However, they could be linearly dependent over L, as we will show in the next example. In general, this means that the dimension of V as an L-vector space is less then or equal to the dimension of V considered as a vector space over K, i.e.,

$$\dim_L V \leq \dim_K V.$$

We illustrate what has just been stated with a concrete example. Let $K = \mathbb{R}$ be the field of real numbers and $L = \mathbb{C}$ the field of complex numbers. The set $\mathbb{C}$ is identified in a natural way with the set $\mathbb{R}^2$, by associating to each complex number $z = x + iy$ the pair of real numbers (x, y). Let $V = \mathbb{C} \cong \mathbb{R}^2$.

[8]This means that L is a field and K is a subfield.

The vector space $V = \mathbb{C}$, as a vector space over the field $\mathbb{C}$, obviously has dimension 1, and a basis is given by the vector $v = 1$. On the other hand, if we consider $\mathbb{C}$ as a vector space over $\mathbb{R}$, it has dimension 2. A basis is given by the vectors $v_1 = 1$ and $v_2 = i$, that correspond, in the identification $\mathbb{C} \cong \mathbb{R}^2$ described above, to the vectors $(1, 0)$ and $(0, 1)$ of the canonical basis of $\mathbb{R}^2$.

Note that the vectors $v_1 = 1$ and $v_2 = i$ are linearly independent over the field $\mathbb{R}$ of real numbers; in fact, if $\alpha v_1 + \beta v_2 = \alpha + i\beta = 0$, with $\alpha, \beta \in \mathbb{R}$, we must have $\alpha = \beta = 0$. On the other hand, they are linearly dependent over the field $\mathbb{C}$: in fact, we have

$$v_1 + iv_2 = 1 + i^2 = 1 - 1 = 0.$$

More generally, if $V = \mathbb{C}^n$, we have $\dim_{\mathbb{C}} V = n$ and $\dim_{\mathbb{R}} V = 2n$.

We end this section by proving a result that relates the dimensions of two vector subspaces of V with the dimensions of their sum and their intersection:

Proposition 1.3.48 (Grassmann's formula). *Let W_1 and W_2 be two vector subspaces of a finitely generated vector space V. We have:*

$$\dim(W_1 + W_2) = \dim W_1 + \dim W_2 - \dim(W_1 \cap W_2).$$

Proof. We set

$$r = \dim W_1, \quad s = \dim W_2, \quad t = \dim(W_1 \cap W_2).$$

Let $\{v_1, \dots, v_t\}$ be a basis of $W_1 \cap W_2$. By Corollary 1.3.44, this set of vectors can be completed to obtain a basis $\{v_1, \dots, v_t, v_{t+1}, \dots, v_r\}$ of W_1 and a basis $\{v_1, \dots, v_t, v'_{t+1}, \dots, v'_s\}$ of W_2.

Since every vector of $W_1 + W_2$ can be written as the sum of a vector of W_1 and a vector of W_2, it can be expressed as a linear combination of the vectors $v_1, \dots, v_t, v_{t+1}, \dots, v_r, v'_{t+1}, \dots, v'_s$. We want to prove that these vectors, besides being generators of $W_1 + W_2$, are also linearly independent.

Suppose that we have

$$\alpha_1 v_1 + \cdots + \alpha_t v_t + \beta_1 v_{t+1} + \cdots + \beta_{r-t} v_r + \gamma_1 v'_{t+1} + \cdots + \gamma_{s-t} v'_s = \mathbf{0}.$$

It follows that

$$\alpha_1 v_1 + \cdots + \alpha_t v_t + \beta_1 v_{t+1} + \cdots + \beta_{r-t} v_r = -(\gamma_1 v'_{t+1} + \cdots + \gamma_{s-t} v'_s).$$

If we denote by w the vector above, we have $w \in W_1 \cap W_2$. Since $\{v_1, \ldots, v_t\}$ is a basis of $W_1 \cap W_2$, the vector w can be uniquely written as

$$w = \lambda_1 v_1 + \cdots + \lambda_t v_t.$$

Hence we have the following equalities:

$$\lambda_1 v_1 + \cdots + \lambda_t v_t = \alpha_1 v_1 + \cdots + \alpha_t v_t + \beta_1 v_{t+1} + \cdots + \beta_{r-t} v_r$$

and

$$\lambda_1 v_1 + \cdots + \lambda_t v_t = -(\gamma_1 v'_{t+1} + \cdots + \gamma_{s-t} v'_s).$$

Since, by hypothesis, the vectors $v_1, \ldots, v_t, v_{t+1}, \ldots, v_r$ are a basis of W_1 and the vectors $v_1, \ldots, v_t, v'_{t+1}, \ldots, v'_s$ are a basis of W_2, from the previous equalities it follows that

$$\begin{aligned} \lambda_1 = \cdots = \lambda_t = 0, &\qquad \alpha_1 = \cdots = \alpha_t = 0, \\ \beta_1 = \cdots = \beta_{r-t} = 0, &\qquad \gamma_1 = \cdots = \gamma_{s-t} = 0. \end{aligned}$$

Hence we have proved that the set of vectors

$$\{v_1, \ldots, v_t, v_{t+1}, \ldots, v_r, v'_{t+1}, \ldots, v'_s\}$$

is a basis of $W_1 + W_2$. It follows that

$$\dim(W_1 + W_2) = r + s - t = \dim W_1 + \dim W_2 - \dim(W_1 \cap W_2). \qquad \square$$

Remark 1.3.49. Let V be a finitely generated vector space. Note that, for every subspace $W \subseteq V$, it is possible to find a subspace W' of V such that $V = W \oplus W'$, i.e., such that we have $V = W + W'$ and $W \cap W' = \{\mathbf{0}\}$. Such a subspace W' is called a *complement* of W.

For this purpose, it is enough to consider a basis $\{w_1, \ldots, w_r\}$ of W and complete it to a basis $\{w_1, \ldots, w_r, v_{r+1}, \ldots, v_n\}$ of V (cf. Corollary 1.3.44). The subspace $W' = \langle v_{r+1}, \ldots, v_n \rangle$, spanned by the vectors $v_{r+1}, \ldots, v_n$, is the subspace we are looking for.

Note that such a subspace W' is not unique. For instance, if $V = K^2$ and W is the subspace spanned by the vector $(1, 0)$, every vector (a, b), with $b \neq 0$, generates a subspace W' such that $V = W \oplus W'$.

Exercises

Exercise 1.3.1. Check if the following sets are vector spaces:

(1) The set of non-negative, continuous real functions defined in the interval $[0,1]$, with the operations of addition and multiplication by a real number.

(2) The set of real functions f defined in $\mathbb{R}$, such that

$$\lim_{x\to+\infty} f(x) = 0,$$

with the operations of addition and multiplication by a real number.

(3) The set of functions from $\mathbb{R}$ to $\mathbb{R}$ that vanish in 1 or in 4.

(4) The set of polynomials of degree equal to n (n a positive integer).

(5) The set of C^2-functions from $\mathbb{R}$ to $\mathbb{R}$ such that

$$f'' + \omega^2 f = 0,$$

with $\omega \in \mathbb{R}$.

(6) The set of continuous real functions $f(x)$ defined in the interval $[0,1]$ such that

$$\int_0^1 f(x)\sin x\,dx = 0.$$

Exercise 1.3.2. Let V be the vector space of polynomials in x of degree ≤ 3, with real coefficients. Verify that the following sets are bases of V:

(1) $\{1, x, x^2, x^3\}$;

(2) $\{1, 1-x, x-x^2, x^2-x^3\}$;

(3) $\{1, 1+x, 1+x+x^2, 1+x+x^2+x^3\}$.

Exercise 1.3.3. In the vector space V of polynomials of degree ≤ 2 consider the polynomials

$$\begin{aligned} p_1(x) &= x^2 + x(1-x) + (1-x)^2 \\ p_2(x) &= x^2 + (1-x)^2 \\ p_3(x) &= x^2 + 1 + (1-x)^2 \\ p_4(x) &= x(1-x). \end{aligned}$$

Is it possible to extract from $\{p_1(x), p_2(x), p_3(x), p_4(x)\}$ a basis of V? If this is possible, find all possible bases.

Exercise 1.3.4. In the vector space of continuous functions from $\mathbb{R}$ to $\mathbb{R}$ consider the functions $f_1(x) = \sin x$, $f_2(x) = \sin 2x$ and $f_3(x) = \sin 3x$. Are these functions linearly independent?

Exercise 1.3.5. In the following cases check if the vectors v_1, v_2, v_3 are a basis of $\mathbb{R}^3$. If not, describe the subspace they span.

(1) $v_1 = (1, 1, 1)$, $v_2 = (3, 0, -1)$, $v_3 = (-1, 1, -1)$;
(2) $v_1 = (1, 2, 3)$, $v_2 = (3, 0, -1)$, $v_3 = (1, 8, 13)$;
(3) $v_1 = (1, 2, -3)$, $v_2 = (1, 0, -1)$, $v_3 = (1, 10, -11)$.

Exercise 1.3.6. In $\mathbb{R}^4$ the following vectors are:
(a) a free set (i.e., a set of linearly independent vectors)? If yes, complete this set to obtain a basis of $\mathbb{R}^4$, otherwise find the linear dependence relations of the vectors and extract a free set from this set of vectors.
(b) a set of generators? If yes, extract a basis of $\mathbb{R}^4$, otherwise find the dimension of the subspace they span.

(1) $v_1 = (1, 1, 1, 1)$, $v_2 = (0, 1, 2, -1)$, $v_3 = (1, 0, -2, 3)$, $v_4 = (2, 1, 0, -1)$, $v_5 = (4, 3, 2, 1)$;
(2) $v_1 = (1, 2, 3, 4)$, $v_2 = (0, 1, 2, -1)$, $v_3 = (3, 4, 5, 16)$;
(3) $v_1 = (1, 2, 3, 4)$, $v_2 = (0, 1, 2, -1)$, $v_3 = (2, 1, 0, 11)$, $v_4 = (3, 4, 5, 14)$.

Exercise 1.3.7. Find a basis of the vector subspace V of $\mathbb{R}^5$ formed by vectors $(x_1, \dots, x_5)$ that are solutions of the following system:

$$\begin{cases} x_1 - 3x_2 + x_4 = 0 \\ x_2 + 3x_3 - x_5 = 0 \\ x_1 + 2x_2 + x_3 - x_4 = 0. \end{cases}$$

Exercise 1.3.8. In $\mathbb{R}^4$ let $v_1 = (1, 2, 3, 4)$ and $v_2 = (1, -2, 3, -4)$. Is it possible to find two real numbers x and y such that $(x, 1, y, 1) \in L\{v_1, v_2\}$? (Recall that $L\{v_1, v_2\}$ denotes the subspace spanned by the vectors v_1 and v_2)

Exercise 1.3.9. Let V be a vector space. Are the following assertions true?

(1) If the vectors v_1, v_2, v_3 are pairwise non-proportional then the set $\{v_1, v_2, v_3\}$ is free.
(2) If none of the vectors $v_1, \dots, v_r$ is a linear combination of the remaining ones then the set $\{v_1, \dots, v_r\}$ is free.

Exercise 1.3.10. In $\mathbb{R}^4$ let $v_1 = (0, 1, -2, 1)$, $v_2 = (1, 0, 2, -1)$, $v_3 = (3, 2, 2, -1)$, $v_4 = (0, 0, 1, 0)$, $v_5 = (0, 0, 0, 1)$. Check whether the following statements are true.

(1) $L\{v_1, v_2, v_3\} = L\{(1, 1, 0, 0), (-1, 1, -4, 2)\}$;
(2) $(1, 1, 0, 0) \in L\{v_1, v_2\} \cap L\{v_2, v_3, v_4\}$;
(3) $\dim(L\{v_1, v_2\} \cap L\{v_2, v_3, v_4\}) = 1$;
(4) $L\{v_1, v_2\} + L\{v_2, v_3, v_4\} = \mathbb{R}^4$;
(5) $L\{v_1, v_2, v_3\} + L\{v_4, v_5\} = \mathbb{R}^4$.

Exercise 1.3.11. Check the linear independence of the following vectors, and find a basis of the subspace they span.

(1) $(1, 0, 1)$, $(0, 2, 2)$, $(3, 7, 1)$, in $\mathbb{R}^3$;

(2) $(1, 0, 0)$, $(0, 1, 1)$, $(1, 1, 1)$, in $\mathbb{R}^3$;

(3) $(1, 2, 1, 2, 1)$, $(2, 1, 2, 1, 2)$, $(1, 0, 1, 1, 0)$, $(0, 1, 0, 0, 1)$, in $\mathbb{R}^5$.

Exercise 1.3.12. Let V be the vector space of polynomials in x of degree $\leq n$, with coefficients in $\mathbb{R}$. Prove that, for every $a \in \mathbb{R}$, the set

$$\{1, x - a, (x - a)^2, \ldots, (x - a)^n\}$$

is a basis of V. Let $f(x) \in V$; write $f(x)$ as a linear combination of the polynomials above. Who are the coefficients of this linear combination?

Exercise 1.3.13. Let $U_t = L\{u_1, u_2\}$ and $V_t = L\{v_1, v_2\}$ be two subspaces of $\mathbb{R}^4$, with $u_1 = (1, t, 2t, 0)$, $u_2 = (t, t, t, t)$, $v_1 = (t - 2, -t, -3t, t)$ and $v_2 = (2, t, 2t, 0)$.

(1) Find if there exists $t \in \mathbb{R}$ such that $U_t + V_t = \mathbb{R}^4$.

(2) For which $t \in \mathbb{R}$ we have $\dim(U_t \cap V_t) = 1$?

(3) Find a basis of $U_1 \cap V_1$ and extend it to a basis of $\mathbb{R}^4$.

Exercise 1.3.14. In $\mathbb{R}^4$ we consider the subspaces $U = L\{v_1, v_2, v_3\}$ and $V = L\{v_4, v_5\}$, where $v_1 = (1, 2, 3, 4)$, $v_2 = (2, 2, 2, 6)$, $v_3 = (0, 2, 4, 4)$, $v_4 = (1, 0, -1, 2)$ and $v_5 = (2, 3, 0, 1)$. Find bases of the subspaces $U \cap V$, U, V and $U + V$.

Exercise 1.3.15. Let U and W be two vector subspaces of a vector space V. Prove that $U \cup W$ is a vector subspace of V if and only if $U \subseteq W$ or $W \subseteq U$.

Exercise 1.3.16. Let U, V and W be three subspaces of the same vector space. Is it true that

$$U \cap (V + W) = (U \cap V) + (U \cap W)?$$

Exercise 1.3.17. Check if the sum of the following subspaces of $\mathbb{R}^4$ is direct: $U = L\{(1, 0, 1, 0), (1, 2, 3, 4)\}$ and $V = L\{(0, 1, 1, 1), (0, 0, 0, 1)\}$.

Exercise 1.3.18. Consider the following subspaces of $\mathbb{R}^4$:

$$U = L\{(1, 0, 1, 0), (0, 1, 1, 1), (0, 0, 0, 1)\}$$

and

$$V = L\{(1, 0, 1, 0), (0, 1, 1, 0)\}.$$

Find a subspace $W \subset \mathbb{R}^4$ such that $U = V \oplus W$. Is such a W unique?

Exercise 1.3.19. Given the following subspaces of $\mathbb{R}^4$,

$$U = L\{(1, 0, 1, 0), (0, 0, 0, 1)\}, \qquad V = L\{(1, 0, 2, 0), (0, 0, 1, 1)\},$$

does there exist $W \subset \mathbb{R}^4$ such that $U \oplus W = V \oplus W = \mathbb{R}^4$? If yes, find such a W. Is it unique?

Exercise 1.3.20. In the vector space V of polynomials of degree ≤ 5 with real coefficients consider the following subspaces:

$$
\begin{aligned}
U_1 &= \{p(x) \in V \mid p(0) = 0\}, \\
U_2 &= \{p(x) \in V \mid p'(1) = 0\}, \\
U_3 &= \{p(x) \in V \mid x^2 + 1 \text{ divides } p(x)\}, \\
U_4 &= \{p(x) \in V \mid p(-x) = p(x), \forall x\}, \\
U_5 &= \{p(x) \in V \mid p(x) = xp'(x), \forall x\}.
\end{aligned}
$$

(1) Find bases of the following subspaces:

$$U_1,\ U_2,\ U_3,\ U_4,\ U_5,\ U_1 \cap U_2,\ U_1 \cap U_3,\ U_1 \cap U_2 \cap U_3,\ U_1 \cap U_2 \cap U_3 \cap U_4.$$

(2) Find subspaces W_1 and W_2 of V such that

$$W_1 \oplus U_4 = W_2 \oplus (U_1 \cap U_3) = V.$$

2

Linear Functions and Matrices

In this chapter we will develop the theory of linear functions between two vector spaces. We will introduce the notion of matrix, we will describe the link between matrices and linear functions and prove the main results of matrix theory. Finally, we will apply these results to the study of systems of linear equations.

2.1 Linear functions

We will now deal with the study of functions between two vector spaces which "respect" the vector space structure, i.e., which are compatible with the operations of vector addition and multiplication of a vector by a scalar.

Definition 2.1.1. Let V and W be two vector spaces over a field K. A function $f\colon V \to W$ is *additive* if

$$f(v_1 + v_2) = f(v_1) + f(v_2),$$

for every $v_1, v_2 \in V$.

Such a function is K*-linear* (or simply *linear*) if it is additive and the following equality holds:

$$f(\lambda v) = \lambda f(v),$$

for every $v \in V$ and any $\lambda \in K$.

A linear function between two vector spaces is also called a *homomorphism* of vector spaces.

Definition 2.1.2. Let $f\colon V \to W$ be a homomorphism of vector spaces. f is an *isomorphism* if there exists a homomorphism $g\colon W \to V$ such that $g \circ f = \mathrm{id}_V$ and $f \circ g = \mathrm{id}_W$.

In other words, saying that f is an isomorphism of vector spaces is equivalent to saying that f is an invertible homomorphism and that its inverse function is linear.

Two vector spaces V and W over K are *isomorphic* if there exists an isomorphism $f: V \to W$. When we want to indicate that V and W are isomorphic without specifying the isomorphism, we simply write $V \cong W$.

From the definition it follows that an isomorphism of vector spaces is a bijective function. We now prove the converse:

Proposition 2.1.3. *Let $f: V \to W$ be a homomorphism of vector spaces. If f is bijective then it is an isomorphism.*

Proof. Since f is bijective it is invertible, hence we have only to prove that the inverse function $f^{-1}: W \to V$ is linear.

Let $w_1, w_2 \in W$ and set $v_1 = f^{-1}(w_1)$ and $v_2 = f^{-1}(w_2)$. From the additivity of f it follows that $f(v_1 + v_2) = f(v_1) + f(v_2) = w_1 + w_2$, hence $f^{-1}(w_1+w_2) = v_1+v_2 = f^{-1}(w_1)+f^{-1}(w_2)$; this proves that f^{-1} is additive.

Let $\lambda \in K$. From the linearity of f it follows that $f(\lambda v_1) = \lambda f(v_1) = \lambda w_1$, hence $f^{-1}(\lambda w_1) = \lambda v_1 = \lambda f^{-1}(w_1)$. This proves that f^{-1} is linear. □

Remark 2.1.4. An injective homomorphism of vector spaces is also called a *monomorphism*, while a surjective homomorphism is called an *epimorphism*. A monomorphism that is also an epimorphism is an isomorphism.

The importance of the notion of isomorphism is due to the fact that it allows to "identify" different vector spaces, provided they are isomorphic. It is thus possible to classify vector spaces, as described in the following result:

Proposition 2.1.5. *Let V be a vector space of dimension n over the field K. Then V is (non canonically) isomorphic to the vector space K^n.*

Proof. Let us choose a basis $\{v_1, v_2, \ldots, v_n\}$ of V. Note that this is always possible, although there is, in general, no 'canonical' choice for such a basis.

Now we can define a function $f: V \to K^n$ that associates to a vector $v \in V$ the unique n-tuple $(\lambda_1, \ldots, \lambda_n) \in K^n$ such that

$$v = \lambda_1 v_1 + \cdots + \lambda_n v_n.$$

It is easy to verify that the function f is linear. Moreover f is bijective because $\{v_1, v_2, \ldots, v_n\}$ is a basis of V. From the Proposition 2.1.3 it follows that f is an isomorphism.

Note that the function f depends on the choice of a basis of V. Since, in general, there is no canonical choice of such a basis, it follows that there is no canonical choice of an isomorphism between V and K^n. □

Corollary 2.1.6. *Two finite dimensional vector spaces over the field K are isomorphic if and only if they have the same dimension.*

2.1.1 Kernel and image

We now introduce two particularly important vector subspaces associated with a linear function:

Definition 2.1.7. Let $f\colon V \to W$ be a linear function between two vector spaces. The *kernel* of f is the set

$$\mathrm{Ker}(f) = \{v \in V \mid f(v) = \mathbf{0}\}.$$

The *image* of f is the set

$$\mathrm{Im}(f) = \{w \in W \mid w = f(v), \text{ for some } v \in V\}.$$

Proposition 2.1.8. *The kernel of a linear function $f\colon V \to W$ is a vector subspace of V, and the image of f is a vector subspace of W.*

Proof. Let $v_1, v_2 \in \mathrm{Ker}(f)$ and let us consider a linear combination $\lambda_1 v_1 + \lambda_2 v_2$, with $\lambda_1, \lambda_2 \in K$. From the linearity of f it follows that

$$f(\lambda_1 v_1 + \lambda_2 v_2) = \lambda_1 f(v_1) + \lambda_2 f(v_2) = \mathbf{0},$$

hence $\lambda_1 v_1 + \lambda_2 v_2 \in \mathrm{Ker}(f)$. This proves that $\mathrm{Ker}(f)$ is a vector subspace of V.

Now let $w_1, w_2 \in \mathrm{Im}(f)$ and $v_1, v_2 \in V$ be such that $w_1 = f(v_1)$ and $w_2 = f(v_2)$. From the linearity of f it follows that

$$f(\lambda_1 v_1 + \lambda_2 v_2) = \lambda_1 f(v_1) + \lambda_2 f(v_2) = \lambda_1 w_1 + \lambda_2 w_2,$$

hence $\lambda_1 w_1 + \lambda_2 w_2 \in \mathrm{Im}(f)$, for every $\lambda_1, \lambda_2 \in K$. This proves that $\mathrm{Im}(f)$ is a vector subspace of W. □

The following result provides a characterization of the monomorphisms in terms of the vanishing of their kernel.

Proposition 2.1.9. *Let $f\colon V \to W$ be a linear function. Then f is injective if and only if* $\mathrm{Ker}(f) = \{\mathbf{0}\}$.

Proof. Let us assume that f is injective. Let $v \in \text{Ker}(f)$: we have $f(v) = \mathbf{0}$. Since $f(\mathbf{0}) = \mathbf{0}$, from the injectivity of f it follows that $v = \mathbf{0}$, hence $\text{Ker}(f) = \{\mathbf{0}\}$.

Conversely, let us suppose that $\text{Ker}(f) = \{\mathbf{0}\}$. Let $v_1, v_2 \in V$ be such that $f(v_1) = f(v_2)$. From the linearity of f we have

$$f(v_1 - v_2) = f(v_1) - f(v_2) = \mathbf{0},$$

hence $v_1 - v_2 \in \text{Ker}(f)$. Since, by hypothesis, $\text{Ker}(f) = \{\mathbf{0}\}$, we have $v_1 - v_2 = \mathbf{0}$, hence $v_1 = v_2$. This proves that f is injective. □

We have seen that the kernel of a linear function $f\colon V \to W$ is always a vector subspace of V. We will now prove that, more generally, the inverse image of any vector $w \in \text{Im}(f)$ is simply obtained by "translating" the kernel of f along any vector $v \in f^{-1}(w)$.

Proposition 2.1.10. *Let $f\colon V \to W$ be a linear function and let $w \in W$. If $w \in \text{Im}(f)$ then*

$$f^{-1}(w) = v + \text{Ker}(f) = \{v + u \mid u \in \text{Ker}(f)\},$$

where v is any vector such that $f(v) = w$. If $w \notin \text{Im}(f)$ then $f^{-1}(w) = \varnothing$.

Proof. Let $v \in V$ be such that $f(v) = w$. For any $u \in \text{Ker}(f)$ we have $f(v+u) = f(v) + f(u) = w + \mathbf{0} = w$. This proves that $v + \text{Ker}(f) \subseteq f^{-1}(w)$.

Conversely, for every $v' \in f^{-1}(w)$ let $u = v' - v$. Then $f(u) = f(v') - f(v) = w - w = \mathbf{0}$, hence $u \in \text{Ker}(f)$. It follows that $v' = v + u \in v + \text{Ker}(f)$, hence we have the inclusion $f^{-1}(w) \subseteq v + \text{Ker}(f)$. The last statement is obvious. □

The dimensions of the kernel and the image of a linear function are related to each other by the following relationship:

Proposition 2.1.11. *Let $f\colon V \to W$ be a linear function. If V has finite dimension then*

$$\dim(V) = \dim \text{Ker}(f) + \dim \text{Im}(f).$$

Proof. Let $n = \dim(V)$ and $r = \dim \text{Ker}(f)$; we must prove that $\dim \text{Im}(f) = n - r$. We choose a basis $\{v_1, v_2, \ldots, v_r\}$ of $\text{Ker}(f)$ and we complete it to a basis $\{v_1, v_2, \ldots, v_r, v_{r+1}, \ldots, v_n\}$ of V. Recall that the images of the vectors of a basis of V form a set of generators of $\text{Im}(f)$. Since $f(v_1) = f(v_2) = \cdots = f(v_r) = \mathbf{0}$, it follows that the vectors

$$w_1 = f(v_{r+1}),\ w_2 = f(v_{r+2}),\ \ldots,\ w_{n-r} = f(v_n)$$

span the image of f. We now prove that these vectors are also linearly independent. Let us consider a linear combination

$$\lambda_1 w_1 + \lambda_2 w_2 + \cdots + \lambda_{n-r} w_{n-r} = \mathbf{0}.$$

Since f is linear, we have

$$\begin{aligned}\mathbf{0} &= \lambda_1 w_1 + \lambda_2 w_2 + \cdots + \lambda_{n-r} w_{n-r} \\ &= \lambda_1 f(v_{r+1}) + \lambda_2 f(v_{r+2}) + \cdots + \lambda_{n-r} f(v_n) \\ &= f(\lambda_1 v_{r+1} + \lambda_2 v_{r+2} + \cdots + \lambda_{n-r} v_n)\end{aligned}$$

hence

$$\lambda_1 v_{r+1} + \lambda_2 v_{r+2} + \cdots + \lambda_{n-r} v_n \in \operatorname{Ker}(f).$$

Since $\{v_1, v_2, \dots, v_r\}$ is a basis of $\operatorname{Ker}(f)$, we have

$$\lambda_1 v_{r+1} + \lambda_2 v_{r+2} + \cdots + \lambda_{n-r} v_n = \mu_1 v_1 + \mu_2 v_2 + \cdots + \mu_r v_r$$

hence

$$\mu_1 v_1 + \mu_2 v_2 + \cdots + \mu_r v_r - \lambda_1 v_{r+1} - \lambda_2 v_{r+2} - \cdots - \lambda_{n-r} v_n = \mathbf{0}.$$

By hypothesis, the vectors $v_1, v_2, \dots, v_n$ are a basis of V, hence

$$\mu_1 = \mu_2 = \cdots = \mu_r = 0, \ \lambda_1 = \lambda_2 = \cdots = \lambda_{n-r} = 0,$$

which proves that the vectors $w_1, w_2, \dots, w_{n-r}$ are linearly independent. It follows that these vectors are a basis of the image of f, hence $\dim \operatorname{Im}(f) = n - r$. □

Remark 2.1.12. Let $f\colon V \to W$ be a linear function between two vector spaces. The dimension of the image of f is called the *rank* of f,

$$\operatorname{rk}(f) = \dim \operatorname{Im}(f).$$

The dimension of the kernel of f is called the *nullity* of f,

$$\operatorname{null}(f) = \dim \operatorname{Ker}(f).$$

The previous result states that for any linear function $f\colon V \to W$ we have

$$\operatorname{rk}(f) + \operatorname{null}(f) = \dim(V).$$

Remark 2.1.13. Let V and W be two vector spaces over K and let us denote by $\mathrm{Hom}(V, W)$ the set of linear functions from V to W. We define the sum of two linear functions $f, g \in \mathrm{Hom}(V, W)$ by setting $(f+g)(v) = f(v)+g(v)$, for every $v \in V$; the resulting function is linear. We define the product of a scalar $\lambda \in K$ and a linear function $f \in \mathrm{Hom}(V, W)$ by setting $(\lambda f)(v) = \lambda(f(v))$, for every $v \in V$. It is easy to verify that, with these two operations, the set $\mathrm{Hom}(V, W)$ is a vector space over K.

We also note that if $W = V$ the composition of two linear functions $f, g \colon V \to V$ is a linear function, hence $g \circ f \in \mathrm{Hom}(V, V)$, for any $f, g \in \mathrm{Hom}(V, V)$. The set $\mathrm{Hom}(V, V)$, with the operations of addition and composition, is a non-commutative (unitary) ring. If we also consider the multiplication of a linear function by a scalar $\lambda \in K$, the resulting structure on the set $\mathrm{Hom}(V, V)$ is called a K-algebra.

Remark 2.1.14. A homomorphism $f \colon V \to V$ is also called an *endomorphism*. If it is invertible it is an *automorphism*. The set of all endomorphisms of a vector space V is denoted by $\mathrm{End}(V)$ and the subset of all automorphisms is denoted by $\mathrm{Aut}(V)$.

Remark 2.1.15. A *diagram* of vector spaces and homomorphisms is *commutative* if, for each pair of vector spaces, all the functions between them that can be obtained as composition of homomorphisms in the diagram are equal. As an example, consider the following diagram:

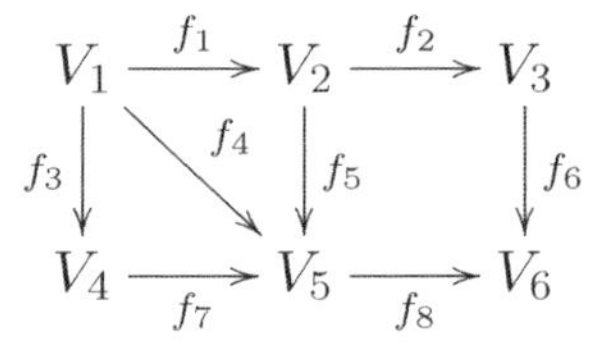

This diagram is commutative if $f_5 \circ f_1 = f_4$, $f_7 \circ f_3 = f_4$, $f_6 \circ f_2 = f_8 \circ f_5$, etc.

2.1.2 Linear functions and bases

We now plan to study the properties of a linear function $f \colon V \to W$ in relation to the choice of bases for the vector spaces V and W.

We begin by proving that a linear function $f \colon V \to W$ is completely determined by knowing the images of the vectors of a basis of V, which can be chosen arbitrarily in W.

Proposition 2.1.16. *Let V and W be two vector spaces over the field K and let $\{v_i\}_{i\in I}$ be a basis (not necessarily finite) of V.*

(*i*) *A homomorphism $f\colon V \to W$ is uniquely determined by knowing the images of the vectors v_i, for any $i \in I$.*

(*ii*) *If we arbitrarily choose vectors $\{w_i\}_{i\in I}$ in W, there exists a unique homomorphism $f\colon V \to W$ such that $f(v_i) = w_i$, for any $i \in I$.*

Proof. (*i*) Let $f\colon V \to W$ be a linear function and suppose we know $f(v_i)$, for every $i \in I$. Since $\{v_i\}_{i\in I}$ is a basis of V, every vector $v \in V$ can be written, in a unique way, as a finite linear combination of the vectors v_i:

$$v = \lambda_1 v_{i_1} + \lambda_2 v_{i_2} + \cdots + \lambda_n v_{i_n}.$$

Since f is linear, we have

$$f(v) = \lambda_1 f(v_{i_1}) + \lambda_2 f(v_{i_2}) + \cdots + \lambda_n f(v_{i_n}), \tag{2.1.1}$$

which proves that if we know $f(v_i)$, for every $i \in I$, we can compute $f(v)$, for every $v \in V$. It follows that if $g\colon V \to W$ is a homomorphism such that $g(v_i) = f(v_i)$, for every $i \in I$, then (2.1.1) implies that $g(v) = f(v)$, for every $v \in V$.

(*ii*) For each $i \in I$ let us arbitrarily choose a vector $w_i \in W$. We define a function $f\colon V \to W$ by setting $f(v_i) = w_i$, for every $i \in I$, and then extending f by linearity to every $v \in V$, i.e., we set

$$f(v) = \lambda_1 f(v_{i_1}) + \lambda_2 f(v_{i_2}) + \cdots + \lambda_n f(v_{i_n}),$$

if $v = \lambda_1 v_{i_1} + \lambda_2 v_{i_2} + \cdots + \lambda_n v_{i_n}$.

It is easy to verify that f is well defined and is linear. The uniqueness of such a function follows from (*i*). □

Corollary 2.1.17. *Let V and W be two vector spaces over the field K, let $\{v_i\}_{i\in I}$ be a basis (not necessarily finite) of V and $f\colon V \to W$ be a linear function.*

(*i*) *f is injective if and only if $\{f(v_i)\}_{i\in I}$ is a free set;*

(*ii*) *f is surjective if and only if $\{f(v_i)\}_{i\in I}$ is a set of generators of W;*

(*iii*) *f is an isomorphism if and only if $\{f(v_i)\}_{i\in I}$ is a basis of W.*

Proof. (*i*) Recall, from Proposition 2.1.9, that f is injective if and only if $\mathrm{Ker}(f) = \{\mathbf{0}\}$. We shall prove the following implication $\mathrm{Ker}(f) = \{\mathbf{0}\} \Rightarrow \{f(v_i)\}_{i \in I}$ is a free set. Let's consider a linear combination

$$\lambda_1 f(v_{i_1}) + \lambda_2 f(v_{i_2}) + \cdots + \lambda_n f(v_{i_n}) = \mathbf{0}.$$

We have

$$\lambda_1 f(v_{i_1}) + \lambda_2 f(v_{i_2}) + \cdots + \lambda_n f(v_{i_n}) = f(\lambda_1 v_{i_1} + \lambda_2 v_{i_2} + \cdots + \lambda_n v_{i_n}),$$

hence

$$\lambda_1 v_{i_1} + \lambda_2 v_{i_2} + \cdots + \lambda_n v_{i_n} \in \mathrm{Ker}(f).$$

Since, by hypothesis, $\mathrm{Ker}(f) = \{\mathbf{0}\}$, we have

$$\lambda_1 v_{i_1} + \lambda_2 v_{i_2} + \cdots + \lambda_n v_{i_n} = \mathbf{0},$$

hence

$$\lambda_1 = \lambda_2 = \cdots = \lambda_n = 0,$$

because the vectors $\{v_i\}_{i \in I}$ are a basis of V.

We now prove the opposite implication. Let $v \in \mathrm{Ker}(f)$ and write

$$v = \lambda_1 v_{i_1} + \lambda_2 v_{i_2} + \cdots + \lambda_n v_{i_n},$$

for some integer n and some $\lambda_1, \ldots, \lambda_n \in K$. Since $f(v) = \mathbf{0}$, from the linearity of f it follows that

$$\lambda_1 f(v_{i_1}) + \lambda_2 f(v_{i_2}) + \cdots + \lambda_n f(v_{i_n}) = \mathbf{0}.$$

Since, by hypothesis, the set $\{f(v_i)\}_{i \in I}$ is free, we have

$$\lambda_1 = \lambda_2 = \cdots = \lambda_n = 0$$

hence $v = \mathbf{0}$, which proves that $\mathrm{Ker}(f) = \{\mathbf{0}\}$.

(*ii*) Recall that f is surjective if and only if $\mathrm{Im}(f) = W$. Note that $\mathrm{Im}(f)$ is spanned by the vectors $f(v_i)$, for $i \in I$. In fact, for every $w \in \mathrm{Im}(f)$ there exists $v \in V$ such that $w = f(v)$. Since $\{v_i\}_{i \in I}$ is a basis of V, the vector v can be written as a linear combination of a finite number of v_i,

$$v = \lambda_1 v_{i_1} + \lambda_2 v_{i_2} + \cdots + \lambda_n v_{i_n}.$$

It follows that

$$w = f(v) = \lambda_1 f(v_{i_1}) + \lambda_2 f(v_{i_2}) + \cdots + \lambda_n f(v_{i_n}),$$

which proves that the set of vectors $\{f(v_i)\}_{i\in I}$ spans the image of f.

It follows that $\mathrm{Im}(f) = W$ if and only if $\{f(v_i)\}_{i\in I}$ is a set of generators of W.

(*iii*) Since f is an isomorphism if and only if it is bijective (cf. Proposition 2.1.3), from (*i*) and (*ii*) it follows that f is an isomorphism if and only if $\{f(v_i)\}_{i\in I}$ is a free set of generators of W, i.e., a basis of W. □

We end this section with some remarks on the invertibility of a linear function.

To state that a linear function $f\colon V \to W$ is invertible means that there exists a linear function $g\colon W \to V$ such that $g \circ f = \mathrm{id}_V$ and $f \circ g = \mathrm{id}_W$ (cf. Definition 2.1.2). It may happen that f is *not* invertible but that there still exists a linear function $g\colon W \to V$ such that $g \circ f = \mathrm{id}_V$ (in this case, however, one cannot also have $f \circ g = \mathrm{id}_W$, otherwise f would be invertible); or it could be that $f \circ g = \mathrm{id}_W$ but not $g \circ f = \mathrm{id}_V$. Such a function g *is not* the inverse of f, but it behaves as such only from the left or only from the right.

Definition 2.1.18. Let $f\colon V \to W$ be a linear function. If there exists a linear function $g\colon W \to V$ such that $g \circ f = \mathrm{id}_V$, then g is called a *left inverse* of f. If we have $f \circ g = \mathrm{id}_W$, then g is called a *right inverse* of f.

It may happen that a linear function has a left inverse but no right inverse, or a right inverse but no left inverse. However, if there is both a left inverse and a right inverse of f they must necessarily be equal and, in this case, f is invertible.

In fact, let us suppose that g_l is a left inverse and that g_r is a right inverse of f. By the associative property of the composition of functions we have

$$(g_l \circ f) \circ g_r = g_l \circ (f \circ g_r)$$

But $g_l \circ f = \mathrm{id}_V$ and $f \circ g_r = \mathrm{id}_W$, because g_l and g_r are, respectively, a left inverse and a right inverse of f. The previous equality then becomes

$$\mathrm{id}_V \circ g_r = g_l \circ \mathrm{id}_W$$

which implies $g_r = g_l$.

The following proposition clarifies under which conditions a linear function admits a left or right inverse.

Proposition 2.1.19. *Let $f\colon V \to W$ be a linear function.*

(*i*) *There exists a left inverse of f if and only if f is injective.*

(*ii*) *There exists a right inverse of f if and only if f is surjective.*

Proof. (*i*) Suppose there is a left inverse g_l of f, i.e., a function $g_l\colon W \to V$ such that $g_l \circ f = \mathrm{id}_V$. To prove that f is injective, it suffices to prove that $\mathrm{Ker}(f) = \{\mathbf{0}\}$. Let $v \in \mathrm{Ker}(f)$. We have:

$$v = \mathrm{id}_V(v) = (g_l \circ f)(v) = g_l(f(v)) = g_l(\mathbf{0}) = \mathbf{0}.$$

This proves that the only vector v that belongs to the kernel of f is the zero vector, hence $\mathrm{Ker}(f) = \{\mathbf{0}\}$ and f is injective.

Conversely, let us assume that f is injective and try to construct a left inverse g_l. We choose a basis $\{v_1, \dots, v_n\}$ of V. From Corollary 2.1.17 we know that the set of vectors $\{w_1 = f(v_1), w_2 = f(v_2), \dots, w_n = f(v_n)\}$ is a free set (because f is injective), hence we can complete it to a basis $\{w_1, \dots, w_n, w_{n+1}, \dots, w_{n+r}\}$ of W. By recalling Proposition 2.1.16, to define the linear function $g_l\colon W \to V$ it suffices to specify the images of the vectors of the basis $\{w_1, \dots, w_n, w_{n+1}, \dots, w_{n+r}\}$ of W. Hence we define g_l by setting $g_l(w_i) = v_i$, for every $i = 1, \dots, n$, and $g_l(w_{n+1}) = g_l(w_{n+2}) = \dots = g_l(w_{n+r}) = \mathbf{0}$. Let us check that this function g_l is a left inverse of f. Let $v = \lambda_1 v_1 + \dots + \lambda_n v_n \in V$. We have:

$$\begin{aligned}(g_l \circ f)(v) = g_l(f(v)) &= g_l\Big(f\Big(\sum_{i=1}^{n} \lambda_i\, v_i\Big)\Big) = g_l\Big(\sum_{i=1}^{n} \lambda_i\, f(v_i)\Big)\\ &= g_l\Big(\sum_{i=1}^{n} \lambda_i\, w_i\Big) = \sum_{i=1}^{n} \lambda_i\, g_l(w_i) = \sum_{i=1}^{n} \lambda_i\, v_i = v.\end{aligned}$$

Since this equality holds for any $v \in V$, this means that $g_l \circ f = \mathrm{id}_V$, which proves that g_l is a left inverse of f.

(*ii*) Suppose there is a right inverse g_r of f, i.e., a function $g_r\colon W \to V$ such that $f \circ g_r = \mathrm{id}_W$. For $w \in W$ we set $v = g_r(w)$. We have:

$$f(v) = f(g_r(w)) = (f \circ g_r)(w) = \mathrm{id}_W(w) = w,$$

which means that $w \in \mathrm{Im}(f)$ hence f is surjective.

Conversely, let us assume that f is surjective and try to construct a right inverse g_r. We choose a basis $\{w_1, \dots, w_m\}$ of W. Since f is surjective, for every w_j ($j = 1, \dots, m$) there exists a vector $v_j \in V$ such that $f(v_j) = w_j$.

Recall that to define a linear function it suffices to specify the images of the vectors of a basis. Hence we can define $g_r\colon W \to V$ by setting $g_r(w_j) = v_j$, for every $j = 1, \ldots, m$. Let us check that this function g_r is a right inverse of f. Let $w = \mu_1 w_1 + \cdots + \mu_m w_m \in W$. We have:

$$\begin{aligned}(f \circ g_r)(w) = f(g_r(w)) &= f\Big(g_r\Big(\sum_{j=1}^{m} \mu_j\, w_j\Big)\Big) = f\Big(\sum_{j=1}^{m} \mu_j\, g_r(w_j)\Big) \\ &= f\Big(\sum_{j=1}^{m} \mu_j\, v_j\Big) = \sum_{j=1}^{m} \mu_j\, f(v_j) = \sum_{j=1}^{m} \mu_j\, w_j = w.\end{aligned}$$

Since this equality holds for any $w \in W$, this means that $f \circ g_r = \mathrm{id}_W$, which proves that g_r is a right inverse of f. □

Exercises

Exercise 2.1.1. Determine if the following functions are linear:

(1) $f\colon \mathbb{R}^2 \to \mathbb{R}^3$, $(x, y) \mapsto (x - y, x + y + 1, 0)$;
(2) $f\colon \mathbb{R}^2 \to \mathbb{R}^2$, $(x, y) \mapsto (2x, x + y)$;
(3) $f\colon \mathbb{R}^2 \to \mathbb{R}$, $(x, y) \mapsto \sin(x - y)$.

Exercise 2.1.2. Find for which values of $t \in \mathbb{R}$ the following function is linear:

$$f\colon \mathbb{R}^3 \to \mathbb{R}^2, \quad (x, y, z) \mapsto (x + ty, tyz).$$

Exercise 2.1.3. Let $f\colon \mathbb{C}^2 \to \mathbb{C}$ be the function given by $f(x, y) = x + \bar{y}$, where $\bar{y}$ denotes the complex conjugate of the complex number y. Determine if f is $\mathbb{C}$-linear.

Exercise 2.1.4. Let $f\colon V \to W$ be a function between two vector spaces. Prove that f is linear if and only if its graph is a vector subspace of $V \times W$.

2.2 Matrices

Let V and W be two vector spaces over the field K, of dimensions n and m, respectively, and let us choose bases $\{v_1, \ldots, v_n\}$ of V and $\{w_1, \ldots, w_m\}$ of W. From Proposition 2.1.16 we know that a linear function $f\colon V \to W$ is uniquely determined by specifying the vectors $f(v_j)$, for every $j = 1, \ldots, n$. Since $\{w_1, \ldots, w_m\}$ is a basis of W, we can write

$$f(v_j) = \sum_{i=1}^{m} a_{ij}\, w_i,$$

for some $a_{ij} \in K$, for every $i = 1, \dots, m$ and $j = 1, \dots, n$.

It follows that a linear function $f\colon V \to W$ is uniquely determined by the datum of $m\,n$ elements a_{ij} of the field K. These elements constitute what goes by the name of *matrix*.

Definition 2.2.1. A *matrix*, with m rows and n columns (or an $m \times n$ matrix) with entries in K is the datum of $m\,n$ elements of K, organized in a rectangular table with m rows and n columns:

$$A = \begin{pmatrix} a_{11} & a_{12} & \dots & a_{1n} \\ a_{21} & a_{22} & \dots & a_{2n} \\ \vdots & \vdots & \ddots & \vdots \\ a_{m1} & a_{m2} & \dots & a_{mn} \end{pmatrix}$$

Such a matrix A will be often denoted simply by

$$A = \big(a_{ij}\big),$$

where $i = 1, \dots, m$ is the *row index* and $j = 1, \dots, n$ is the *column index*.

To every linear function $f\colon V \to W$ we can associate an $m \times n$ matrix with entries in K. Obviously this matrix depends not only on the function f but also on the choice of the bases of V and W. When it is necessary to explicitly indicate the dependence on the bases, we will write $A = M_{\mathbf{v}}^{\mathbf{w}}(f)$ to denote the matrix of $f\colon V \to W$ with respect to the bases $\mathbf{v} = \{v_1, \dots, v_n\}$ of V and $\mathbf{w} = \{w_1, \dots, w_m\}$ of W.

Remark 2.2.2. Recall that if a vector $v \in V$ is written as a linear combination

$$v = \lambda_1 v_1 + \lambda_2 v_2 + \dots + \lambda_n v_n$$

of the vectors of a basis $\mathbf{v} = \{v_1, \dots, v_n\}$ of V, the coefficients $\lambda_1, \dots, \lambda_n$ that appear in this expression are called the *coordinates* of v in the given basis. Also in this case, when it is appropriate to explicitly indicate the basis, we will write $(\lambda_1, \dots, \lambda_n)^{\mathbf{v}}$.

We can then observe that, from the definition of the matrix $M_{\mathbf{v}}^{\mathbf{w}}(f)$ associated with a homomorphism $f\colon V \to W$, with respect to the bases $\mathbf{v} = \{v_1, \dots, v_n\}$ of V and $\mathbf{w} = \{w_1, \dots, w_m\}$ of W, it follows that the coordinates of the vector $f(v_j)$ in the basis of W are written in the j-th column of the matrix $M_{\mathbf{v}}^{\mathbf{w}}(f)$. This observation is useful when it is necessary to explicitly write the matrix associated with a given linear function.

Remark 2.2.3. Throughout this section we will always assume that all vector spaces have finite dimension. However, we point out that many results can be extended, with suitable modifications, also to vector spaces of infinite dimension.

We now consider two linear functions $f, g\colon V \to W$ and denote by $A = M_{\mathbf{v}}^{\mathbf{w}}(f) = (a_{ij})$ and $B = M_{\mathbf{v}}^{\mathbf{w}}(g) = (b_{ij})$ the corresponding matrices. The sum of f and g is the linear function defined by $(f+g)(v) = f(v) + g(v)$, for any $v \in V$. In particular, for each vector v_j of the basis of V, we have

$$\begin{aligned}(f+g)(v_j) = f(v_j) + g(v_j) &= \sum_{i=1}^{m} a_{ij}\, w_i + \sum_{i=1}^{m} b_{ij}\, w_i \\ &= \sum_{i=1}^{m} (a_{ij} + b_{ij})\, w_i.\end{aligned}$$

This means that the elements of the matrix $M_{\mathbf{v}}^{\mathbf{w}}(f+g)$, associated with the function $f+g$, are the sums $a_{ij} + b_{ij}$ of the elements of the matrices $A = M_{\mathbf{v}}^{\mathbf{w}}(f)$ and $B = M_{\mathbf{v}}^{\mathbf{w}}(g)$, associated with f and g, respectively. This result motivates the following definition:

Definition 2.2.4. Let $A = (a_{ij})$ and $B = (b_{ij})$ be two $m \times n$ matrices with entries in K. Their sum is the matrix

$$A + B = (a_{ij} + b_{ij}),$$

obtained by adding the elements of A and B that are in the same positions.

The previous result can therefore be expressed as follows:

$$M_{\mathbf{v}}^{\mathbf{w}}(f+g) = M_{\mathbf{v}}^{\mathbf{w}}(f) + M_{\mathbf{v}}^{\mathbf{w}}(g).$$

Given $\lambda \in K$, let's consider the function λf defined by $(\lambda f)(v) = \lambda f(v)$, for every $v \in V$. Evaluating this function on the basis vectors of V, we have

$$(\lambda f)(v_j) = \lambda f(v_j) = \lambda \sum_{i=1}^{m} a_{ij}\, w_i = \sum_{i=1}^{m} (\lambda a_{ij})\, w_i.$$

It follows that the elements of the matrix $M_{\mathbf{v}}^{\mathbf{w}}(\lambda f)$, associated with the function λf, are given by the product of λ and the elements of $M_{\mathbf{v}}^{\mathbf{w}}(f)$.

Definition 2.2.5. The product of a scalar $\lambda \in K$ and a matrix $A = (a_{ij})$ with entries in K is the matrix

$$\lambda A = (\lambda a_{ij}).$$

This definition allows us to express the previous result as follows:

$$M_{\mathbf{v}}^{\mathbf{w}}(\lambda f) = \lambda M_{\mathbf{v}}^{\mathbf{w}}(f).$$

We denote by $M_{m,n}(K)$ the set of matrices with m rows and n columns with entries in K. From what we have seen above, it follows that there is a bijection between the sets $\mathrm{Hom}(V, W)$ and $M_{m,n}(K)$:

$$\mathrm{Hom}(V, W) \to M_{m,n}(K) \qquad f \mapsto M_{\mathbf{v}}^{\mathbf{w}}(f).$$

The sets $\mathrm{Hom}(V, W)$, with the operations of addition of functions and multiplication of a function by a scalar, and $M_{m,n}(K)$, with the two operations previously defined, are vector spaces over K. The equalities

$$M_{\mathbf{v}}^{\mathbf{w}}(f+g) = M_{\mathbf{v}}^{\mathbf{w}}(f) + M_{\mathbf{v}}^{\mathbf{w}}(g), \quad M_{\mathbf{v}}^{\mathbf{w}}(\lambda f) = \lambda M_{\mathbf{v}}^{\mathbf{w}}(f)$$

mean that the bijective function $f \mapsto M_{\mathbf{v}}^{\mathbf{w}}(f)$ is linear, hence it is an isomorphism between the vector spaces $\mathrm{Hom}(V, W)$ and $M_{m,n}(K)$.

By analogy with the definition of the canonical basis of K^n, we define the matrices E_{ij}, for $i = 1, \dots, m$ and $j = 1, \dots, n$, whose elements are all zero except the one in place (i, j) (i.e., the one on the intersection of the i-th row and the j-th column) that is 1. It is easy to check that the $m\,n$ matrices E_{ij} are a basis of the vector space $M_{m,n}(K)$. In fact, every matrix $A = (a_{ij})$ can be uniquely written as follows:

$$A = \sum_{i,j} a_{ij} E_{ij}.$$

We can summarize what we have just seen in the following result:

Proposition 2.2.6. *$M_{m,n}(K)$ is a vector space over K of dimension $m\,n$. If V and W are K-vector spaces of dimensions n and m respectively, there is an isomorphism $\mathrm{Hom}(V, W) \cong M_{m,n}(K)$. This isomorphism is not canonical because it depends on the choice of bases of V and W.*

In particular, if $V = W$ (hence $m = n$), the vector space $\mathrm{End}(V) = \mathrm{Hom}(V, V)$ is isomorphic to $M_n(K) = M_{n,n}(K)$ and its dimension is n^2.

Let us now see which operation between matrices corresponds to the composition of two linear functions. Let U, V and W be three vector spaces of dimensions r, n and m, respectively, and choose bases $\mathbf{u} = \{u_1, \dots, u_r\}$, $\mathbf{v} = \{v_1, \dots, v_n\}$ and $\mathbf{w} = \{w_1, \dots, w_m\}$. Let $f\colon V \to W$ and $g\colon U \to V$ be two linear functions and denote by $A = M_{\mathbf{v}}^{\mathbf{w}}(f)$ the matrix of f, by

$B = M^{\mathbf{v}}_{\mathbf{u}}(g)$ the matrix of g and by $C = M^{\mathbf{w}}_{\mathbf{u}}(f \circ g)$ that of $f \circ g \colon U \to W$. We note that A is an $m \times n$ matrix, B is $n \times r$, and C is $m \times r$.

For each vector u_j of the basis of U, we have:

$$\begin{aligned}(f \circ g)(u_j) = f(g(u_j)) &= f\Big(\sum_{h=1}^{n} b_{hj}\, v_h\Big)\\ &= \sum_{h=1}^{n} b_{hj}\, f(v_h) = \sum_{h=1}^{n} b_{hj}\Big(\sum_{i=1}^{m} a_{ih}\, w_i\Big)\\ &= \sum_{i=1}^{m}\Big(\sum_{h=1}^{n} a_{ih}\, b_{hj}\Big)\, w_i.\end{aligned}$$

Since $C = (c_{ij})$ is the matrix of $f \circ g$, we also have

$$(f \circ g)(u_j) = \sum_{i=1}^{m} c_{ij}\, w_i.$$

From the equality of these two expressions (and the fact that the vectors $\{w_1, \ldots, w_m\}$ are a basis of W) it follows that

$$c_{ij} = \sum_{h=1}^{n} a_{ih}\, b_{hj},$$

for every $i = 1, \ldots, m$ and $j = 1, \ldots, r$. We will therefore use this formula to define a multiplication of matrices, so that the product of the matrices A and B associated with the homomorphisms f and g gives precisely the matrix associated with the composition $f \circ g$.

Definition 2.2.7. Given two matrices $A \in M_{m,n}(K)$ and $B \in M_{n,r}(K)$, their *product* is the matrix $C \in M_{m,r}(K)$ whose elements are

$$c_{ij} = \sum_{h=1}^{n} a_{ih}\, b_{hj}, \tag{2.2.1}$$

for every $i = 1, \ldots, m$ and $j = 1, \ldots, r$.

With this definition we can write

$$M^{\mathbf{w}}_{\mathbf{u}}(f \circ g) = M^{\mathbf{w}}_{\mathbf{v}}(f) \cdot M^{\mathbf{v}}_{\mathbf{u}}(g)$$

Let's see in more detail how such a product of matrices is computed. Let A and B be two matrices as above; we want to compute the product $C = AB$. To compute the element c_{ij}, that is on the i-th row and on the j-th column of the matrix C, we have to select the i-th row of the matrix A and the j-th column of the matrix B:

$$(a_{i1}, a_{i2}, \dots, a_{in}) \qquad \begin{pmatrix} b_{1j} \\ b_{2j} \\ \vdots \\ b_{nj} \end{pmatrix}$$

Then we have to "multiply" this row and this column as indicated in the formula (2.2.1), i.e., we must add the products of the components of the two vectors:

$$c_{ij} = a_{i1}\, b_{1j} + a_{i2}\, b_{2j} + \cdots + a_{in}\, b_{nj}.$$

We observe that to do this it is essential that the length of the rows of A coincide with the length of the columns of B. The matrix resulting from the product of A and B will have as many rows as the matrix A and as many columns as the matrix B.

A particular case of product between matrices occurs when the matrix B has only one column, i.e., when B is reduced to a vector (written in column): in this way we obtain the product of a matrix and a vector, the result of which is again a vector. More precisely, given a matrix $A \in M_{m,n}(K)$ and a vector $v = (x_1, x_2, \dots, x_n) \in K^n$ (that we will write in column), the product Av is the vector $w = (y_1, y_2, \dots, y_m) \in K^m$ given by

$$\begin{pmatrix} y_1 \\ y_2 \\ \vdots \\ y_m \end{pmatrix} = \begin{pmatrix} a_{11} & a_{12} & \dots & a_{1n} \\ a_{21} & a_{22} & \dots & a_{2n} \\ \vdots & \vdots & \ddots & \vdots \\ a_{m1} & a_{m2} & \dots & a_{mn} \end{pmatrix} \begin{pmatrix} x_1 \\ x_2 \\ \vdots \\ x_n \end{pmatrix}$$

We obtain a linear function

$$F_A \colon K^n \to K^m, \qquad v \mapsto w = F_A(v) = Av$$

and the matrix associated with this function (with respect to the canonical bases of K^n and K^m) is precisely the matrix A.

Remark 2.2.8. In a completely equivalent way we can consider the particular case of the product of A and B, when the matrix A reduces to a vector (now written as a row vector). Let's consider a matrix $B \in M_{n,r}(K)$ and a vector $v = (x_1, x_2, \ldots, x_n) \in K^n$ (that we write as a row vector). The product vB is the vector $w = (y_1, y_2, \ldots, y_r) \in K^r$ given by

$$(y_1, y_2, \ldots, y_r) = (x_1, x_2, \ldots, x_n) \begin{pmatrix} b_{11} & b_{12} & \ldots & b_{1r} \\ b_{21} & b_{22} & \ldots & b_{2r} \\ \vdots & \vdots & \ddots & \vdots \\ b_{n1} & b_{n2} & \ldots & b_{nr} \end{pmatrix}$$

Again, a linear function is obtained

$$G_B \colon K^n \to K^r, \qquad v \mapsto w = G_B(v) = vB.$$

We therefore conclude that a homomorphism between two vector spaces such as K^n and K^m can be described both by the product of a *row* vector and a given matrix, and by the product of another matrix and a *column* vector. Obviously one can pass from one description to the other simply by exchanging the role of the rows with that of the columns. The operation that transform an $m \times n$ matrix into an $n \times m$ matrix swapping rows and columns is called *transposition*.

Definition 2.2.9. Let $A = (a_{ij}) \in M_{m,n}(K)$. The *transpose* of A is the matrix ${}^tA \in M_{n,m}(K)$ whose element in the position (i, j) is a_{ji}, i.e., it is the element in the position (j, i) of the matrix A.

The transpose of a column vector is the same vector written as a row, and vice versa. For notation convenience, from now on the vectors of K^n will always be thought of as column vectors:

$$v = \begin{pmatrix} x_1 \\ x_2 \\ \vdots \\ x_n \end{pmatrix}$$

To denote that a vector is considered as a row vector, we will write ${}^tv$:

$${}^tv = (x_1, x_2, \ldots, x_n).$$

As a last special case of the product of two matrices, let's see what happens when both A and B reduce to vectors (written the first as a row and the second as a column). In this case the result of the product is a scalar, i.e., an element of K:

$$(a_1, a_2, \dots, a_n) \begin{pmatrix} b_1 \\ b_2 \\ \vdots \\ b_n \end{pmatrix} = a_1 b_1 + a_2 b_2 + \cdots + a_n b_n \in K.$$

In this way we obtain a product between two vectors of K^n, the result of which is a scalar: this is the so-called *scalar product* of two vectors.

Definition 2.2.10. Let $v = {}^t(x_1, x_2, \dots, x_n)$ and $w = {}^t(y_1, y_2, \dots, y_n)$ be two elements of K^n. Their *scalar product*, denoted by $v \cdot w$ (or, sometimes, by $\langle v, w \rangle$) is defined as follows:

$$v \cdot w = {}^t v\, w = \sum_{i=1}^{n} x_i\, y_i.$$

We will deal with this notion of scalar product and its generalizations later.

Let us now look at some properties of the transposition.

Proposition 2.2.11. *Let $A, B \in M_{m,n}(K)$ and $\lambda \in K$. We have:*

(*i*) ${}^t({}^tA) = A$;

(*ii*) ${}^t(A + B) = {}^tA + {}^tB$;

(*iii*) ${}^t(\lambda A) = \lambda\, {}^tA$.

If $A \in M_{m,n}(K)$ and $B \in M_{n,r}(K)$, we have

(*iv*) ${}^t(AB) = {}^tB\, {}^tA$.

Proof. The first three properties are obvious, so let's prove the fourth. Let us denote by a_{ij} the elements of A and by $\tilde{a}_{ij}$ the elements of ${}^tA$, hence $\tilde{a}_{ij} = a_{ji}$. Similarly we denote by b_{ij} and $\tilde{b}_{ij}$ the elements of B and ${}^tB$, respectively, by c_{ij} the elements of the product AB and by $\tilde{c}_{ij}$ the elements of ${}^t(AB)$. Finally, let's denote by d_{ij} the elements of ${}^tB\, {}^tA$. Recalling the definition of the product of two matrices, we have:

$$\tilde{c}_{ij} = c_{ji} = \sum_h a_{jh}\, b_{hi},$$

and

$$d_{ij} = \sum_h \tilde{b}_{ih}\, \tilde{a}_{hj} = \sum_h a_{jh}\, b_{hi}.$$

It follows that $d_{ij} = \tilde{c}_{ij}$, for every i and j. □

Let us now return to the product of matrices and study some of its properties in more detail.

Proposition 2.2.12. *Let A, B, C be three matrices and $\lambda, \mu \in K$. Whenever the sums and the products indicated are defined, we have:*

(i) $(AB)C = A(BC)$;

(ii) $(A+B)C = AC + BC$;

(iii) $A(B+C) = AB + AC$;

(iv) $\lambda(AB) = (\lambda A)B = A(\lambda B)$;

(v) $(\lambda + \mu)A = \lambda A + \mu A$;

(vi) $(\lambda\mu)A = \lambda(\mu A)$.

Proof. All these properties derive from the analogous properties of operations defined on linear functions: for example, the associative property of the product of matrices $(AB)C = A(BC)$ is equivalent to the associative property of the composition of functions $(f \circ g) \circ h = f \circ (g \circ h)$. In any case, they can be proved directly by means of a simple calculation. As an example, we prove the first.

Let us denote by a_{ij} the elements of the matrix A, by b_{ij} those of B and by c_{ij} the elements of C. We denote by d_{ij} the elements of the product of A and B and by e_{ij} those of the product $(AB)C$. From the definition of the product of two matrices we have:

$$e_{ij} = \sum_h d_{ih}\, c_{hj} = \sum_h \Big(\sum_k a_{ik}\, b_{kh}\Big)\, c_{hj} = \sum_{h,k} a_{ik}\, b_{kh}\, c_{hj}.$$

Now it suffices to observe that if we compute the elements of the product $A(BC)$ in an analogous way, we find exactly the same expression. □

Let $f\colon V \to W$ be a linear function between two vector spaces of dimensions n and m, respectively. We have already observed that the choice of a basis $\mathbf{v} = \{v_1, \ldots, v_n\}$ of V determines an isomorphism $\alpha_{\mathbf{v}}\colon V \xrightarrow{\sim} K^n$ that associates to each vector $v \in V$ the n-tuple $(\lambda_1, \ldots, \lambda_n)^{\mathbf{v}}$ of its coordinates in the basis $\mathbf{v}$. Similarly the choice of a basis $\mathbf{w} = \{w_1, \ldots, w_m\}$ of W determines an isomorphism $\beta_{\mathbf{w}}\colon W \xrightarrow{\sim} K^m$ that associates to every vector $w \in W$ the m-tuple $(\mu_1, \ldots, \mu_m)^{\mathbf{w}}$ of its coordinates in the basis $\mathbf{w}$.

Let $A = M_{\mathbf{v}}^{\mathbf{w}}(f)$ be the matrix of f with respect to the chosen bases. A determines a linear function $F\colon K^n \to K^m$, defined by

$$F: \begin{pmatrix} \lambda_1 \\ \vdots \\ \lambda_n \end{pmatrix}^{\mathbf{v}} \mapsto \begin{pmatrix} \mu_1 \\ \vdots \\ \mu_m \end{pmatrix}^{\mathbf{w}} = M_{\mathbf{v}}^{\mathbf{w}}(f) \begin{pmatrix} \lambda_1 \\ \vdots \\ \lambda_n \end{pmatrix}^{\mathbf{v}}.$$

Proposition 2.2.13. *With the previous notations, the diagram*

$$\begin{array}{ccc} V & \xrightarrow{\;f\;} & W \\ {\scriptstyle \alpha_{\mathbf{v}}}\big\downarrow\wr & & \wr\big\downarrow{\scriptstyle \beta_{\mathbf{w}}} \\ K^n & \xrightarrow[\;F\;]{} & K^m \end{array} \tag{2.2.2}$$

is commutative.

Proof. We must prove that $\beta_{\mathbf{w}} \circ f = F \circ \alpha_{\mathbf{v}}$. Let $v \in V$ and let's write v as a linear combination of the vectors of the basis $\mathbf{v}$:

$$v = \lambda_1 v_1 + \lambda_2 v_2 + \cdots + \lambda_n v_n.$$

According to the definition of the function $\alpha_{\mathbf{v}}$, we have $\alpha_{\mathbf{v}}(v) = {}^t(\lambda_1, \dots, \lambda_n)^{\mathbf{v}}$ (remember that we decided to write the elements of K^n as column vectors). If we compute $F(\alpha_{\mathbf{v}}(v))$ we obtain the vector

$$M_{\mathbf{v}}^{\mathbf{w}}(f) \begin{pmatrix} \lambda_1 \\ \vdots \\ \lambda_n \end{pmatrix}^{\mathbf{v}},$$

whose i-th component is

$$a_{i1}\lambda_1 + a_{i2}\lambda_2 + \cdots + a_{in}\lambda_n = \sum_{j=1}^{n} a_{ij}\lambda_j. \tag{2.2.3}$$

Now we compute $f(v)$. From the linearity of f and the definition of the matrix $A = M_{\mathbf{v}}^{\mathbf{w}}(f) = (a_{ij})$ associated with f, we have:

$$\begin{aligned} f(v) &= f(\lambda_1 v_1 + \lambda_2 v_2 + \cdots + \lambda_n v_n) = \sum_{j=1}^{n} \lambda_j f(v_j) \\ &= \sum_{j=1}^{n} \lambda_j \Big(\sum_{i=1}^{m} a_{ij} w_i \Big) = \sum_{i=1}^{m} \Big(\sum_{j=1}^{n} a_{ij}\lambda_j \Big) w_i. \end{aligned}$$

We set

$$\mu_i = \sum_{j=1}^{n} a_{ij}\lambda_j, \tag{2.2.4}$$

for $i = 1, \dots, m$, so that we have

$$f(v) = \sum_{i=1}^{m} \mu_i w_i.$$

The m-tuple $(\mu_1, \dots, \mu_m)$ gives the coordinates of the vector $f(v)$ in the basis $\mathbf{w}$, hence we have $\beta_{\mathbf{w}}(f(v)) = {}^t(\mu_1, \dots, \mu_m)^{\mathbf{w}}$. Now it suffices to note that the expression of μ_i in (2.2.4) coincides with the expression (2.2.3) of the i-th component of the vector $F(\alpha_{\mathbf{v}}(v))$. We have thus proved that $F(\alpha_{\mathbf{v}}(v)) = \beta_{\mathbf{w}}(f(v))$, for every $v \in V$. □

Remark 2.2.14. This result provides a direct method for computing the image via $f\colon V \to W$ of any vector $v \in V$, given the matrix of f with respect to some fixed bases of the two vector spaces V and W.

First we determine the coordinates $(\lambda_1, \dots, \lambda_n)^{\mathbf{v}}$ of the vector v in the basis $\mathbf{v}$ of V, then we multiply the matrix $A = M_{\mathbf{v}}^{\mathbf{w}}(f)$ associated with f by the vector $(\lambda_1, \dots, \lambda_n)^{\mathbf{v}}$, written as a column vector. The resulting vector gives the coordinates of $f(v)$ with respect to the basis $\mathbf{w}$ of W.

It follows from the commutative diagram (2.2.2) that the kernels of f and F are isomorphic (the isomorphism been induced by $\alpha_{\mathbf{v}}$). Similarly the isomorphism $\beta_{\mathbf{w}}$ induces an isomorphism of $\operatorname{Im}(f)$ and $\operatorname{Im}(F)$. In particular these vector spaces have the same dimension, hence we have

$$\operatorname{null}(f) = \operatorname{null}(F) \quad \text{and} \quad \operatorname{rk}(f) = \operatorname{rk}(F).$$

Since the linear function $F\colon K^n \to K^m$ is given by the multiplication by the matrix $A = M_{\mathbf{v}}^{\mathbf{w}}(f)$, we give the following definition:

Definition 2.2.15. Let A be an $m \times n$ matrix with entries in K and $F\colon K^n \to K^m$ be the linear function defined by left multiplying a column vector by the matrix A. We define the *rank* and the *nullity* of the matrix A by setting

$$\begin{aligned} \operatorname{rk}(A) &= \operatorname{rk}(F) = \dim \operatorname{Im}(F), \\ \operatorname{null}(A) &= \operatorname{null}(F) = \dim \operatorname{Ker}(F). \end{aligned}$$

Note that the subspace $\operatorname{Im}(F)$ of K^m is spanned by the *columns* of A (we can remark that in the isomorphism $\beta_{\mathbf{w}}\colon W \xrightarrow{\sim} K^m$ the columns of the matrix

$A = M_{\mathbf{v}}^{\mathbf{w}}(f)$ correspond to the images by $f\colon V \to W$ of the basis vectors of V, and these images span the subspace $\mathrm{Im}(f)$ of W). It follows that the dimension of $\mathrm{Im}(F)$, i.e., the rank of F, is equal to the maximum number of linearly independent columns of the matrix A. We have thus proved the following result:

Proposition 2.2.16. *The rank of a matrix A is the maximum number of linearly independent columns of A.*

Remark 2.2.17. The previous result is often used as a definition of the rank of a matrix. It is called *column rank*, to distinguish it from the *row rank*, defined as the maximum number of linearly independent rows. We will see later that, in fact, these two notions of rank always coincide, i.e., in any matrix the maximum number of linearly independent columns is always equal to the maximum number of linearly independent rows.

We will now prove that, given a linear function $f\colon V \to W$, it is always possible to choose a basis $\mathbf{v} = \{v_1, \dots, v_n\}$ of V and a basis $\mathbf{w} = \{w_1, \dots, w_m\}$ of W such that the matrix of f with respect to these bases has a particularly simple form.

Theorem 2.2.18. *Let V and W be two vector spaces, with* $\dim V = n$ *and* $\dim W = m$*, and let $f\colon V \to W$ be a linear function. Let* $r = \mathrm{rk}(f) = \dim(\mathrm{Im}\, f)$*. Then it is possible to choose a basis $\mathbf{v} = \{v_1, \dots, v_n\}$ of V and a basis $\mathbf{w} = \{w_1, \dots, w_m\}$ of W such that the matrix of f with respect to these bases is*

$$D_r = \begin{pmatrix} \mathbf{1}_r & \mathbf{0} \\ \mathbf{0} & \mathbf{0} \end{pmatrix}$$

where $\mathbf{1}_r$ is the $r \times r$ identity matrix and $\mathbf{0}$ are blocks made up entirely of zeros.

Proof. Since $\dim(\mathrm{Im}\, f) = r$ we have $\dim(\mathrm{Ker}\, f) = n - r$. Let us choose a basis $\{u_1, \dots, u_{n-r}\}$ of the kernel of f. The vectors $u_1, \dots, u_{n-r}$ are linearly independent, hence we can complete the basis of $\mathrm{Ker}\, f$ to a basis of V, denoted by $\mathbf{v} = \{v_1, \dots, v_r, u_1, \dots, u_{n-r}\}$.

Now we set $w_1 = f(v_1), w_2 = f(v_2), \dots, w_r = f(v_r)$ and note that, since $f(u_1) = \mathbf{0}, \dots, f(u_{n-r}) = \mathbf{0}$, the vectors $w_1, \dots, w_r$ span the image of f. Let's verify that $w_1, \dots, w_r$ are linearly independent.

Consider a linear combination $\alpha_1 w_1 + \alpha_2 w_2 + \cdots + \alpha_r w_r = \mathbf{0}$. Since $w_i = f(v_i)$, we have $\alpha_1 f(v_1) + \alpha_2 f(v_2) + \cdots + \alpha_r f(v_r) = \mathbf{0}$. Since f is linear we can rewrite this expression as follows

$$f(\alpha_1 v_1 + \alpha_2 v_2 + \cdots + \alpha_r v_r) = \mathbf{0}$$

from which it follows that $\alpha_1 v_1 + \alpha_2 v_2 + \cdots + \alpha_r v_r \in \operatorname{Ker} f$. Recalling that a basis of $\operatorname{Ker} f$ is $u_1, \ldots, u_{n-r}$, we must have

$$\alpha_1 v_1 + \alpha_2 v_2 + \cdots + \alpha_r v_r = \beta_1 u_1 + \cdots + \beta_{n-r} u_{n-r}$$

or, equivalently,

$$\alpha_1 v_1 + \cdots + \alpha_r v_r - \beta_1 u_1 - \cdots - \beta_{n-r} u_{n-r} = \mathbf{0}.$$

Since the vectors $v_1, \ldots, v_r, u_1, \ldots, u_{n-r}$ are a basis of V, it follows that all the coefficients α_i and β_j must vanish, hence the vectors $w_1, \ldots, w_r$ are linearly independent. Hence we can complete this set of vectors to a basis $\mathbf{w} = \{w_1, \ldots, w_r, w_{r+1}, \ldots, w_m\}$ of W.

At this point it suffices to recall how the matrix of a linear function with respect to the assigned bases is constructed to conclude that the matrix of f with respect to the bases $\mathbf{v}$ and $\mathbf{w}$ is precisely the matrix D_r. □

2.2.1 Square matrices

We have seen that the product of two matrices (as well as the composition of two functions) is not always defined: in order for the product AB to be defined it is necessary (and sufficient) that the number of columns of the matrix A is equal to the number of rows of B. If we limit ourselves to considering only matrices of type $n \times n$, these problems disappear and the product of two matrices is always defined.

Definition 2.2.19. A matrix with entries in K is a *square matrix of order* n if it has n rows and n columns. The set of square matrices of order n is denoted by $M_n(K)$ instead of $M_{n,n}(K)$.

If V is a vector space of dimension n over K and if we have chosen a basis $\mathbf{v} = \{v_1, \ldots, v_n\}$ of V, to every endomorphism $f\colon V \to V$ there corresponds a square matrix $A = M^{\mathbf{v}}_{\mathbf{v}}(f) \in M_n(K)$. This correspondence gives a bijection between $\operatorname{End}(V)$ and $M_n(K)$.

The set $\operatorname{End}(V)$, with the operations of addition of functions, multiplication of a function by a scalar and composition of two functions, and the set $M_n(K)$, with the operations of addition of matrices, multiplication of a matrix by a scalar and multiplication of two matrices, are K-algebras. Recalling the equalities $M^{\mathbf{v}}_{\mathbf{v}}(f+g) = M^{\mathbf{v}}_{\mathbf{v}}(f) + M^{\mathbf{v}}_{\mathbf{v}}(g)$, $M^{\mathbf{v}}_{\mathbf{v}}(\lambda f) = \lambda M^{\mathbf{v}}_{\mathbf{v}}(f)$, $M^{\mathbf{v}}_{\mathbf{v}}(f \circ g) = M^{\mathbf{v}}_{\mathbf{v}}(f) \cdot M^{\mathbf{v}}_{\mathbf{v}}(g)$, we deduce that the bijection

$$\operatorname{End}(V) \xrightarrow{\sim} M_n(K), \quad f \mapsto M^{\mathbf{v}}_{\mathbf{v}}(f),$$

is an isomorphism of K-algebras.

We note that the neutral element for the addition operation is the *null matrix*

$$\mathbf{0} = \mathbf{0}_n = \begin{pmatrix} 0 & 0 & \cdots & 0 \\ 0 & 0 & \cdots & 0 \\ \vdots & \vdots & \ddots & \vdots \\ 0 & 0 & \cdots & 0 \end{pmatrix}$$

(that corresponds to the null function $f\colon V \to V$, $f(v) = \mathbf{0}$, for every $v \in V$), while the neutral element for the operation of multiplication of matrices is the *identity matrix*

$$\mathbf{1} = \mathbf{1}_n = \begin{pmatrix} 1 & 0 & \cdots & 0 \\ 0 & 1 & \cdots & 0 \\ \vdots & \vdots & \ddots & \vdots \\ 0 & 0 & \cdots & 1 \end{pmatrix}$$

(that corresponds to the identity $\mathrm{id}\colon V \to V$), i.e., it is the matrix whose entries on the *main diagonal* are 1, while all the other entries are 0. In fact, it is easy to check that for any matrix $A \in M_n(K)$ we have

$$\mathbf{1}_n A = A\,\mathbf{1}_n = A.$$

Finally we note that the product of matrices does not have the commutative property: if A and B are two matrices in $M_n(K)$ we have, in general,

$$AB \neq BA.$$

This should not be surprising as it simply reflects the fact that the composition of two linear functions $f, g\colon V \to V$ is not, in general, commutative, i.e., $f \circ g \neq g \circ f$.

Given $\lambda \in K$, the matrix $\lambda\mathbf{1}_n$, i.e.,

$$\begin{pmatrix} \lambda & 0 & \cdots & 0 \\ 0 & \lambda & \cdots & 0 \\ \vdots & \vdots & \ddots & \vdots \\ 0 & 0 & \cdots & \lambda \end{pmatrix}$$

is called a *scalar matrix*. It corresponds to the homomorphism $f\colon V \to V$ defined by $f(v) = \lambda v$. It is easy to verify that a scalar matrix commutes with any other matrix, i.e.,

$$(\lambda\mathbf{1}_n)A = A(\lambda\mathbf{1}_n),$$

for every $A \in M_n(K)$.

More generally, a matrix like

$$\begin{pmatrix} \lambda_1 & 0 & \cdots & 0 \\ 0 & \lambda_2 & \cdots & 0 \\ \vdots & \vdots & \ddots & \vdots \\ 0 & 0 & \cdots & \lambda_n \end{pmatrix}$$

i.e., a matrix in which all elements are zero, except at most those on the main diagonal, is called a *diagonal matrix.* Note that, in general, a diagonal matrix *does not* commute with other matrices. However diagonal matrices commute with each other.

An *upper triangular* matrix is a matrix in which all elements below the main diagonal are zero, i.e., a matrix like

$$\begin{pmatrix} a_{11} & a_{12} & a_{13} & \cdots & a_{1n} \\ 0 & a_{22} & a_{23} & \cdots & a_{2n} \\ 0 & 0 & a_{33} & \cdots & a_{3n} \\ \vdots & \vdots & \vdots & \ddots & \vdots \\ 0 & 0 & 0 & \cdots & a_{nn} \end{pmatrix}$$

Similarly we define a *lower triangular* matrix as a matrix in which all elements above the main diagonal are zero, i.e., a matrix like

$$\begin{pmatrix} a_{11} & 0 & 0 & \cdots & 0 \\ a_{21} & a_{22} & 0 & \cdots & 0 \\ a_{31} & a_{32} & a_{33} & \cdots & 0 \\ \vdots & \vdots & \vdots & \ddots & \vdots \\ a_{n1} & a_{n2} & a_{n3} & \cdots & a_{nn} \end{pmatrix}$$

Note that the sum and product of two upper (respectively, lower) triangular matrices is a matrix of the same type.

Remark 2.2.20. Let us consider, for example, the case of square matrices of order 2, with rational elements. Let

$$A = \begin{pmatrix} 2 & -1 \\ -4 & 2 \end{pmatrix}, \quad B = \begin{pmatrix} 1 & 3 \\ 2 & 6 \end{pmatrix}.$$

It is immediately verified that the product AB is the null matrix, however neither A nor B is zero. This shows that, in general, in the ring $M_n(K)$

of square matrices there are elements other than zero, with the property that their product is equal to zero (these elements are called *zero divisors*): therefore the so-called "cancellation law" (according to which the product of two factors is zero if and only if at least one of the two factors is zero) does not hold.

Let us consider the matrix

$$C = \begin{pmatrix} 0 & 1 \\ 0 & 0 \end{pmatrix}$$

This is not the null matrix, but $C^2 = CC = \mathbf{0}$. More generally, in the ring $M_n(K)$ there are matrices $C \neq \mathbf{0}$ such that $C^r = \mathbf{0}$, for some $r > 1$. These matrices are called *nilpotent*.

We now come to the problem of the invertibility of a matrix in $M_n(K)$. Since the neutral element for the multiplication is the identity matrix $\mathbf{1}_n$, the inverse of $A \in M_n(K)$ is a matrix $B \in M_n(K)$ such that

$$AB = BA = \mathbf{1}_n.$$

Of course the existence in $M_n(K)$ of zero divisors prevents the existence of inverses of all non-zero matrices. In fact, if $A \in M_n(K)$ is a zero divisor and B is a non-zero matrix such that $AB = \mathbf{0}$, then, if the inverse matrix A^{-1} of A existed, we would have

$$B = \mathbf{1}_n B = (A^{-1}A)B = A^{-1}(AB) = A^{-1}\mathbf{0} = \mathbf{0},$$

against the assumption that $B \neq \mathbf{0}$.

On the other hand, considering the isomorphism of K-algebras between $\mathrm{End}(V)$ and $M_n(K)$, where V is a vector space of dimension n with a chosen basis, we note that stating that a matrix A is invertible is equivalent to stating that the corresponding linear function $f\colon V \to V$ is invertible, but this is true if and only if f is bijective, i.e., if and only if f is an isomorphism.

We recall that the subset of $\mathrm{End}(V)$ of all invertible linear functions (i.e., the isomorphisms) has been denoted by $\mathrm{Aut}(V)$. The corresponding subset of $M_n(K)$, consisting of the matrices associated with the elements of $\mathrm{Aut}(V)$, i.e., the invertible matrices, will be denoted by $\mathrm{GL}(n, K)$ and called the *general linear group* of order n with elements in K. It is a non-commutative group for the operation of multiplication of matrices.

2.2.2 Change of basis

We have repeatedly pointed out that the matrix associated with a linear function $f\colon V \to W$ depends on the choice of a basis of the vector space V and a basis of W: changing the bases also changes the matrix associated with f. In this section we want to find out how the matrix of f changes if we modify our choice of bases of V and W.

Let V and W be two vector spaces over the field K, of dimensions n e m respectively, and let $f\colon V \to W$ be a linear function. Let $\mathbf{v} = \{v_1, \dots, v_n\}$ and $\mathbf{v}' = \{v'_1, \dots, v'_n\}$ be two bases of V and $\mathbf{w} = \{w_1, \dots, w_m\}$ and $\mathbf{w}' = \{w'_1, \dots, w'_m\}$ be two bases of W. Let us denote by $A = M^{\mathbf{w}}_{\mathbf{v}}(f) = (a_{ij})$ the matrix of f with respect to the bases $\mathbf{v}$ and $\mathbf{w}$, and by $A' = M^{\mathbf{w}'}_{\mathbf{v}'}(f) = (a'_{ij})$ the matrix of f with respect to the bases $\mathbf{v}'$ and $\mathbf{w}'$. Recall that this means that

$$f(v_j) = \sum_{i=1}^{m} a_{ij}\, w_i, \quad \text{and} \quad f(v'_j) = \sum_{i=1}^{m} a'_{ij}\, w'_i,$$

for every $j = 1, \dots, n$.

We denote by $\alpha_{\mathbf{v}}\colon V \xrightarrow{\sim} K^n$ the isomorphism that associates to each vector $v \in V$ the n-tuple $(\lambda_1, \dots, \lambda_n)^{\mathbf{v}}$ of its coordinates in the basis $\mathbf{v}$ and by $\alpha_{\mathbf{v}'}\colon V \xrightarrow{\sim} K^n$ the isomorphism that associates to each $v \in V$ the n-tuple $(\lambda'_1, \dots, \lambda'_n)^{\mathbf{v}'}$ of its coordinates in the basis $\mathbf{v}'$.

Similarly we denote by $\beta_{\mathbf{w}}\colon W \xrightarrow{\sim} K^m$ the isomorphism that associates to each vector $w \in W$ the m-tuple $(\mu_1, \dots, \mu_m)^{\mathbf{w}}$ of its coordinates in the basis $\mathbf{w}$ and by $\beta_{\mathbf{w}'}\colon W \xrightarrow{\sim} K^m$ the isomorphism that associates to each $w \in W$ the m-tuple $(\mu'_1, \dots, \mu'_m)^{\mathbf{w}'}$ of its coordinates in the basis $\mathbf{w}'$.

By composing $\alpha_{\mathbf{v}}$ with the inverse of the isomorphism $\alpha_{\mathbf{v}'}$ we obtain an automorphism of K^n that corresponds to the multiplication by some matrix $P \in M_n(K)$. We denote this isomorphism by $F_P\colon K^n \xrightarrow{\sim} K^n$. We get the following commutative diagram:

$$\begin{array}{ccccc} & & V & & \\ & {\scriptstyle \alpha_{\mathbf{v}'}} \swarrow & & \searrow {\scriptstyle \alpha_{\mathbf{v}}} & \\ K^n & & \xrightarrow[F_P]{\hspace{3em}} & & K^n \end{array} \tag{2.2.5}$$

Similarly by composing $\beta_{\mathbf{w}}$ with the inverse of the isomorphism $\beta_{\mathbf{w}'}$ we obtain an automorphism of K^m that corresponds to the multiplication by some other matrix $Q \in M_m(K)$. We denote this isomorphism by $F_Q\colon K^m \xrightarrow{\sim} K^m$. We obtain the following commutative diagram:

(2.2.6)

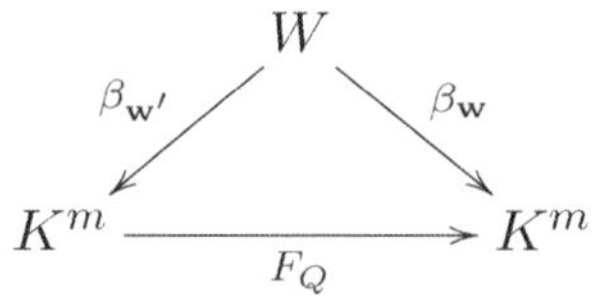

Note that the matrices P and Q are invertible, because the corresponding linear functions F_P and F_Q are isomorphisms.

Let's now get a more explicit description of the matrices P and Q. We start with the matrix P, which corresponds to the isomorphism

$$F_P = \alpha_{\mathbf{v}} \circ \alpha_{\mathbf{v}'}^{-1} \colon K^n \xrightarrow{\sim} K^n.$$

We have already observed that the columns of P are given by the images of the vectors of the canonical basis of K^n. Let $e_j = {}^t(0, \dots, 0, 1, 0, \dots, 0)$ be the j-th vector of the canonical basis of K^n (all coordinates are 0 except the j-th which is 1). Through the isomorphism $\alpha_{\mathbf{v}'}$ the vector $e_j \in K^n$ corresponds to the j-th vector v'_j of the basis $\mathbf{v}'$ of V. It follows that

$$F_P(e_j) = \alpha_{\mathbf{v}}(\alpha_{\mathbf{v}'}^{-1}(e_j)) = \alpha_{\mathbf{v}}(v'_j),$$

where $\alpha_{\mathbf{v}}(v'_j) \in K^n$ is the vector of the coordinates of v'_j in the basis $\mathbf{v}$; this is the j-th column of P.

In conclusion, we can state that the *columns* of the matrix P are the coordinates of the vectors $v'_1, \dots, v'_n$ of the basis $\mathbf{v}'$ of V computed with respect to the first basis $\mathbf{v}$. In other words, this means that P is the matrix associated with the identity function $\mathrm{id}_V \colon V \to V$, with respect to the basis $\mathbf{v}'$ of the domain and the basis $\mathbf{v}$ of the target space (the codomain), $P = M^{\mathbf{v}}_{\mathbf{v}'}(\mathrm{id}_V)$. From the commutativity of the diagram (2.2.5) it follows that

$$\begin{pmatrix} \lambda_1 \\ \vdots \\ \lambda_n \end{pmatrix}^{\mathbf{v}} = M^{\mathbf{v}}_{\mathbf{v}'}(\mathrm{id}_V) \begin{pmatrix} \lambda'_1 \\ \vdots \\ \lambda'_n \end{pmatrix}^{\mathbf{v}'}$$

The matrix $P = M^{\mathbf{v}}_{\mathbf{v}'}(\mathrm{id}_V)$ allows to pass from the coordinates $(\lambda'_1, \dots, \lambda'_n)^{\mathbf{v}'}$ of a vector in the basis $\mathbf{v}'$ to the coordinates $(\lambda_1, \dots, \lambda_n)^{\mathbf{v}}$ of the same vector, in the basis $\mathbf{v}$. For this reason the matrix $M^{\mathbf{v}}_{\mathbf{v}'}(\mathrm{id}_V)$ is called the *change of basis matrix* (in the vector space V).

With a similar reasoning, exchanging the roles of the two bases, it can be shown that the columns of the inverse matrix P^{-1} are precisely the coordinates of the vectors $v_1, \dots, v_n$ of the basis $\mathbf{v}$ of V computed with respect to the second basis $\mathbf{v}'$, i.e., $P^{-1} = M_{\mathbf{v}}^{\mathbf{v}'}(\mathrm{id}_V)$.

In a completely analogous way it can be proved that the j-th column of the matrix Q is formed by the coordinates of the j-th vector w'_j of the basis $\mathbf{w}'$ computed with respect to the basis $\mathbf{w}$. In other words, Q is the matrix whose *columns* are formed by the coordinates of the vectors $w'_1, \dots, w'_m$ of the basis $\mathbf{w}'$ of W computed with respect to the first basis $\mathbf{w}$. This means that Q is the matrix associated with the identity $\mathrm{id}_W : W \to W$, with respect to the basis $\mathbf{w}'$ of the domain and the basis $\mathbf{w}$ of the codomain, $Q = M_{\mathbf{w}'}^{\mathbf{w}}(\mathrm{id}_W)$. From the commutativity of the diagram (2.2.6) it follows that

$$\begin{pmatrix} \mu_1 \\ \vdots \\ \mu_m \end{pmatrix}^{\mathbf{w}} = M_{\mathbf{w}'}^{\mathbf{w}}(\mathrm{id}_W) \begin{pmatrix} \mu'_1 \\ \vdots \\ \mu'_m \end{pmatrix}^{\mathbf{w}'}$$

The matrix $Q = M_{\mathbf{w}'}^{\mathbf{w}}(\mathrm{id}_W)$ allows us to pass from the coordinates of a vector in the basis $\mathbf{w}'$ to the coordinates of the same vector in the basis $\mathbf{w}$. $M_{\mathbf{w}'}^{\mathbf{w}}(\mathrm{id}_W)$ is then the change of basis matrix in the vector space W.

Similarly it can be shown that the columns of the inverse matrix Q^{-1} are the coordinates of the vectors $w_1, \dots, w_m$ of the basis $\mathbf{w}$ of W computed with respect to the second basis $\mathbf{w}'$, i.e., $Q^{-1} = M_{\mathbf{w}}^{\mathbf{w}'}(\mathrm{id}_W)$.

Recalling the result stated in the Proposition 2.2.13, we can summarize what has been said so far in the following commutative diagram

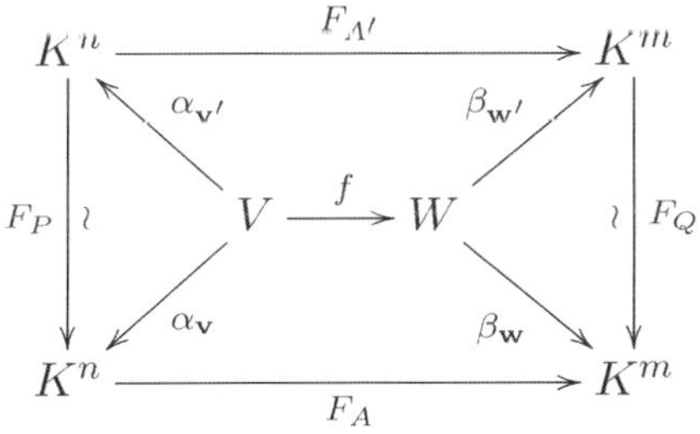

where F_A and $F_{A'}$ are the linear functions defined by the multiplication by $A = M_{\mathbf{v}}^{\mathbf{w}}(f)$ and by $A' = M_{\mathbf{v}'}^{\mathbf{w}'}(f)$, respectively.

From the commutativity of this diagram it follows that

$$F_A \circ F_P = F_Q \circ F_{A'},$$

which is equivalent to the following equality between products of matrices

$$AP = QA'.$$

From this we deduce that

$$A = QA'P^{-1} \quad \text{and} \quad A' = Q^{-1}AP,$$

or, with a more expressive notation,

$$\begin{aligned} M^{\mathbf{w}}_{\mathbf{v}}(f) &= M^{\mathbf{w}}_{\mathbf{w}'}(\mathrm{id}_W) \cdot M^{\mathbf{w}'}_{\mathbf{v}'}(f) \cdot M^{\mathbf{v}'}_{\mathbf{v}}(\mathrm{id}_V) \\ M^{\mathbf{w}'}_{\mathbf{v}'}(f) &= M^{\mathbf{w}'}_{\mathbf{w}}(\mathrm{id}_W) \cdot M^{\mathbf{w}}_{\mathbf{v}}(f) \cdot M^{\mathbf{v}}_{\mathbf{v}'}(\mathrm{id}_V) \end{aligned} \tag{2.2.7}$$

These two equivalent expressions allow us to determine the matrix $A' = M^{\mathbf{w}'}_{\mathbf{v}'}(f)$ of a linear function $f\colon V \to W$ with respect to the bases $\mathbf{v}'$ of V and $\mathbf{w}'$ of W when we know the matrix $A = M^{\mathbf{w}}_{\mathbf{v}}(f)$ of f with respect to the bases $\mathbf{v}$ and $\mathbf{w}$ and the change of basis matrices $P = M^{\mathbf{v}}_{\mathbf{v}'}(\mathrm{id}_V)$ and $Q = M^{\mathbf{w}}_{\mathbf{w}'}(\mathrm{id}_W)$.

In the special case where $W = V$, i.e., when f is an endomorphism of a vector space V, the previous commutative diagram reduces to the following

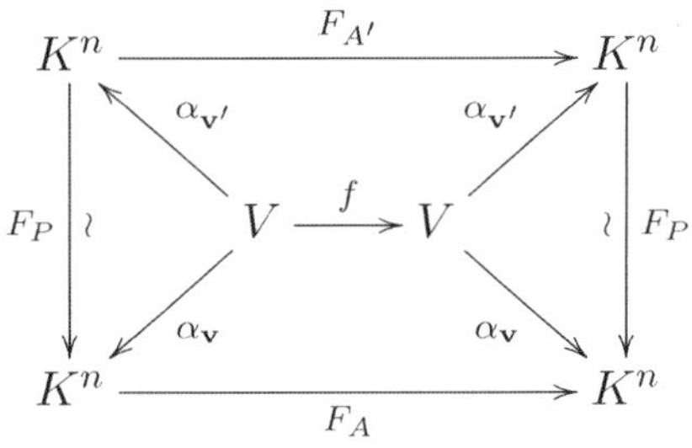

and the equalities (2.2.7) become

$$A = PA'P^{-1} \quad \text{e} \quad A' = P^{-1}AP,$$

or, with a different notation,

$$\begin{aligned} M^{\mathbf{v}}_{\mathbf{v}}(f) &= M^{\mathbf{v}}_{\mathbf{v}'}(\mathrm{id}_V) \cdot M^{\mathbf{v}'}_{\mathbf{v}'}(f) \cdot M^{\mathbf{v}'}_{\mathbf{v}}(\mathrm{id}_V) \\ M^{\mathbf{v}'}_{\mathbf{v}'}(f) &= M^{\mathbf{v}'}_{\mathbf{v}}(\mathrm{id}_V) \cdot M^{\mathbf{v}}_{\mathbf{v}}(f) \cdot M^{\mathbf{v}}_{\mathbf{v}'}(\mathrm{id}_V) \end{aligned} \tag{2.2.8}$$

We now give the following definition:

Definition 2.2.21. Two square matrices A, B of order n with entries in K are *similar* if there exists an invertible matrix $S \in M_n(K)$ (i.e., $S \in \mathrm{GL}_n(K)$) such that

$$A = SBS^{-1}$$

or, equivalently,

$$B = S^{-1}AS.$$

From the above, the following result follows:

Corollary 2.2.22. *Two matrices $A, B \in M_n(K)$ represent the same endomorphism f of a vector space V of dimension n over K, with respect to different bases, if and only if they are similar.*

Remark 2.2.23. Note that the relation of "similarity" is an equivalence relation in the set $M_n(K)$ of square matrices of order n with entries in K.

More generally, for matrices that are not necessarily square, we can give the following definition:

Definition 2.2.24. Two $m \times n$ matrices A and B with entries in K are *equivalent* if they represent the same linear function $f\colon K^n \to K^m$ with respect to different bases, i.e., if there exist invertible matrices $P \in \mathrm{GL}_n(K)$ and $Q \in \mathrm{GL}_m(K)$ such that $B = Q^{-1}AP$ (or, equivalently, by exchanging the roles of Q and Q^{-1}, such that $B = QAP$).

Recalling the Theorem 2.2.18 we can conclude that every matrix $A \in M_{m,n}(K)$, of rank r, is equivalent to the matrix

$$D_r = \begin{pmatrix} \mathbf{1}_r & \mathbf{0} \\ \mathbf{0} & \mathbf{0} \end{pmatrix}$$

where $\mathbf{1}_r$ is the identity matrix of order r and $\mathbf{0}$ are blocks made up entirely of zeros. It follows that two matrices $A, B \in M_{m,n}(K)$ are equivalent if and only if they have the same rank.

Exercises

Exercise 2.2.1. Let $f\colon V \to W$ be a linear function between two vector spaces. Let $\{v_1, v_2, v_3\}$ be a basis of V and $\{w_1, w_2, w_3, w_4\}$ be a basis of W, and suppose that f is defined by $f(v_1) = 2w_1 - 3w_2 + w_4$, $f(v_2) = w_2 - 2w_3 + 3w_4$ and $f(v_3) = w_1 + w_2 + w_3 - 3w_4$. Write the matrix of f in the given bases.

Exercise 2.2.2. Let V and W be two vector spaces with bases $\{v_1, v_2, v_3\}$ and $\{w_1, w_2\}$, respectively. Let $f\colon V \to W$ be the linear function whose matrix (in the given bases) is

$$A = \begin{pmatrix} 2 & -1 & 1 \\ 3 & 2 & -3 \end{pmatrix}$$

(1) We choose for V the new basis $v_1' = v_2 + v_3$, $v_2' = v_1 + v_3$, $v_3' = v_1 + v_2$. Write the matrix A' of f with respect to the bases $\{v_1', v_2', v_3'\}$ and $\{w_1, w_2\}$.

(2) We choose for W the new basis $w'_1 = \frac{1}{2}(w_1 + w_2)$ and $w'_2 = \frac{1}{2}(w_1 - w_2)$. Write the matrix A'' of f with respect to the bases $\{v'_1, v'_2, v'_3\}$ and $\{w'_1, w'_2\}$.

Exercise 2.2.3. Let V be the subspace of $C^\infty(\mathbb{R})$ generated by the functions $f_1(x) = e^{2x} + \cos x$, $f_2(x) = \cos x + \sin x$ and $f_3(x) = \sin x$. Prove that f_1, f_2 and f_3 are linearly independent and write the matrix (with respect to the basis $\{f_1, f_2, f_3\}$) of the endomorphism of V that sends a function to its derivative.

Exercise 2.2.4. Find the matrices, with respect to the canonical bases, of all the linear functions $f\colon \mathbb{R}^3 \to \mathbb{R}^4$ such that $f(1,2,-1) = (0,1,0,1)$, $f(3,-1,2) = (1,2,0,-1)$ and $f(-1,5,-4) = (2,0,3,2)$.

Exercise 2.2.5. Find the matrices, with respect to the canonical bases, of all the linear functions $f\colon \mathbb{R}^3 \to \mathbb{R}^2$ such that $f(0,-2,1) = (3,-1)$, $f(1,1,-2) = (1,2)$ and $f(2,-4,-1) = (11,1)$.

Exercise 2.2.6. Let V be the set of all polynomial functions of degree ≤ 4 with real coefficients that vanish in 0 and 1, and W be the set of all polynomial functions of degree ≤ 3 with real coefficients whose integral between 0 and 1 is zero.

(1) Prove that V and W are vector spaces and write bases of V and W.

(2) Let $D\colon V \to W$ be the linear function that sends a function to its derivative. Prove that D is well defined and write the matrix of D with respect to the bases previously found.

Exercise 2.2.7. Let V and W be two vector spaces with bases $\{v_1, v_2, v_3, v_4\}$ and $\{w_1, w_2, w_3\}$, respectively. Write the matrix (with respect to these bases) of the linear function $\phi\colon V \to W$ defined by $\phi(v_1) = w_1 - w_2$, $\phi(v_2) = 2w_2 - 6w_3$, $\phi(v_3) = -2w_1 + 2w_2$, $\phi(v_4) = w_2 - 3w_3$. Compute the dimensions of $\operatorname{Ker}\phi$ and $\operatorname{Im}\phi$ and write bases of these subspaces. Check whether $w_1 + w_2 + w_3 \in \operatorname{Im}\phi$.

Exercise 2.2.8. Let $\phi_\lambda\colon \mathbb{R}^3 \to \mathbb{R}^4$ be the homomorphism whose matrix (with respect to the canonical bases) is

$$A_\lambda = \begin{pmatrix} 1 & \lambda & 0 \\ 0 & \lambda & 0 \\ 1 & 0 & 1 \\ 0 & 0 & 0 \end{pmatrix}$$

(1) Is it true that, for every $\lambda \in \mathbb{R}$, there exists a homomorphism $\psi\colon \mathbb{R}^4 \to \mathbb{R}^3$ such that $\psi \circ \phi_\lambda$ is surjective?

(2) Find the values of λ for which there exist $x, y, z \in \mathbb{R}$ such that $BA_\lambda = \mathbf{1}$, where

$$B = \begin{pmatrix} 1 & x & 0 & 0 \\ 0 & y & 0 & 0 \\ -1 & z & 1 & 0 \end{pmatrix}$$

Exercise 2.2.9. Find out whether the endomorphism of $\mathbb{R}^3$ defined by

$$f(x, y, z) = (x + 2y, y + z, 2z - x)$$

is injective or surjective. Write bases of $\operatorname{Ker} f$ and $\operatorname{Im} f$ and determine if the sum of the kernel and the image of f is direct.

Exercise 2.2.10. Let $f\colon \mathbb{R}^3 \to \mathbb{R}^3$ be the endomorphism defined by

$$\begin{aligned} f(1,0,0) &= (2,-1,0) \\ f(0,1,0) &= (1,-1,1) \\ f(0,1,-1) &= (0,2,2). \end{aligned}$$

Write the matrix of f with respect to the canonical basis of $\mathbb{R}^3$. Compute the dimensions of the kernel and the image of f and write bases of these subspaces.

Exercise 2.2.11. Let $f\colon \mathbb{R}^3 \to \mathbb{R}^3$ be the endomorphism given by the matrix (with respect to the canonical basis)

$$A = \begin{pmatrix} 1 & 2 & 3 \\ 1 & 1 & 1 \\ 1 & 1 & 1 \end{pmatrix}$$

Compute the rank of f and write bases of $\operatorname{Ker} f$ and $\operatorname{Im} f$.

Exercise 2.2.12. Let V be a finite dimensional vector space. State under which conditions on V there exists an endomorphism $\phi\colon V \to V$ such that $\operatorname{Ker} \phi = \operatorname{Im} \phi$.

Exercise 2.2.13. Compute the rank of the matrix

$$A = \begin{pmatrix} 0 & 1 & 2 & 1 \\ 1 & 1 & 1 & 0 \\ 0 & -1 & 1 & 1 \\ 1 & 1 & 4 & 2 \end{pmatrix}$$

Exercise 2.2.14. Determine, as $a \in \mathbb{R}$ varies, the rank of the matrix

$$A = \begin{pmatrix} 0 & 1 & 2 & 1 & 0 \\ 1 & 2 & 2 & 1 & 1 \\ 1 & 1 & a & 0 & 1 \\ 0 & a & 2a & a^2 & 0 \end{pmatrix}$$

2.3 Linear systems

We now resume the study of systems of linear equations in the light of what we have learned about linear functions and matrices.

Given a system of m linear equations with n unknowns

$$S : \begin{cases} a_{11}\, x_1 + a_{12}\, x_2 + \cdots + a_{1n}\, x_n = b_1 \\ a_{21}\, x_1 + a_{22}\, x_2 + \cdots + a_{2n}\, x_n = b_2 \\ \cdot \quad \cdot \quad \cdot \quad \cdot \quad \cdot \quad \cdot \quad \cdot \quad \cdot \quad \cdot \quad \cdot \\ a_{m1}\, x_1 + a_{m2}\, x_2 + \cdots + a_{mn}\, x_n = b_m \end{cases}$$

we denote by $A = (a_{ij})$ the matrix of the coefficients, by $X = {}^t(x_1, \ldots, x_n)$ the *column* vector of the n unknowns and by $B = {}^t(b_1, \ldots, b_m)$ the *column* vector of the constant terms. Recalling the definition of the product of a matrix and a (column) vector, it is immediate to verify that the system S is equivalent to the following equation:

$$AX = B,$$

or, more explicitly,

$$\begin{pmatrix} a_{11} & a_{12} & \ldots & a_{1n} \\ a_{21} & a_{22} & \ldots & a_{2n} \\ \vdots & \vdots & \ddots & \vdots \\ a_{m1} & a_{m2} & \ldots & a_{mn} \end{pmatrix} \begin{pmatrix} x_1 \\ x_2 \\ \vdots \\ x_n \end{pmatrix} = \begin{pmatrix} b_1 \\ b_2 \\ \vdots \\ b_m \end{pmatrix}$$

If we denote by $F_A \colon K^n \to K^m$ the linear function defined by $F_A(X) = AX$, for every $X \in K^n$ (F_A is the linear function whose matrix, with respect to the canonical bases of K^n and K^m is A), the equation $AX = B$ is equivalent to $F_A(X) = B$. It follows that the set of solutions of the system S is the inverse image by F_A of the vector $B \in K^m$:

$$F_A^{-1}(B) = \{X \in K^n \mid AX = B\}.$$

Remark 2.3.1. More generally, given a linear function $f \colon V \to W$ between two vector spaces V and W of dimensions n and m, respectively, we can consider the following problem: given a vector $w \in W$, determine its inverse image $f^{-1}(w) \subseteq V$, i.e., determine all vectors $v \in V$ such that $f(v) = w$.

By choosing bases $\mathbf{v} = \{v_1, \ldots, v_n\}$ of V and $\mathbf{w} = \{w_1, \ldots, w_m\}$ of W, to the homomorphism f we can associate a matrix $A = (a_{ij})$, with m rows and n columns, such that for every vector

$$v = \lambda_1 v_1 + \cdots + \lambda_n v_n \in V,$$

if we write $f(v)$ as a linear combination of the vectors of the basis $\mathbf{w}$

$$f(v) = \mu_1 w_1 + \cdots + \mu_m w_m,$$

then we have

$$\begin{pmatrix} a_{11} & a_{12} & \dots & a_{1n} \\ a_{21} & a_{22} & \dots & a_{2n} \\ \vdots & \vdots & \ddots & \vdots \\ a_{m1} & a_{m2} & \dots & a_{mn} \end{pmatrix} \begin{pmatrix} \lambda_1 \\ \lambda_2 \\ \vdots \\ \lambda_n \end{pmatrix} = \begin{pmatrix} \mu_1 \\ \mu_2 \\ \vdots \\ \mu_m \end{pmatrix}$$

Hence if we denote by $(b_1, \dots, b_m) \in K^m$ the vector of the coordinates of w in the basis $\mathbf{w}$ of W and we denote by $(x_1, \dots, x_n) \in K^n$ the coordinates of a generic vector $v \in V$ (in the basis $\mathbf{v}$ of V), the problem of finding the vectors v such that $f(v) = w$ translates into the problem of finding the solutions of the following system of linear equations:

$$\begin{pmatrix} a_{11} & a_{12} & \dots & a_{1n} \\ a_{21} & a_{22} & \dots & a_{2n} \\ \vdots & \vdots & \ddots & \vdots \\ a_{m1} & a_{m2} & \dots & a_{mn} \end{pmatrix} \begin{pmatrix} x_1 \\ x_2 \\ \vdots \\ x_n \end{pmatrix} = \begin{pmatrix} b_1 \\ b_2 \\ \vdots \\ b_m \end{pmatrix}$$

Recall that a linear system $AX = B$ is homogeneous if $B = \mathbf{0}$.

Proposition 2.3.2. *The set of solutions of a homogeneous linear system of m equations with n unknowns with coefficients in K is a vector subspace of K^n*

Proof. It suffices to note that the set of solutions of the system $AX = \mathbf{0}$ is the kernel of the linear function $F_A \colon K^n \to K^m$, which is a vector subspace of K^n (see Proposition 2.1.8). However, we can also provide a direct proof.

Let X_1 and X_2 be two solutions of the system $AX = \mathbf{0}$. For every $\lambda_1, \lambda_2 \in K$, we have

$$A(\lambda_1 X_1 + \lambda_2 X_2) = \lambda_1 A X_1 + \lambda_2 A X_2 = \lambda_1 \mathbf{0} + \lambda_2 \mathbf{0} = \mathbf{0},$$

hence $\lambda_1 X_1 + \lambda_2 X_2$ is a solution of the same system. This means that the set of solutions of the system $AX = \mathbf{0}$ is a vector subspace of K^n. □

For non-homogeneous linear systems we have:

Proposition 2.3.3. *Every solution of the non-homogeneous linear systems*

$$S : AX = B$$

can be written as the sum of a fixed solution of S and a solution of the corresponding homogeneous system. In other words, if we denote by Σ_B the set of solutions of S and by $\Sigma_{\mathbf{0}}$ the set of solutions of the corresponding homogeneous system $AX = \mathbf{0}$, we have $\Sigma_B = \varnothing$ (if S has no solutions), or

$$\Sigma_B = \overline{X} + \Sigma_{\mathbf{0}} = \{\overline{X} + Y \mid Y \in \Sigma_{\mathbf{0}}\}, \tag{2.3.1}$$

where $\overline{X}$ is any solution of S.

Proof. Let $F_A \colon K^n \to K^m$ be the linear function whose matrix is A (with respect to the canonical bases of K^n and K^m). Then we have $\Sigma_B = F_A^{-1}(B)$ and $\Sigma_{\mathbf{0}} = F_A^{-1}(\mathbf{0}) = \operatorname{Ker} F_A$. The equality (2.3.1) follows from Proposition 2.1.10. □

We are now able to determine conditions which guarantee the existence of solutions of a system of linear equations.

Proposition 2.3.4. *Let $S : AX = B$ be a system of m linear equations with n unknowns. Let $F_A \colon K^n \to K^m$ be the linear function given by $F_A(X) = AX$, for every $X \in K^n$. The following conditions are equivalent:*

- (*i*) *The system S has solutions;*
- (*ii*) $B \in \operatorname{Im}(F_A)$*;*
- (*iii*) *The vector B is a linear combination of the columns of A;*
- (*iv*) *The rank of the matrix A is equal to the rank of the augmented matrix*[1] *$(A \mid B)$, where this is the matrix obtained by adding to A the column B of the constant terms.*

Proof. Since the set of solutions of S is equal to $F_A^{-1}(B)$, the equivalence of (*i*) and (*ii*) follows from the fact that $F_A^{-1}(B) \neq \varnothing$ if and only if $B \in \operatorname{Im}(F_A)$.

To prove the equivalence of (*ii*) and (*iii*) it suffices to recall that the subspace $\operatorname{Im}(F_A)$ of K^m is spanned by the columns of A. Hence $B \in \operatorname{Im}(F_A)$ if and only if B is a linear combination of the columns of A.

[1] The augmented matrix of the linear system $AX = B$ is the matrix $(A \mid B)$ obtained by adding to the matrix A the column B of the constant terms.

Now we prove that (*iii*) $\Rightarrow$ (*iv*). For this it suffices to recall that the rank of the matrix A (that is equal to the dimension of $\mathrm{Im}(F_A)$) is the maximum number of linearly independent columns of A. Hence if the column B is a linear combination of the columns of A, by adding B to the matrix A the rank of A does not change. It follows that $\mathrm{rk}(A) = \mathrm{rk}(A \mid B)$.

Conversely, the fact that the matrices A and $(A \mid B)$ have the same rank means that adding the column B to the matrix A did not change its rank, hence B must be a linear combination of the columns of A. This proves that (*iv*) $\Rightarrow$ (*iii*). □

What we have seen so far allows us to prove the following theorem:

Theorem 2.3.5 (ROUCHÉ–CAPELLI). *Let $S : AX = B$ be a system of m linear equations with n unknowns. S has solutions if and only if $\mathrm{rk}(A) = \mathrm{rk}(A \mid B)$. In this case, if we denote by r the common value of the ranks of the two matrices, we have:*

(*i*) *if $r = n$ the system has only one solution;*

(*ii*) *if $r < n$ the system has infinite solutions that depend on $n - r$ free parameters (it is also customary to say that S has ∞^{n-r} solutions).*

Proof. That equality between the ranks of the matrices A and $(A \mid B)$ is a necessary and sufficient condition for the solvability of the system S has been proved in the previous proposition. Moreover, in the Proposition 2.3.3 we proved that the set of solutions of S is

$$\overline{X} + \Sigma_{\mathbf{0}} = \{\overline{X} + Y \mid Y \in \Sigma_{\mathbf{0}}\},$$

where $\overline{X}$ is any solution of S and $\Sigma_{\mathbf{0}}$ is the set of solutions of the homogeneous system $AX = \mathbf{0}$ corresponding to S. This means that $\Sigma_{\mathbf{0}} = \mathrm{Ker}(F_A)$, where $F_A \colon K^n \to K^m$ is the linear function defined by $F_A(X) = AX$. Since $r = \mathrm{rk}(A) = \dim(\mathrm{Im}\, F_A)$, from the Proposition 2.1.11 it follows that

$$\dim \Sigma_{\mathbf{0}} = n - r.$$

If $r = n$ we have $\Sigma_{\mathbf{0}} = \{\mathbf{0}\}$ hence S has only one solution $\overline{X}$. If $r < n$ the vector subspace $\Sigma_{\mathbf{0}}$ has dimension $n - r$. This means that its elements can be written as linear combinations of $n - r$ basis vectors. In such a linear combination, therefore, there are $n - r$ coefficients which can assume any value in the field K. □

2.3.1 Solving a linear system: the elimination method

The method for solving a linear system we will now describe, known as *Gauss' elimination method*, is based on the observation that certain algebraic manipulations, such as exchanging two equations with each other, multiplying both sides of an equation by the same non-zero constant, adding or subtracting two equations or, more generally, adding to an equation a multiple of another, transform a given linear system into an equivalent one, i.e., into a new system having the same solutions as the previous one.

The idea is therefore to use the operations described above (also known as *elementary operations*) to transform a system of linear equations into increasingly simpler equivalent systems. We will now try to briefly describe how this idea can actually be implemented.

Let us consider a system of m linear equations with n unknowns

$$S : \begin{cases} a_{11}\,x_1 + a_{12}\,x_2 + \cdots + a_{1n}\,x_n = b_1 \\ a_{21}\,x_1 + a_{22}\,x_2 + \cdots + a_{2n}\,x_n = b_2 \\ \cdot \quad \cdot \quad \cdot \quad \cdot \quad \cdot \quad \cdot \quad \cdot \quad \cdot \quad \cdot \quad \cdot \\ a_{m1}\,x_1 + a_{m2}\,x_2 + \cdots + a_{mn}\,x_n = b_m \end{cases}$$

We denote by $A = (a_{ij})$ the matrix of the coefficients and by $(A \,|\, B)$ the augmented matrix of the system, obtained by adding to A the column B of constant terms.

If the first column of the matrix A is entirely zero, the unknown x_1 does not actually appear in the system S. In this case we move on to the next column (i.e., to the next unknown). Otherwise we choose a row of A in which the coefficient of the unknown x_1 is different from zero. Suppose it is the i-th row: therefore $a_{i1} \neq 0$. We can then divide both sides of the i-th equation by a_{i1}, and then swap the i-th equation with the first. Thus we obtain a new system, equivalent to the given one, in which the augmented matrix is

$$\left(\begin{array}{cccc|c} 1 & a'_{12} & \cdots & a'_{1n} & b'_1 \\ a'_{21} & a'_{22} & \cdots & a'_{2n} & b'_2 \\ \vdots & \vdots & \ddots & \vdots & \vdots \\ a'_{m1} & a'_{m2} & \cdots & a'_{mn} & b'_m \end{array}\right)$$

Now, for each $i \geq 2$, we replace the i-th row of this matrix (i.e., the i-th equation of the system) with the sum of this row and the first row multiplied by $-a'_{i1}$, thus obtaining a new system whose augmented matrix is

$$\left(\begin{array}{cccc|c} 1 & a'_{12} & \dots & a'_{1n} & b'_1 \\ 0 & a''_{22} & \dots & a''_{2n} & b''_2 \\ \vdots & \vdots & \ddots & \vdots & \vdots \\ 0 & a''_{m2} & \dots & a''_{mn} & b''_m \end{array}\right)$$

We denote by S' the subsystem obtained by neglecting the first equation. In the system S' the unknown x_1 no longer appears. We can now repeat the above procedure for the system S'.

At the end of this procedure we will obtain a system, equivalent to the initial one, whose augmented matrix is in the so-called "echelon form," i.e., a form in which, in each row, the first non-zero element (in our specific case this element is equal to 1) is located to the right of the first non-zero element of the previous row.

Now the system can be easily solved starting from the last equation through a simple backward substitution, analogous to the one used in the substitution method.

Let us now clarify the algorithm just described by means of some concrete examples.

Example 1. (System with no solutions) Let us consider the following system of linear equations, with coefficients in $\mathbb{Q}$.

$$S : \begin{cases} 2x_1 - 4x_2 = -4 \\ 3x_1 - 6x_2 + 3x_3 = -3 \\ x_1 - 2x_2 - x_3 = -2 \end{cases}$$

The augmented matrix is

$$\left(\begin{array}{ccc|c} 2 & -4 & 0 & -4 \\ 3 & -6 & 3 & -3 \\ 1 & -2 & -1 & -2 \end{array}\right)$$

To make an element equal to 1 appear in the position $(1, 1)$ of the matrix there are three different possibilities: divide the first row by 2, divide the second row by 3 and swap it with the first, or swap the first and third row. We choose the latter, obtaining the matrix

$$\left(\begin{array}{ccc|c} 1 & -2 & -1 & -2 \\ 3 & -6 & 3 & -3 \\ 2 & -4 & 0 & -4 \end{array}\right)$$

Now we need to make zeros appear in the first column, below the first element. To do this we add the first row multiplied by -3 to the second row, and then we add the first row multiplied by -2 to the third row. The resulting matrix is

$$\left(\begin{array}{ccc|c} 1 & -2 & -1 & -2 \\ 0 & 0 & 6 & 3 \\ 0 & 0 & 2 & 0 \end{array}\right)$$

Now we start again on the second row, dividing it by 6 so that its first nonzero element is 1.

$$\left(\begin{array}{ccc|c} 1 & -2 & -1 & -2 \\ 0 & 0 & 1 & 1/2 \\ 0 & 0 & 2 & 0 \end{array}\right)$$

To the third row we add the second row multiplied by -2, obtaining the matrix

$$\left(\begin{array}{ccc|c} 1 & -2 & -1 & -2 \\ 0 & 0 & 1 & 1/2 \\ 0 & 0 & 0 & -1 \end{array}\right)$$

This matrix is finally in the form we want (the echelon form). It corresponds to the system

$$S' : \begin{cases} x_1 - 2x_2 - x_3 = -2 \\ x_3 = 1/2 \\ 0 = -1 \end{cases}$$

which is equivalent to the original system. Since the third equation has been reduced to the equality $0 = -1$, which is false, we conclude that this system has no solutions.

Example 2. (System with one solution) Let us consider the following system of linear equations, with coefficients in $\mathbb{Q}$.

$$S : \begin{cases} 2x_1 - 2x_2 + 8x_3 = 5 \\ 2x_2 + 6x_3 = 1 \\ x_1 - 2x_2 + 4x_3 = -1 \\ x_1 + 10x_3 = 0 \end{cases}$$

The augmented matrix is

$$\left(\begin{array}{ccc|c} 2 & -2 & 8 & 5 \\ 0 & 2 & 6 & 1 \\ 1 & -2 & 4 & -1 \\ 1 & 0 & 10 & 0 \end{array}\right)$$

Let's swap the first and third rows:

$$\left(\begin{array}{ccc|c} 1 & -2 & 4 & -1 \\ 0 & 2 & 6 & 1 \\ 2 & -2 & 8 & 5 \\ 1 & 0 & 10 & 0 \end{array}\right)$$

To the third row we add the first row multiplied by -2 and to the fourth row we subtract the first row:

$$\left(\begin{array}{ccc|c} 1 & -2 & 4 & -1 \\ 0 & 2 & 6 & 1 \\ 0 & 2 & 0 & 7 \\ 0 & 2 & 6 & 1 \end{array}\right)$$

Now we divide the second row by 2:

$$\left(\begin{array}{ccc|c} 1 & -2 & 4 & -1 \\ 0 & 1 & 3 & 1/2 \\ 0 & 2 & 0 & 7 \\ 0 & 2 & 6 & 1 \end{array}\right)$$

We continue by adding the second row multiplied by -2 to the third row and by adding the second row multiplied by -2 to the fourth row. We get:

$$\left(\begin{array}{ccc|c} 1 & -2 & 4 & -1 \\ 0 & 1 & 3 & 1/2 \\ 0 & 0 & -6 & 6 \\ 0 & 0 & 0 & 0 \end{array}\right)$$

Now we divide the third row by -6:

$$\left(\begin{array}{ccc|c} 1 & -2 & 4 & -1 \\ 0 & 1 & 3 & 1/2 \\ 0 & 0 & 1 & -1 \\ 0 & 0 & 0 & 0 \end{array}\right)$$

This matrix is in the echelon form. It corresponds to the system

$$S' : \begin{cases} x_1 - 2x_2 + 4x_3 = -1 \\ \qquad\quad x_2 + 3x_3 = 1/2 \\ \qquad\qquad\quad x_3 = -1 \\ \qquad\qquad\quad 0 = 0 \end{cases}$$

which is equivalent to the original system. The solution of this system can be easily obtained, starting from the last equation, by means of a backward substitution: from the third equation we obtain $x_3 = -1$, then by replacing this value in the second equation we get $x_2 = 1/2 - 3x_3 = 7/2$. Finally, by replacing in the first equation the values of x_2 and x_3, we find $x_1 = -1 + 2x_2 - 4x_3 = 10$. We conclude that the solution of the system is

$$\begin{cases} x_1 = 10 \\ x_2 = 7/2 \\ x_3 = -1 \end{cases}$$

Example 3. (System with infinite solutions) Let us consider the following system of linear equations, with coefficients in $\mathbb{Q}$.

$$S : \begin{cases} 2x_1 + 5x_3 = 1 \\ 4x_1 - 3x_2 + 4x_3 = 5 \\ 2x_1 - x_2 + 3x_3 = 2 \end{cases}$$

The augmented matrix is

$$\left(\begin{array}{ccc|c} 2 & 0 & 5 & 1 \\ 4 & -3 & 4 & 5 \\ 2 & -1 & 3 & 2 \end{array}\right)$$

We divide the first row by 2:

$$\left(\begin{array}{ccc|c} 1 & 0 & 5/2 & 1/2 \\ 4 & -3 & 4 & 5 \\ 2 & -1 & 3 & 2 \end{array}\right)$$

To the second row we add the first row multiplied by -4 and to the third row we add the first row multiplied by -2:

$$\left(\begin{array}{ccc|c} 1 & 0 & 5/2 & 1/2 \\ 0 & -3 & -6 & 3 \\ 0 & -1 & -2 & 1 \end{array}\right)$$

Now we divide the second row by -3:

$$\left(\begin{array}{ccc|c} 1 & 0 & 5/2 & 1/2 \\ 0 & 1 & 2 & -1 \\ 0 & -1 & -2 & 1 \end{array}\right)$$

Finally we add the second row to the third row:

$$\left(\begin{array}{ccc|c} 1 & 0 & 5/2 & 1/2 \\ 0 & 1 & 2 & -1 \\ 0 & 0 & 0 & 0 \end{array}\right)$$

This matrix is in the echelon form. It corresponds to the system

$$S' : \begin{cases} x_1 + 5x_3/2 = 1/2 \\ x_2 + 2x_3 = -1 \\ 0 = 0 \end{cases}$$

which is equivalent to the original system. We note that this system has infinite solutions. In fact, from the second equation we obtain $x_2 = -1 - 2x_3$ and from the first equation we get $x_1 = \frac{1}{2} - \frac{5}{2}x_3$. The unknown x_3 is indeterminate and can have any value. In conclusion, the given system admits infinitely many solutions, depending on one parameter (∞^1 solutions), which can be expressed as follows:

$$\begin{cases} x_1 = 1/2 - 5x_3/2 \\ x_2 = -1 - 2x_3 \\ x_3 \text{ any value.} \end{cases}$$

2.3.2 Computation of the rank of a matrix

Let A be an $m \times n$ matrix with entries in the field K. Recall that the rank of A has been defined as the maximum number of linearly independent columns of the matrix A (see Definition 2.2.15). This number is the dimension of the image of the linear function $F_A\colon K^n \to K^m$ whose matrix, with respect to the canonical bases of K^n and K^m, is A.

In a completely analogous way, we can define the *row rank* of A as the maximum number of linearly independent rows, i.e., as the dimension of the vector subspace of K^n generated by the rows of A. At first glance it might seem that there is no reason why the column rank of a matrix should match its row rank. However, these two numbers always turn out to be equal, as we will prove later (see Chapter 3, Theorem 3.3.1).

We will now see how the Gauss' elimination method provides a very useful tool for the computation of the row rank of a matrix. This follows from the fact that the elementary operations on the rows of a matrix used in Gauss' method do not alter the number of linearly independent rows. Consequently, if we apply Gaussian elimination to transform a matrix A into a matrix A' in echelon form, the row rank of A' will necessarily be equal to the row rank of A. At this point, the computation of the rank is immediate. In fact, the following result holds:

Proposition 2.3.6. *If a matrix A is in the echelon form, its row rank is equal to the number of nonzero rows.*

Proof. Suppose that A has m rows, of which the first r are nonzero. Let $v_1, v_2, \ldots, v_r$ be the nonzero rows of A, where $v_i = (a_{i1}, a_{i2}, \ldots, a_{in})$. Since A is in the echelon form, the first nonzero element of each row v_i is located to the right of the first nonzero element of the previous row v_{i-1}. Hence, if a_{1h_1} is the first nonzero element of the row v_1, all the other elements of the column h_1 of the matrix A, i.e., the elements a_{jh_1} with $j \geq 2$, are zero. We now consider a linear combination of the vectors $v_1, v_2, \ldots, v_r$:

$$\lambda_1 v_1 + \lambda_2 v_2 + \cdots + \lambda_r v_r = \mathbf{0}.$$

If we focus our attention on the components in the position h_1, we find

$$\lambda_1 a_{1h_1} + \lambda_2 0 + \cdots + \lambda_r 0 = 0$$

hence $\lambda_1 = 0$.

Similarly we denote by a_{2h_2} the first nonzero element of the row v_2. Then all the elements of the column h_2, from the third row on, i.e., the elements a_{jh_2} with $j \geq 3$, are zero. From the equality

$$\lambda_2 v_2 + \cdots + \lambda_r v_r = \mathbf{0}$$

by looking only at the components in the position h_2, we get

$$\lambda_2 a_{2h_2} + \lambda_3 0 + \cdots + \lambda_r 0 = 0$$

hence $\lambda_2 = 0$. By repeating the reasoning described above, it can thus be shown that all the coefficients $\lambda_1, \lambda_2, \ldots, \lambda_r$ are zero. This proves that the vectors $v_1, v_2, \ldots, v_r$ are linearly independent, hence the number r of nonzero rows of A is equal to the number of linearly independent rows, i.e., the rank of A. □

Example 2.3.7. We illustrate the method just described on a concrete example. We want to compute the (row) rank of the following matrix:

$$A = \begin{pmatrix} 0 & 2 & -1 & 6 & 1 \\ 3 & -4 & 14 & 3 & 2 \\ 1 & -2 & 5 & 3 & 3 \\ 1 & 0 & 4 & 3 & 0 \end{pmatrix}$$

Let's swap the first row with the third:

$$\begin{pmatrix} 1 & -2 & 5 & 3 & 3 \\ 3 & -4 & 14 & 3 & 2 \\ 0 & 2 & -1 & 6 & 1 \\ 1 & 0 & 4 & 3 & 0 \end{pmatrix}$$

To the second row we add the first multiplied by -3 and to the fourth row we subtract the first:

$$\begin{pmatrix} 1 & -2 & 5 & 3 & 3 \\ 0 & 2 & -1 & -6 & -7 \\ 0 & 2 & -1 & 6 & 1 \\ 0 & 2 & -1 & 0 & -3 \end{pmatrix}$$

From the third row we subtract the second and from the fourth row we subtract the second:

$$\begin{pmatrix} 1 & -2 & 5 & 3 & 3 \\ 0 & 2 & -1 & -6 & -7 \\ 0 & 0 & 0 & 12 & 8 \\ 0 & 0 & 0 & 6 & 4 \end{pmatrix}$$

Finally, to the fourth row we add the third multiplied by $-1/2$:

$$\begin{pmatrix} 1 & -2 & 5 & 3 & 3 \\ 0 & 2 & -1 & -6 & -7 \\ 0 & 0 & 0 & 12 & 8 \\ 0 & 0 & 0 & 0 & 0 \end{pmatrix}$$

This matrix is in the echelon form. Since there are three nonzero rows, we conclude that its (row) rank is 3. This is then also the rank of the matrix A we started from.

As a further exercise, the reader is advised to apply the algorithm just described to the transpose matrix of A, in order to verify that the column rank is also equal to 3.

Remark 2.3.8. To compute the column rank of the matrix A, it suffices to apply the Gaussian elimination method to the transpose of A. Alternatively, Gauss' method can be modified so as to perform the elementary operations on the columns of A rather than on its rows. The goal is to bring the given matrix into a suitable "echelon form," which is the transpose of the echelon form described above. That is, it is an echelon form in which the first nonzero element of each *column* is located below the first nonzero element of the previous column.

2.3.3 Computation of the inverse matrix

We now want to describe an algorithm, derived from the Gaussian elimination, for computing the inverse of a square matrix. This algorithm is based on the observation that performing elementary operations on the rows of a matrix A is equivalent to multiplying A, on the left, by a suitable invertible matrix.

Let us consider, for example, the operation which consists in exchanging two rows of the matrix A. For every $i, j = 1, \ldots, n$, with $i \neq j$, we denote by $P(i, j)$ the matrix whose elements p_{hk} $(1 \leq h, k \leq n)$ are:

$$p_{hk} = \begin{cases} 1 & \text{if } h = k,\ h \neq i,\ h \neq j, \\ 1 & \text{if } h = i \text{ and } k = j \text{ or if } h = j \text{ and } k = i, \\ 0 & \text{otherwise.} \end{cases}$$

If A is an $n \times n$ matrix and we set $A' = P(i, j)A$, it is easy to verify that the matrix A' is obtained from A by swapping the i-th and j-th rows. For example, if $n = 4$, $i = 1$, $j = 3$, we have

$$P(1,3) = \begin{pmatrix} 0 & 0 & 1 & 0 \\ 0 & 1 & 0 & 0 \\ 1 & 0 & 0 & 0 \\ 0 & 0 & 0 & 1 \end{pmatrix}$$

and

$$\begin{pmatrix} 0 & 0 & 1 & 0 \\ 0 & 1 & 0 & 0 \\ 1 & 0 & 0 & 0 \\ 0 & 0 & 0 & 1 \end{pmatrix} \cdot \begin{pmatrix} a_{11} & a_{12} & a_{13} & a_{14} \\ a_{21} & a_{22} & a_{23} & a_{24} \\ a_{31} & a_{32} & a_{33} & a_{34} \\ a_{41} & a_{42} & a_{43} & a_{44} \end{pmatrix} = \begin{pmatrix} a_{31} & a_{32} & a_{33} & a_{34} \\ a_{21} & a_{22} & a_{23} & a_{24} \\ a_{11} & a_{12} & a_{13} & a_{14} \\ a_{41} & a_{42} & a_{43} & a_{44} \end{pmatrix}$$

From what has just been said, it follows immediately that

$$P(i, j)^2 = P(i, j)P(i, j) = \mathbf{1},$$

hence all the matrices $P(i, j)$ are invertible.

Now let us consider the matrix $M(i; \lambda) = (m_{hk})$ defined, for every $i = 1, \ldots, n$ and any $\lambda \in K$, $\lambda \neq 0$, by setting

$$m_{hk} = \begin{cases} 1 & \text{if } h = k,\ h \neq i, \\ \lambda & \text{if } h = k = i, \\ 0 & \text{otherwise.} \end{cases}$$

The multiplication (on the left) of $M(i;\lambda)$ and another matrix A has the effect of multiplying the i-th row of A by λ. For example, if $n=4$ and $i=2$, we have

$$M(2;\lambda)=\begin{pmatrix}1&0&0&0\\0&\lambda&0&0\\0&0&1&0\\0&0&0&1\end{pmatrix}$$

and

$$\begin{pmatrix}1&0&0&0\\0&\lambda&0&0\\0&0&1&0\\0&0&0&1\end{pmatrix}\cdot\begin{pmatrix}a_{11}&a_{12}&a_{13}&a_{14}\\a_{21}&a_{22}&a_{23}&a_{24}\\a_{31}&a_{32}&a_{33}&a_{34}\\a_{41}&a_{42}&a_{43}&a_{44}\end{pmatrix}=\begin{pmatrix}a_{11}&a_{12}&a_{13}&a_{14}\\\lambda a_{21}&\lambda a_{22}&\lambda a_{23}&\lambda a_{24}\\a_{31}&a_{32}&a_{33}&a_{34}\\a_{41}&a_{42}&a_{43}&a_{44}\end{pmatrix}$$

Note that, since we assumed $\lambda\neq 0$, we have $M(i;\lambda)M(i;\lambda^{-1})=\mathbf{1}$, hence the matrices $M(i;\lambda)$ are invertible.

Finally, for every $i,j=1,\ldots,n$, with $i\neq j$, and any $\alpha\in K$, we define the matrix $S(i,j;\alpha)=(s_{hk})$ by setting

$$s_{hk}=\begin{cases}1 & \text{if } h=k,\\ \alpha & \text{if } h=i \text{ and } k=j,\\ 0 & \text{otherwise.}\end{cases}$$

If A is an $n\times n$ matrix and we set $A'=S(i,j;\alpha)A$, it is easy to verify that the matrix A' is obtained from A by adding to the i-th row the j-th row multiplied by α. For example, if $n=4$, $i=2$, $j=4$, we have

$$S(2,4;\alpha)=\begin{pmatrix}1&0&0&0\\0&1&0&\alpha\\0&0&1&0\\0&0&0&1\end{pmatrix}$$

and

$$\begin{pmatrix}1&0&0&0\\0&1&0&\alpha\\0&0&1&0\\0&0&0&1\end{pmatrix}\cdot\begin{pmatrix}a_{11}&a_{12}&a_{13}&a_{14}\\a_{21}&a_{22}&a_{23}&a_{24}\\a_{31}&a_{32}&a_{33}&a_{34}\\a_{41}&a_{42}&a_{43}&a_{44}\end{pmatrix}=$$

$$=\begin{pmatrix}a_{11}&a_{12}&a_{13}&a_{14}\\a_{21}+\alpha a_{41}&a_{22}+\alpha a_{42}&a_{23}+\alpha a_{43}&a_{24}+\alpha a_{44}\\a_{31}&a_{32}&a_{33}&a_{34}\\a_{41}&a_{42}&a_{43}&a_{44}\end{pmatrix}$$

It follows that

$$S(i,j;\alpha)S(i,j;-\alpha) = \mathbf{1},$$

hence all the matrices $S(i,j;\alpha)$ are invertible.

A sequence of elementary operations on the rows of a matrix A is therefore equivalent to a sequence of multiplications, on the left, by invertible matrices of the types previously described. Since the product of any number of invertible matrices is still an invertible matrix, we conclude that the effect of any number of elementary operations on the rows of a matrix A can be obtained simply by multiplying the matrix A, on the left, by a suitable invertible matrix.

Remark 2.3.9. It is easy to verify that multiplying a matrix A *on the right* by the matrices $P(i,j)$, $M(i;\lambda)$ and $S(i,j;\alpha)$ previously described is equivalent to perform elementary operations on the *columns* of A. More precisely, if we set $A' = AP(i,j)$, the matrix A' is obtained by swapping the i-th and j-th columns of A, if $A' = AM(i;\lambda)$ then A' is obtained by multiplying the i-th column of A by λ, and if $A' = AS(i,j;\alpha)$ then A' is obtained from A by adding to the j-th column the i-th column multiplied by α. Therefore, the effect of any number of elementary operations on the columns of A can be obtained by multiplying the matrix A, on the right, by a suitable invertible matrix.

Suppose now that the square matrix A, of order n, is invertible. Using elementary operations on the rows it is possible to transform A into a matrix in the echelon form, in which the first nonzero element of each row can be made equal to 1. Since A is invertible,[2] its rank must be n, hence in the echelon form there are no rows consisting only of zeros. Since A is a square matrix of order n, this is equivalent to saying that the echelon form we obtain after applying Gauss' elimination algorithm is an upper triangular matrix A', with all elements on the main diagonal equal to 1.

At this point it is easy to convince oneself that, by means of suitable elementary operations on the rows of the matrix A', it is possible to transform this matrix into the identity matrix $\mathbf{1}$. If we denote by B the matrix that represents the effect of all the elementary operations on the rows that we have performed to transform the matrix A into the identity matrix, we have $BA = \mathbf{1}$. The matrix B is therefore the inverse of the matrix A. It can be

[2]Recall that A is invertible if and only if the corresponding linear function $F_A \colon K^n \to K^n$ is an isomorphism, which means that $\operatorname{Im} F_A = K^n$, hence $\operatorname{rk}(A) = \operatorname{rk}(F_A) = n$.

determined by scrupulously writing down all the matrices corresponding to the elementary operations on the rows which have been carried out.

A much more effective method to determine the matrix B is the following. Next to the matrix A, we write the identity matrix $\mathbf{1}$, so as to obtain a matrix with n rows and $2n$ columns which we will denote by $(A\,|\,\mathbf{1})$. In this way, all the operations that we perform on the rows of A must also be performed on the rows of the matrix $\mathbf{1}$. The effect of these elementary operations is equivalent to the multiplication on the left by the (unknown) matrix B. We therefore have:

$$B(A\,|\,\mathbf{1}) = (BA\,|\,B\mathbf{1}) = (\mathbf{1}\,|\,B).$$

This means that when we have transformed the matrix A into the identity matrix, the matrix $\mathbf{1}$ written to the right of A will have been automatically transformed into the matrix B, which is the inverse of A.

As an example, let us apply the algorithm just described to determine the inverse of the matrix

$$A = \begin{pmatrix} 1 & 2 & 1 \\ 1 & 3 & 0 \\ 1 & 2 & 2 \end{pmatrix}$$

First we write the identity matrix next to the matrix A, obtaining

$$\left(\begin{array}{ccc|ccc} 1 & 2 & 1 & 1 & 0 & 0 \\ 1 & 3 & 0 & 0 & 1 & 0 \\ 1 & 2 & 2 & 0 & 0 & 1 \end{array}\right)$$

Now, using elementary operations on the rows, we try to transform the matrix A into the identity matrix. If we succeed in doing this, the matrix that we will find on the right will be the inverse of A.

In detail, the operations to be performed are, for example, the following: we subtract the first row from the second row, and the first row from the third row, obtaining

$$\left(\begin{array}{ccc|ccc} 1 & 2 & 1 & 1 & 0 & 0 \\ 0 & 1 & -1 & -1 & 1 & 0 \\ 0 & 0 & 1 & -1 & 0 & 1 \end{array}\right)$$

Now we add the third row to the second row, and we subtract the third row from the first row, obtaining

$$\left(\begin{array}{ccc|ccc} 1 & 2 & 0 & 2 & 0 & -1 \\ 0 & 1 & 0 & -2 & 1 & 1 \\ 0 & 0 & 1 & -1 & 0 & 1 \end{array}\right)$$

Now we add to the first row the second row multiplied by -2, obtaining

$$\left(\begin{array}{ccc|ccc} 1 & 0 & 0 & 6 & -2 & -3 \\ 0 & 1 & 0 & -2 & 1 & 1 \\ 0 & 0 & 1 & -1 & 0 & 1 \end{array}\right)$$

Hence we have

$$A^{-1} = \begin{pmatrix} 6 & -2 & -3 \\ -2 & 1 & 1 \\ -1 & 0 & 1 \end{pmatrix}$$

As an exercise, verify that the matrix just found is actually the inverse of A, i.e., that $AA^{-1} = \mathbf{1}$.

Remark 2.3.10. If the matrix A, whose inverse we want to compute, were not invertible it would have rank strictly smaller than n, therefore its echelon form A', obtained in the first part of the algorithm, would have at least one row containing only zeros. We would thus realize that A is not invertible.

Exercises

Exercise 2.3.1. Solve the exercises proposed at the end of Section 1.1 of Chap. 1 by using Gauss' elimination method.

Exercise 2.3.2. Compute the inverse of

$$A = \begin{pmatrix} 2 & -3 & -1 \\ 2 & -1 & -3 \\ 1 & -3 & -1 \end{pmatrix}$$

Exercise 2.3.3. Compute the inverse of

$$A = \begin{pmatrix} 2 & 1 & -1 & 0 \\ -1 & 0 & 3 & 2 \\ 1 & 1 & -3 & -1 \\ 0 & 3 & 2 & 4 \end{pmatrix}$$

3

Determinants

In this chapter we will define the notion of determinant of a square matrix with coefficients in a field and we will study its main properties. Before doing that, we need to recall some elementary properties of permutations of a finite set of elements.

3.1 Permutations

Definition 3.1.1. Let $A = \{a_1, a_2, \ldots, a_n\}$ be a set of n elements. A *permutation* of the elements of A is a bijective function $\sigma \colon A \to A$.

The permutation σ can be conveniently represented with the following table

$$\begin{pmatrix} a_1 & a_2 & a_3 & \ldots & a_n \\ a_{j_1} & a_{j_2} & a_{j_3} & \ldots & a_{j_n} \end{pmatrix}$$

where we set $\sigma(a_i) = a_{j_i}$, for $i = 1, \ldots, n$.

For example, if $n = 4$ and σ is the permutation defined by $\sigma(a_1) = a_3$, $\sigma(a_2) = a_2$, $\sigma(a_3) = a_4$ and $\sigma(a_4) = a_1$, the corresponding table is

$$\begin{pmatrix} a_1 & a_2 & a_3 & a_4 \\ a_3 & a_2 & a_4 & a_1 \end{pmatrix}$$

In such a representation the presence of the symbol a is, in fact, completely superfluous; it is more convenient to represent σ using the table

$$\begin{pmatrix} 1 & 2 & 3 & \ldots & n \\ j_1 & j_2 & j_3 & \ldots & j_n \end{pmatrix}.$$

We denote by $\mathfrak{S}_n$ the set of all permutations of n elements. We observe that the composition of two permutations is still a permutation and the same is true for the inverse of a permutation. It is straightforward to verify that the composition of permutations defines a group structure on $\mathfrak{S}_n$. We also note that, if $n \geq 3$ and $\sigma, \tau \in \mathfrak{S}_n$, in general we have $\sigma \circ \tau \neq \tau \circ \sigma$, hence $\mathfrak{S}_n$ is a non-commutative group.

Proposition 3.1.2. *The cardinality of $\mathfrak{S}_n$ is $n!$.*

Proof. To define a permutation $\sigma \in \mathfrak{S}_n$ we must specify $\sigma(i)$ for every $i = 1, \ldots, n$. For $\sigma(1)$ there are n possible values, while for $\sigma(2)$ there are only $n-1$ possible values, because $\sigma(2) \neq \sigma(1)$. Similarly for $\sigma(3)$ there are only $n-2$ possible values because $\sigma(3) \neq \sigma(1)$ and $\sigma(3) \neq \sigma(2)$. We deduce that the possible values of $\sigma(j)$ are $n - j + 1$, for every $j = 1, \ldots, n$. It follows that the number of possible permutations is given by the product $n(n-1)(n-2)\cdots 2 \cdot 1 = n!$. □

Definition 3.1.3. Let $\sigma \in \mathfrak{S}_n$. We say that in σ there is an *inversion* whenever we have $i < j$ but $\sigma(i) > \sigma(j)$.

For example, in the permutation

$$\sigma = \begin{pmatrix} 1 & 2 & 3 & 4 & 5 \\ 3 & 2 & 5 & 1 & 4 \end{pmatrix}$$

there are 5 inversions: in fact we have $1 < 2$ but $\sigma(1) = 3 > \sigma(2) = 2$, $1 < 4$ but $\sigma(1) = 3 > \sigma(4) = 1$, $2 < 4$ but $\sigma(2) = 2 > \sigma(4) = 1$, $3 < 4$ but $\sigma(3) = 5 > \sigma(4) = 1$ and $3 < 5$ but $\sigma(3) = 5 > \sigma(5) = 4$. Note that to count the number of inversions in σ it suffices to observe that in the second row of the table the number 3 appears before the numbers 2 and 1 (two inversions), 2 appears before 1 (another inversion) and finally 5 appears before 1 and 4 (two more inversions).

Definition 3.1.4. We say that $\sigma \in \mathfrak{S}_n$ is an *even* (resp., *odd*) permutation if the number of inversions in σ is even (resp., odd). We define the *sign* of σ, denoted by $\operatorname{sgn}(\sigma)$, by setting

$$\operatorname{sgn}(\sigma) = \begin{cases} 1 & \text{if } \sigma \text{ is even,} \\ -1 & \text{if } \sigma \text{ is odd.} \end{cases}$$

We can also write

$$\operatorname{sgn}(\sigma) = (-1)^{i(\sigma)},$$

where $i(\sigma)$ is the number of inversions in σ.

It is a completely elementary fact that any permutation $\sigma \in \mathfrak{S}_n$ can be written as the product (i.e., as the composition) of a finite number of disjoint cycles. For example, the permutation

$$\sigma = \begin{pmatrix} 1 & 2 & 3 & 4 & 5 & 6 & 7 & 8 \\ 5 & 6 & 8 & 4 & 3 & 7 & 2 & 1 \end{pmatrix}$$

can be written as follows:

$$\sigma = (1\,5\,3\,8)(2\,6\,7)(4).$$

To express a given permutation σ as a product of disjoint cycles, it suffices to see where σ sends the number 1, then where it sends $i_1 = \sigma(1)$, then where it sends $i_2 = \sigma(i_1)$, etc.

Note that a cycle of length 1 (like (4) in the previous example) corresponds to the identical permutation, so it can be omitted. We conclude that every permutation can be written as a product of disjoint cycles of length ≥ 2.

Definition 3.1.7. A cycle of length 2 (or 2-cycle) is called a *transposition*.

It is easy to verify that each r-cycle can be written (not necessarily uniquely) as a product of $r - 1$ transpositions. In fact we have

$$(i_1\, i_2\, \dots\, i_r) = (i_1\, i_r)(i_1\, i_{r-1}) \cdots (i_1\, i_3)(i_1\, i_2), \tag{3.1.2}$$

or also

$$(i_1\, i_2\, \dots\, i_r) = (i_{r-1}\, i_r)(i_{r-2}\, i_r) \cdots (i_2\, i_r)(i_1\, i_r). \tag{3.1.3}$$

It should be remembered here that, in the composition product of two or more permutations, the following convention is used: $(\sigma \circ \tau)(i) = \sigma(\tau(i))$.

Since every permutation is a product of a finite number of disjoint cycles, and since every cycle is a finite product of transpositions, it follows that every permutation can always be written as a product of a finite number of transpositions.

Let us consider now the transposition $\sigma = (i\,j) \in \mathfrak{S}_n$, with $i < j$. It is easy to verify that in this permutation there are $2(j-i)-1$ inversions (this is left as an exercise), hence its sign is

$$\operatorname{sgn}(\sigma) = (-1)^{2(j-i)-1} = -1.$$

Each transposition has the sign -1, therefore if a permutation is expressed as a product of r transpositions, its sign is $(-1)^r$. From (3.1.2) and (3.1.3) we deduce the following result:

Proposition 3.1.8. *If σ is an r-cycle then* $\operatorname{sgn}(\sigma) = (-1)^{r-1}$.

Proof. We have seen that an r-cycle $(i_1\, i_2\, \dots\, i_r)$ can be written as the product of $r-1$ transpositions. □

Corollary 3.1.9. *If a permutation σ is written as the product of s disjoint cycles of length r_1, r_2, ..., r_s, respectively, its sign is given by*

$$\operatorname{sgn}(\sigma) = (-1)^{r_1+r_2+\cdots+r_s-s}.$$

Proof. Since the sign of a cycle of length r is $(-1)^{r-1}$, the sign of σ is given by the following product:

$$(-1)^{r_1-1}(-1)^{r_2-1}\cdots(-1)^{r_s-1} = (-1)^{r_1+r_2+\cdots+r_s-s}.$$ □

Example 3.1.10. Consider the following permutation $\xi \in \mathfrak{S}_n$:

$$\xi = \begin{pmatrix} 1 & 2 & 3 & \dots & h & h+1 & \dots & n \\ h & 1 & 2 & \dots & h-1 & h+1 & \dots & n \end{pmatrix}$$

This permutation is a cycle of length h, in fact we have

$$\xi = (h \;\; h-1 \;\; h-2 \;\; \dots \;\; 2 \;\; 1).$$

It follows that $\operatorname{sgn}(\xi) = (-1)^{h-1}$.

Let us now consider a particular type of transposition.

Definition 3.1.11. We say that a permutation $\sigma \in \mathfrak{S}_n$ is an *exchange of contiguous elements* if $\sigma(i) = i+1$ and $\sigma(i+1) = i$, for some $i = 1, \dots, n-1$, and $\sigma(j) = j$, for every $j \neq i, i+1$. Such a permutation swaps two contiguous elements with each other and leaves all others fixed.

We conclude this section by proving the following result:

Proposition 3.1.12. *Every permutation $\sigma \in \mathfrak{S}_n$ can be obtained as the composition of a finite number (eventually zero)* of exchanges of contiguous elements.

Proof. Let σ be the permutation

$$\sigma = \begin{pmatrix} 1 & 2 & \dots & n-1 & n \\ j_1 & j_2 & \dots & j_{n-1} & j_n \end{pmatrix}$$

Starting from the identity

$$\begin{pmatrix} 1 & 2 & \dots & n-1 & n \\ 1 & 2 & \dots & n-1 & n \end{pmatrix}$$

and repeatedly swapping the number j_n with the next one, it is possible to put j_n in the last position, thus obtaining the permutation

$$\begin{pmatrix} 1 & 2 & \dots & n-1 & n \\ j'_1 & j'_2 & \dots & j'_{n-1} & j_n \end{pmatrix}$$

Now that the last element has been arranged, we can repeatedly swap the number j_{n-1} with the next one to get the permutation

$$\begin{pmatrix} 1 & 2 & \dots & n-1 & n \\ j''_1 & j''_2 & \dots & j_{n-1} & j_n \end{pmatrix}$$

Continuing in this way it is possible to arrange in the correct position the numbers j_{n-2}, j_{n-3}, etc., up to the number j_1, thus obtaining the permutation σ. □

Remark 3.1.13. Since the sign of an exchange of contiguous elements is -1, if a permutation σ is obtained as a composition of k exchanges, we have

$$\operatorname{sgn}(\sigma) = (-1)^k.$$

Thus a permutation is even (resp., odd) if and only if it can be obtained as a composition of an even (resp., odd) number of exchanges of contiguous elements.

Exercises

Exercise 3.1.1. Let's consider the permutations

$$\sigma = \begin{pmatrix} 1 & 2 & 3 & 4 & 5 \\ 3 & 5 & 1 & 4 & 2 \end{pmatrix}, \quad \tau = \begin{pmatrix} 1 & 2 & 3 & 4 & 5 \\ 5 & 2 & 1 & 4 & 3 \end{pmatrix}$$

Compute $\sigma \circ \tau$, $\tau \circ \sigma$, σ^{-1} and τ^{-1}.

Exercise 3.1.2. Determine the sign of the following permutations:

$$\sigma_1 = \begin{pmatrix} 1 & 2 & 3 \\ 3 & 2 & 1 \end{pmatrix}, \quad \sigma_2 = \begin{pmatrix} 1 & 2 & 3 & 4 \\ 4 & 2 & 1 & 3 \end{pmatrix}, \quad \sigma_3 = \begin{pmatrix} 1 & 2 & 3 & 4 & 5 \\ 3 & 2 & 5 & 4 & 1 \end{pmatrix}$$

Exercise 3.1.3. Write the permutation

$$\sigma = \begin{pmatrix} 1 & 2 & 3 & 4 & 5 & 6 & 7 & 8 \\ 5 & 3 & 8 & 2 & 6 & 1 & 7 & 4 \end{pmatrix}$$

as a product of disjoint cycles and as a product of transpositions, and compute its sign.

Exercise 3.1.4. Write the following permutations as a product of disjoint cycles:

$$\begin{pmatrix} 1 & 2 & 3 & 4 & 5 & 6 \\ 4 & 6 & 5 & 1 & 3 & 2 \end{pmatrix}, \quad \begin{pmatrix} 1 & 2 & 3 & 4 & 5 & 6 \\ 5 & 3 & 2 & 6 & 4 & 1 \end{pmatrix}, \quad \begin{pmatrix} 1 & 2 & 3 & 4 & 5 & 6 \\ 3 & 5 & 6 & 4 & 1 & 2 \end{pmatrix}$$

Exercise 3.1.5. We consider the permutations represented by the following products of cycles:

$$\begin{aligned} \sigma_1 &= (1\,2\,3\,4)(5\,6\,7)(2\,6\,1)(4\,7) \\ \sigma_2 &= (1\,2\,3\,4\,5)(6\,7)(1\,3\,5\,7)(1\,6\,3) \\ \sigma_3 &= (1\,4)(1\,2\,3)(4\,5)(1\,4) \end{aligned}$$

Write σ_1, σ_2 and σ_3 as a product of disjoint cycles.

Exercise 3.1.6. Write the permutation

$$\sigma = \begin{pmatrix} 1 & 2 & 3 & 4 \\ 2 & 4 & 1 & 3 \end{pmatrix}$$

as a composition of exchanges of contiguous elements and determine its sign.

3.2 The determinant of a square matrix

In this section we will define the determinant of a square matrix with entries in a field K and study its main properties.

Definition 3.2.1. Let $A = (a_{ij}) \in M_n(K)$ be a square matrix of order n with entries in a field K. The *determinant* of A is

$$\det A = \sum_{\sigma \in \mathfrak{S}_n} \operatorname{sgn}(\sigma)\, a_{1\sigma(1)} a_{2\sigma(2)} \cdots a_{n\sigma(n)},$$

where the sum is performed over all permutations of n elements $\sigma \in \mathfrak{S}_n$. The determinant of a matrix A is often denoted by $|A|$.

Remark 3.2.2. We note that the definition of determinant also applies to the case in which A is a square matrix with entries in a commutative ring R; in this case the determinant of A is an element of the ring R.

As an example, we apply this definition to explicitly compute the determinant of a matrix for small values of n.

If $n = 1$, i.e., if the matrix A reduces to a single element, $A = a_{11}$, we have $\det A = a_{11}$ (the only permutation of a single element is the identity, whose sign 1).

If $n = 2$, i.e., if

$$A = \begin{pmatrix} a_{11} & a_{12} \\ a_{21} & a_{22} \end{pmatrix}$$

there are two permutations of two elements,

$$\sigma_1 = \begin{pmatrix} 1 & 2 \\ 1 & 2 \end{pmatrix}, \qquad \sigma_2 = \begin{pmatrix} 1 & 2 \\ 2 & 1 \end{pmatrix}.$$

We have $\operatorname{sgn}(\sigma_1) = 1$ and $\operatorname{sgn}(\sigma_2) = -1$, hence

$$\det A = a_{11}a_{22} - a_{12}a_{21}.$$

Finally we consider the case $n = 3$. There are six permutations of three elements:

$$\sigma_1 = \begin{pmatrix} 1 & 2 & 3 \\ 1 & 2 & 3 \end{pmatrix}, \qquad \sigma_2 = \begin{pmatrix} 1 & 2 & 3 \\ 1 & 3 & 2 \end{pmatrix}, \qquad \sigma_3 = \begin{pmatrix} 1 & 2 & 3 \\ 2 & 3 & 1 \end{pmatrix},$$
$$\sigma_4 = \begin{pmatrix} 1 & 2 & 3 \\ 2 & 1 & 3 \end{pmatrix}, \qquad \sigma_5 = \begin{pmatrix} 1 & 2 & 3 \\ 3 & 1 & 2 \end{pmatrix}, \qquad \sigma_6 = \begin{pmatrix} 1 & 2 & 3 \\ 3 & 2 & 1 \end{pmatrix}.$$

It is easy to check that $\operatorname{sgn}(\sigma_1) = \operatorname{sgn}(\sigma_3) = \operatorname{sgn}(\sigma_5) = 1$, and $\operatorname{sgn}(\sigma_2) = \operatorname{sgn}(\sigma_4) = \operatorname{sgn}(\sigma_6) = -1$, hence

$$\begin{aligned} \det A = {} & a_{11}a_{22}a_{33} - a_{11}a_{23}a_{32} + a_{12}a_{23}a_{31} \\ & - a_{12}a_{21}a_{33} + a_{13}a_{21}a_{32} - a_{13}a_{22}a_{31}. \end{aligned}$$

Remark 3.2.3. A practical method to remember the previous formula for computing the determinant of a matrix of order 3 is the so-called *Sarrus rule.* We copy the first and second column of A to the right of the third column, as follows:

$$\begin{pmatrix} a_{11} & a_{12} & a_{13} & a_{11} & a_{12} \\ a_{21} & a_{22} & a_{23} & a_{21} & a_{22} \\ a_{31} & a_{32} & a_{33} & a_{31} & a_{32} \end{pmatrix}$$

Now we need to consider the sum of the products of the elements located along the three diagonals oriented from north-west to south-east

$$a_{11}a_{22}a_{33} + a_{12}a_{23}a_{31} + a_{13}a_{21}a_{32}$$

and subtract the sum of the products of the elements located along the three diagonals oriented from north-east to south-west

$$a_{13}a_{22}a_{31} + a_{11}a_{23}a_{32} + a_{12}a_{21}a_{33}.$$

We obtain

$$a_{11}a_{22}a_{33} + a_{12}a_{23}a_{31} + a_{13}a_{21}a_{32} - a_{13}a_{22}a_{31} - a_{11}a_{23}a_{32} - a_{12}a_{21}a_{33},$$

which is the determinant of A.

It should be noted that there is no analogue of the *Sarrus rule* for computing the determinant of a matrix of order greater than 3.

In general, remembering that the number of permutations of n elements is $n!$, the computation of the determinant of a square matrix A of order n consists of a sum of $n!$ addends, each of which is a product of n elements of A, taken one from each row and each column, with a sign given by the sign of the corresponding permutation.

Example 3.2.4. Let us try to determine how many "operations" are needed to compute the determinant of a matrix of order 50. The expansion of such a determinant consists of a sum of 50! addends, each of which is a product of 50 elements of the matrix (taken with the appropriate sign). Hence there are $50 \times 50!$ products, followed by 50! sums, that gives a total of $50 \times 50! + 50! = 50!\,(50 + 1) = 51!$ "operations" (and where we have neglected all the operations necessary to determine the sign of each of the 50! permutations). We note that $51! \approx 1.55 \times 10^{66}$.

Even if we had a computer capable of performing a trillion such "operations" per second, the time required to compute such a determinant would be approximately 1.55×10^{54} seconds, which equals approximately 4.9×10^{46} years, which corresponds to more than 3×10^{36} times the entire life of the universe!

For a matrix of order 30, the same reasoning leads to a time necessary for the computation of the determinant equal to about 17300 times the lifetime of the universe.

This example shows that the formula used to define the determinant is essentially unusable for the actual computation, except in cases where n is very small. However, this does not mean at all that such a formula is "useless." Indeed, it allows us to prove many remarkable properties of determinants.

Proposition 3.2.5. *Let*

$$A = \begin{pmatrix} a_{11} & 0 & 0 & \dots & 0 \\ 0 & a_{22} & 0 & \dots & 0 \\ 0 & 0 & a_{33} & \dots & 0 \\ \vdots & \vdots & \vdots & \ddots & \vdots \\ 0 & 0 & 0 & \dots & a_{nn} \end{pmatrix}$$

be a diagonal matrix. Then $\det A = a_{11}a_{22}\cdots a_{nn}$.

Proof. Since all elements outside the main diagonal are zero, the only nonzero product that we find in the formula for the determinant is $a_{11}a_{22}\cdots a_{nn}$, which corresponds to the identity permutation whose sign is 1. □

More generally, the following result holds:

Proposition 3.2.6. *Let*

$$A = \begin{pmatrix} a_{11} & a_{12} & a_{13} & \dots & a_{1n} \\ 0 & a_{22} & a_{23} & \dots & a_{2n} \\ 0 & 0 & a_{33} & \dots & a_{3n} \\ \vdots & \vdots & \vdots & \ddots & \vdots \\ 0 & 0 & 0 & \dots & a_{nn} \end{pmatrix}$$

be an upper triangular matrix. Then $\det A = a_{11}a_{22}\cdots a_{nn}$.

Proof. Since $A = (a_{ij})$ is upper triangular, we have $a_{ij} = 0$ if $i > j$. We note that for each permutation σ of $\{1, 2, \dots, n\}$, different from the identity, there exists at least one index i such that $i > \sigma(i)$. It follows that the product $a_{1\sigma(1)}a_{2\sigma(2)}\cdots a_{n\sigma(n)}$ is zero, because at least one of the factors is zero. This implies that the only permutation that gives a nonzero contribution to the computation of the determinant is the identity, hence we have $\det A = a_{11}a_{22}\cdots a_{nn}$. □

Of course, an analogous result (with an analogous proof) holds also for lower triangular matrices. This also follows from the following result:

Proposition 3.2.7. *Let* $A \in M_n(K)$ *and let* ${}^tA$ *be its transpose. Then*

$$\det({}^tA) = \det(A).$$

Proof. Let $\sigma \in \mathfrak{S}_n$ and let σ^{-1} be its inverse. Note that as σ varies among the elements of $\mathfrak{S}_n$ the same happens to its inverse. Let us denote by a_{ij} the elements of the matrix A and by $\tilde{a}_{ij}$ the elements of ${}^tA$. Recall that $\tilde{a}_{ij} = a_{ji}$. From the definition of determinant, we have

$$\begin{aligned}\det({}^tA) &= \sum_{\sigma \in \mathfrak{S}_n} \operatorname{sgn}(\sigma)\, \tilde{a}_{1\sigma(1)} \tilde{a}_{2\sigma(2)} \cdots \tilde{a}_{n\sigma(n)} \\ &= \sum_{\sigma \in \mathfrak{S}_n} \operatorname{sgn}(\sigma)\, a_{\sigma(1)1} a_{\sigma(2)2} \cdots a_{\sigma(n)n}.\end{aligned}$$

Now we observe that in each product $a_{\sigma(1)1} a_{\sigma(2)2} \cdots a_{\sigma(n)n}$ there are n elements, taken one from each row and each column of A. It follows that this product can also be written as $a_{1\sigma^{-1}(1)} a_{2\sigma^{-1}(2)} \cdots a_{n\sigma^{-1}(n)}$, i.e., we have

$$a_{\sigma(1)1} a_{\sigma(2)2} \cdots a_{\sigma(n)n} = a_{1\sigma^{-1}(1)} a_{2\sigma^{-1}(2)} \cdots a_{n\sigma^{-1}(n)},$$

for every $\sigma \in \mathfrak{S}_n$.

An example may serve to clarify what has just been stated. Let $n = 4$ and consider the permutation

$$\sigma = \begin{pmatrix} 1 & 2 & 3 & 4 \\ 3 & 1 & 4 & 2 \end{pmatrix}$$

whose inverse is

$$\sigma^{-1} = \begin{pmatrix} 1 & 2 & 3 & 4 \\ 2 & 4 & 1 & 3 \end{pmatrix}$$

Then we have

$$a_{\sigma(1)1} a_{\sigma(2)2} a_{\sigma(3)3} a_{\sigma(4)4} = a_{31} a_{12} a_{43} a_{24}$$

and

$$a_{1\sigma^{-1}(1)} a_{2\sigma^{-1}(2)} a_{3\sigma^{-1}(3)} a_{4\sigma^{-1}(4)} = a_{12} a_{24} a_{31} a_{43}$$

which are obviously the same.

If we also recall that $\operatorname{sgn}(\sigma^{-1}) = \operatorname{sgn}(\sigma)$, we can write:

$$\begin{aligned}\det({}^tA) &= \sum_{\sigma \in \mathfrak{S}_n} \operatorname{sgn}(\sigma)\, a_{\sigma(1)1} a_{\sigma(2)2} \cdots a_{\sigma(n)n} \\ &= \sum_{\sigma \in \mathfrak{S}_n} \operatorname{sgn}(\sigma)\, a_{1\sigma^{-1}(1)} a_{2\sigma^{-1}(2)} \cdots a_{n\sigma^{-1}(n)} \\ &= \sum_{\sigma \in \mathfrak{S}_n} \operatorname{sgn}(\sigma^{-1})\, a_{1\sigma^{-1}(1)} a_{2\sigma^{-1}(2)} \cdots a_{n\sigma^{-1}(n)} \\ &= \sum_{\tau \in \mathfrak{S}_n} \operatorname{sgn}(\tau)\, a_{1\tau(1)} a_{2\tau(2)} \cdots a_{n\tau(n)} = \det A,\end{aligned}$$

where we have set $\tau = \sigma^{-1}$. □

Given a matrix $A \in M_n(K)$ we denote by $A^{(1)}$, $A^{(2)}$, ..., $A^{(n)}$ its rows

$$A^{(i)} = (a_{i1}, a_{i2}, \dots, a_{in}),$$

and by $A_{(1)}$, $A_{(2)}$, ..., $A_{(n)}$ its columns

$$A_{(j)} = \begin{pmatrix} a_{1j} \\ a_{2j} \\ \vdots \\ a_{nj} \end{pmatrix}$$

Hence we can write

$$A = (A_{(1)}, A_{(2)}, \dots, A_{(n)}) = \begin{pmatrix} A^{(1)} \\ A^{(2)} \\ \vdots \\ A^{(n)} \end{pmatrix}$$

We can now state and prove the following property:

Proposition 3.2.8. *If the i-th row of a matrix A is a linear combination of two row vectors*

$$A^{(i)} = \alpha V + \alpha' V',$$

with $V = (v_1, v_2, \dots, v_n)$ and $V' = (v'_1, v'_2, \dots, v'_n)$, then

$$\det A = \alpha \det \begin{pmatrix} A^{(1)} \\ \vdots \\ A^{(i-1)} \\ V \\ A^{(i+1)} \\ \vdots \\ A^{(n)} \end{pmatrix} + \alpha' \det \begin{pmatrix} A^{(1)} \\ \vdots \\ A^{(i-1)} \\ V' \\ A^{(i+1)} \\ \vdots \\ A^{(n)} \end{pmatrix}$$

Similarly if the j-th column of A is a linear combination of two column vectors

$$A_{(j)} = \beta W + \beta' W',$$

with $W = {}^t(w_1, w_2, \dots, w_n)$ and $W' = {}^t(w'_1, w'_2, \dots, w'_n)$, then

$$\begin{aligned} \det A = {} & \beta \det(A_{(1)}, \dots, A_{(j-1)}, W, A_{(j+1)}, \dots, A_{(n)}) \\ & + \beta' \det(A_{(1)}, \dots, A_{(j-1)}, W', A_{(j+1)}, \dots, A_{(n)}). \end{aligned}$$

Proof. Since the determinant of a matrix coincides with that of its transpose, it suffices to prove the statement about the rows. Let us suppose that $A^{(i)} = \alpha V + \alpha' V'$, i.e., that $a_{ij} = \alpha v_j + \alpha' v'_j$, for $j = 1, \dots, n$. We have:

$$\begin{aligned}
\det A &= \sum_{\sigma \in \mathfrak{S}_n} \operatorname{sgn}(\sigma)\, a_{1\sigma(1)} \cdots a_{i\sigma(i)} \cdots a_{n\sigma(n)} \\
&= \sum_{\sigma \in \mathfrak{S}_n} \operatorname{sgn}(\sigma)\, a_{1\sigma(1)} \cdots (\alpha v_{\sigma(i)} + \alpha' v'_{\sigma(i)}) \cdots a_{n\sigma(n)} \\
&= \alpha \sum_{\sigma \in \mathfrak{S}_n} \operatorname{sgn}(\sigma)\, a_{1\sigma(1)} \cdots v_{\sigma(i)} \cdots a_{n\sigma(n)} \\
&\qquad + \alpha' \sum_{\sigma \in \mathfrak{S}_n} \operatorname{sgn}(\sigma)\, a_{1\sigma(1)} \cdots v'_{\sigma(i)} \cdots a_{n\sigma(n)} \\
&= \alpha \det \begin{pmatrix} A^{(1)} \\ \vdots \\ V \\ \vdots \\ A^{(n)} \end{pmatrix} + \alpha' \det \begin{pmatrix} A^{(1)} \\ \vdots \\ V' \\ \vdots \\ A^{(n)} \end{pmatrix}
\end{aligned}$$

□

Proposition 3.2.9. *Let A' be the matrix obtained by swapping two rows (or two columns) of A. Then*

$$\det A' = -\det A.$$

Proof. Since $\det({}^tA) = \det(A)$, it suffices to prove the statement about the rows. Let $A' = (a'_{ij})$ be the matrix obtained by swapping the h-th and k-th rows of A, with $1 \le h < k \le n$. We have:

$$\begin{aligned}
\det A' &= \sum_{\sigma \in \mathfrak{S}_n} \operatorname{sgn}(\sigma)\, a'_{1\sigma(1)} \cdots a'_{h\sigma(h)} \cdots a'_{k\sigma(k)} \cdots a'_{n\sigma(n)} \\
&= \sum_{\sigma \in \mathfrak{S}_n} \operatorname{sgn}(\sigma)\, a_{1\sigma(1)} \cdots a_{k\sigma(h)} \cdots a_{h\sigma(k)} \cdots a_{n\sigma(n)}.
\end{aligned}$$

Let us denote by τ the permutation that swaps h and k fixing all other elements, and set $\eta = \sigma \circ \tau$. We have

$$\det A' = \sum_{\sigma \in \mathfrak{S}_n} \operatorname{sgn}(\sigma)\, a_{1\eta(1)} \cdots a_{h\eta(h)} \cdots a_{k\eta(k)} \cdots a_{n\eta(n)}.$$

Note that as σ varies in the set $\mathfrak{S}_n$ the same happens for η, moreover we have $\operatorname{sgn}(\eta) = \operatorname{sgn}(\sigma)\operatorname{sgn}(\tau) = -\operatorname{sgn}(\sigma)$, because the sign of a transposition is -1. Hence we have

$$\begin{aligned}
\det A' &= \sum_{\sigma\in\mathfrak{S}_n} \operatorname{sgn}(\sigma)\, a_{1\eta(1)} \cdots a_{h\eta(h)} \cdots a_{k\eta(k)} \cdots a_{n\eta(n)} \\
&= \sum_{\eta\in\mathfrak{S}_n} -\operatorname{sgn}(\eta)\, a_{1\eta(1)} \cdots a_{h\eta(h)} \cdots a_{k\eta(k)} \cdots a_{n\eta(n)} \\
&= -\det A.
\end{aligned}$$

□

Remark 3.2.10. The determinant of a matrix can be considered as a function of its n rows

$$\det\colon K^n \times \cdots \times K^n \to K, \quad (A^{(1)}, \ldots, A^{(n)}) \mapsto \det \begin{pmatrix} A^{(1)} \\ \vdots \\ A^{(n)} \end{pmatrix}$$

or as a function of its n columns

$$\det\colon K^n \times \cdots \times K^n \to K, \quad (A_{(1)}, \ldots, A_{(n)}) \mapsto \det(A_{(1)}, \ldots, A_{(n)}).$$

The Proposition 3.2.8 states that both these functions are *multilinear*, i.e., they are linear in each of their n variables. Then the Proposition 3.2.9 states that both these functions are *alternating*, i.e., they change sign whenever two of their variables are swapped with each other. The determinant thus provides an example of an alternating multilinear function.

Corollary 3.2.11. *Let $A = (a_{ij})$ be a square matrix of order n. For every permutation $\sigma \in \mathfrak{S}_n$ we have:*

$$\det \begin{pmatrix} A^{(\sigma(1))} \\ A^{(\sigma(2))} \\ \vdots \\ A^{(\sigma(n))} \end{pmatrix} = \operatorname{sgn}(\sigma) \det \begin{pmatrix} A^{(1)} \\ A^{(2)} \\ \vdots \\ A^{(n)} \end{pmatrix} = \operatorname{sgn}(\sigma) \det A$$

and

$$\det(A_{(\sigma(1))}, A_{(\sigma(2))}, \ldots, A_{(\sigma(n))}) = \operatorname{sgn}(\sigma) \det(A_{(1)}, A_{(2)}, \ldots, A_{(n)}) = \det A.$$

Proof. It suffices to recall that every permutation can be written as a product of a finite number of transpositions and that, if σ is a product of r transpositions, we have $\operatorname{sgn}(\sigma) = (-1)^r$. □

Corollary 3.2.12. *If a square matrix A has two equal rows (or columns) then* $\det A = 0$.

Proof. By swapping the two equal rows (or the two equal columns) the matrix A does not change, but its determinant must change sign. Therefore we have $\det A = -\det A$, hence $2\det A = 0$. If the characteristic of the field K is not 2, it follows that $\det A = 0$. If K is a field of characteristic 2 we can use the definition of the determinant. Suppose that the h-th and k-th rows of A are equal, i.e., that $a_{hj} = a_{kj}$ for every $j = 1, \ldots, n$. Since $\operatorname{char}(K) = 2$, we have $\operatorname{sgn}(\sigma) = 1$ for every $\sigma \in \mathfrak{S}_n$, hence

$$\det A = \sum_{\sigma \in \mathfrak{S}_n} a_{1\sigma(1)} \cdots a_{h\sigma(h)} \cdots a_{k\sigma(k)} \cdots a_{n\sigma(n)}.$$

For every permutation $\sigma \in \mathfrak{S}_n$ let us denote by σ' the composition of σ with the transposition that swaps h and k. In the formula for the determinant of A, the addend corresponding to σ' is equal to the one corresponding to σ, in fact:

$$\begin{aligned} a_{1\sigma'(1)} \cdots a_{h\sigma'(h)} \cdots a_{k\sigma'(k)} \cdots a_{n\sigma'(n)} &= a_{1\sigma(1)} \cdots a_{h\sigma(k)} \cdots a_{k\sigma(h)} \cdots a_{n\sigma(n)} \\ &= a_{1\sigma(1)} \cdots a_{k\sigma(k)} \cdots a_{h\sigma(h)} \cdots a_{n\sigma(n)} \end{aligned}$$

where in the last equality we used the fact that the h-th and k-th rows are equal. This means that the addends that appear in the summation over all permutations $\sigma \in \mathfrak{S}_n$ are pairwise equal. Since the characteristic of K is 2, this implies that $\det A = 0$. □

Proposition 3.2.13. *If a matrix $A \in M_n(K)$ has a row (or a column) consisting only of zeros, then its determinant is zero.*

Proof. Since $\det(A) = \det({}^tA)$, it suffices to consider the case of a zero row. Let us suppose that the i-th row of A is zero, i.e., that $a_{ij} = 0$ for every $j = 1, \ldots, n$. For each permutation $\sigma \in \mathfrak{S}_n$, the product

$$a_{1\sigma(1)} \cdots a_{i\sigma(i)} \cdots a_{n\sigma(n)}$$

contains the factor $a_{i\sigma(i)} = 0$, hence this product is zero. From the definition of the determinant it follows that $\det A = 0$. □

Proposition 3.2.14. *The determinant of a matrix $A \in M_n(K)$ does not change if to a row (resp., to a column) of A we add a linear combination of the other rows (resp., of the other columns).*

Proof. Once more, it suffices to prove the statement about the rows. Let B be the matrix obtained from A by replacing its i-th row $A^{(i)}$ with the row

$$B^{(i)} = A^{(i)} + \sum_{j \neq i} \alpha_j A^{(j)},$$

with $\alpha_j \in K$. Recalling the multilinearity of the determinant, we have:

$$\det B = \det \begin{pmatrix} A^{(1)} \\ \vdots \\ A^{(i-1)} \\ A^{(i)} \\ A^{(i+1)} \\ \vdots \\ A^{(n)} \end{pmatrix} + \sum_{j \neq i} \alpha_j \det \begin{pmatrix} A^{(1)} \\ \vdots \\ A^{(i-1)} \\ A^{(j)} \\ A^{(i+1)} \\ \vdots \\ A^{(n)} \end{pmatrix}$$

Since $j \neq i$, in the matrix

$$\begin{pmatrix} A^{(1)} \\ \vdots \\ A^{(i-1)} \\ A^{(j)} \\ A^{(i+1)} \\ \vdots \\ A^{(n)} \end{pmatrix}$$

there are two equal rows, hence the determinant of this matrix is zero. Therefore we have $\det B = \det A$. □

Proposition 3.2.15. *If $A \in M_n(K)$ and $\alpha \in K$, then*

$$\det(\alpha A) = \alpha^n \det(A).$$

Proof. From the multilinearity of the determinant it follows that

$$\det(\alpha A) = \det \begin{pmatrix} \alpha A^{(1)} \\ \alpha A^{(2)} \\ \vdots \\ \alpha A^{(n)} \end{pmatrix} = \alpha^n \det \begin{pmatrix} A^{(1)} \\ A^{(2)} \\ \vdots \\ A^{(n)} \end{pmatrix} = \alpha^n \det A.$$

□

We are now able to prove the following result:

Theorem 3.2.16 (Binet theorem). *For any matrices $A, B \in M_n(K)$, we have:*

$$\det(AB) = \det(A)\det(B).$$

Proof. Let $A = (a_{ij})$ and $B = (b_{ij})$, and write B as follows:

$$B = \begin{pmatrix} B^{(1)} \\ \vdots \\ B^{(n)} \end{pmatrix}$$

Computing the product of A and B, we obtain

$$\det(AB) = \det \begin{pmatrix} a_{11}B^{(1)} + \cdots + a_{1n}B^{(n)} \\ a_{21}B^{(1)} + \cdots + a_{2n}B^{(n)} \\ \vdots \\ a_{n1}B^{(1)} + \cdots + a_{nn}B^{(n)} \end{pmatrix}$$

By using the multilinearity of the determinant, we have

$$\det(AB) = \sum_{j_1}\sum_{j_2}\cdots\sum_{j_n} a_{1j_1}a_{2j_2}\cdots a_{nj_n} \det \begin{pmatrix} B^{(j_1)} \\ \vdots \\ B^{(j_n)} \end{pmatrix}$$

Since the determinant of a matrix with two equal rows is zero, the above expression becomes:

$$\det(AB) = \sum_{\sigma \in \mathfrak{S}_n} a_{1\sigma(1)}a_{2\sigma(2)}\cdots a_{n\sigma(n)} \det \begin{pmatrix} B^{(\sigma(1))} \\ \vdots \\ B^{(\sigma(n))} \end{pmatrix}$$

Finally, recalling that the determinant is an alternating function (Corollary 3.2.11), we have:

$$\begin{aligned} \det(AB) &= \sum_{\sigma \in \mathfrak{S}_n} a_{1\sigma(1)}a_{2\sigma(2)}\cdots a_{n\sigma(n)} \operatorname{sgn}(\sigma) \det \begin{pmatrix} B^{(1)} \\ \vdots \\ B^{(n)} \end{pmatrix} \\ &= \Big(\sum_{\sigma \in \mathfrak{S}_n} \operatorname{sgn}(\sigma) a_{1\sigma(1)}a_{2\sigma(2)}\cdots a_{n\sigma(n)} \Big) \det B \\ &= \det(A)\det(B). \end{aligned}$$

□

Corollary 3.2.17. *If $A \in M_n(K)$ is an invertible matrix, then*

$$\det(A^{-1}) = (\det A)^{-1}.$$

Proof. If A is invertible, we have $AA^{-1} = \mathbf{1}_n$. By Binet theorem we get:

$$1 = \det(\mathbf{1}_n) = \det(AA^{-1}) = \det(A)\det(A^{-1}),$$

hence $\det(A^{-1}) = (\det A)^{-1}$. □

Binet theorem admits an important generalization that we now describe. Let A be a matrix with m rows and n columns and suppose that $m \leq n$. Given m indices $i_1, i_2, \dots, i_m \in \{1, 2, \dots, n\}$, we denote by

$$A_{(i_1,i_2,\dots,i_m)} = \left(A_{(i_1)}, A_{(i_2)}, \dots, A_{(i_m)}\right)$$

the square matrix of order m formed by the columns of A that correspond to the given indices.

Similarly if B is a matrix with n rows and m columns, with $m \leq n$, given m indices $i_1, i_2, \dots, i_m \in \{1, 2, \dots, n\}$, we denote by

$$B^{(i_1,i_2,\dots,i_m)} = \begin{pmatrix} B^{(i_1)} \\ B^{(i_2)} \\ \vdots \\ B^{(i_m)} \end{pmatrix}$$

the square matrix of order m formed by the rows of B that correspond to the given indices. Note that, with these hypotheses, the product AB is a square matrix of order m

Theorem 3.2.18 (Cauchy–Binet formula). *For any two matrices $A \in M_{m,n}(K)$ and $B \in M_{n,m}(K)$, with $m < n$, we have:*

$$\det(AB) = \sum_{1 \leq i_1 < i_2 < \cdots < i_m \leq n} \det(A_{(i_1,i_2,\dots,i_m)}) \det(B^{(i_1,i_2,\dots,i_m)})$$

Proof. The proof of this result is obtained by suitably adapting the proof of Binet theorem.

Let $A = (a_{ij})$ be an $m \times n$ matrix and $B = (b_{ij})$ an $n \times m$ matrix, and let us write B as follows:

$$B = \begin{pmatrix} B^{(1)} \\ \vdots \\ B^{(n)} \end{pmatrix}$$

By computing the product of A and B we obtain

$$\det(AB) = \det \begin{pmatrix} a_{11}B^{(1)} + \cdots + a_{1n}B^{(n)} \\ a_{21}B^{(1)} + \cdots + a_{2n}B^{(n)} \\ \vdots \\ a_{m1}B^{(1)} + \cdots + a_{mn}B^{(n)} \end{pmatrix}$$

Since the determinant is multilinear, we have

$$\begin{aligned} \det(AB) &= \sum_{i_1=1}^{n} a_{1i_1} \sum_{i_2=1}^{n} a_{2i_2} \cdots \sum_{i_m=1}^{n} a_{mi_m} \det \begin{pmatrix} B^{(i_1)} \\ \vdots \\ B^{(i_m)} \end{pmatrix} \\ &= \sum_{i_1,\dots,i_m=1}^{n} a_{1i_1} a_{2i_2} \cdots a_{mi_m} \det \begin{pmatrix} B^{(i_1)} \\ \vdots \\ B^{(i_m)} \end{pmatrix} \end{aligned}$$

Since the determinant of a matrix with two equal rows is zero, the above expression becomes:

$$\det(AB) = \sum_{i_1,\dots,i_m}^{*} a_{1i_1} a_{2i_2} \cdots a_{mi_m} \det \begin{pmatrix} B^{(i_1)} \\ \vdots \\ B^{(i_m)} \end{pmatrix}$$

where the summation is performed over all choices of pairwise distinct indices $i_1, \dots, i_m \in \{1, 2, \dots, n\}$.

We can rearrange this sum by first summing over strictly ascending indices, $1 \le i_1 < i_2 < \cdots < i_m \le n$, and then summing over all possible permutations of those m indices, thus obtaining the following formula:

$$\det(AB) = \sum_{1 \le i_1 < i_2 < \cdots < i_m \le n} \sum_{\sigma \in \mathfrak{S}_m} a_{1i_{\sigma(1)}} a_{2i_{\sigma(2)}} \cdots a_{mi_{\sigma(m)}} \det \begin{pmatrix} B^{(i_{\sigma(1)})} \\ \vdots \\ B^{(i_{\sigma(m)})} \end{pmatrix}$$

Recalling that the determinant is an alternating function, we have

$$\det \begin{pmatrix} B^{(i_{\sigma(1)})} \\ \vdots \\ B^{(i_{\sigma(m)})} \end{pmatrix} = \operatorname{sgn}(\sigma) \det \begin{pmatrix} B^{(i_1)} \\ \vdots \\ B^{(i_m)} \end{pmatrix} = \operatorname{sgn}(\sigma) \det(B^{(i_1,\dots,i_m)}).$$

The above formula becomes:

$$\det(AB) = \sum_{1\le i_1<\cdots<i_m\le n} \Big(\sum_{\sigma\in\mathfrak{S}_m} \operatorname{sgn}(\sigma)\, a_{1i_{\sigma(1)}} a_{2i_{\sigma(2)}} \cdots a_{mi_{\sigma(m)}} \Big) \det(B^{(i_1,\ldots,i_m)}).$$

Now it suffices to note that, from the definition of the determinant, we have

$$\det(A_{(i_1,i_2,\ldots,i_m)}) = \sum_{\sigma\in\mathfrak{S}_m} \operatorname{sgn}(\sigma)\, a_{1i_{\sigma(1)}} a_{2i_{\sigma(2)}} \cdots a_{mi_{\sigma(m)}}.$$

Hence we obtain

$$\det(AB) = \sum_{1\le i_1<\cdots<i_m\le n} \det(A_{(i_1,\ldots,i_m)}) \det(B^{(i_1,\ldots,i_m)}). \qquad \square$$

Remark 3.2.19. We note that the Cauchy–Binet formula trivially holds also if $m > n$. In fact in this case we have $\det(AB) = 0$, as we will now see. Let $A \in M_{m,n}(K)$ and $B \in M_{n,m}(K)$, with $m > n$. The product AB is the following $m \times m$ matrix:

$$AB = \begin{pmatrix} a_{11}B^{(1)} + \cdots + a_{1n}B^{(n)} \\ a_{21}B^{(1)} + \cdots + a_{2n}B^{(n)} \\ \vdots \\ a_{m1}B^{(1)} + \cdots + a_{mn}B^{(n)} \end{pmatrix}$$

The m rows of this matrix are linear combinations of the n vectors $B^{(1)}$, $B^{(2)}$, $\ldots$, $B^{(n)}$. Since $m > n$, the m rows of the matrix AB are linearly dependent, hence we have $\det(AB) = 0$. On the other hand, in the summation there is no choice of indices for which the condition $1 < i_1 < \cdots < i_m \le n$ hold, hence we get an empty summation, whose result is 0.

Remark 3.2.20. We note that the summation over all possible choices of indices such that $1 \le i_1 < \cdots < i_m \le n$ corresponds to the summation over all subsets $S = \{i_1, i_2, \ldots, i_m\}$ of the set $\{1, 2, \ldots, n\}$; hence this summation has $\binom{n}{m} = \frac{n!}{m!\,(n-m)!}$ addends.

In the particular case where $m = n$, i.e., if A and B are two square matrices of order n, a single addend appears in the summation, corresponding to the only possible choice of indices, given by $i_1 = 1$, $i_2 = 2$, $\ldots$, $i_n = n$. In this case the Cauchy–Binet formula reduces to $\det(AB) = \det(A)\det(B)$.

Corollary 3.2.21. *Let $A \in M_{m,n}(K)$. If $m \le n$, we have*

$$\det(A\,{}^tA) = \sum_{1\le i_1<\cdots<i_m\le n} \big(\det A_{(i_1,\ldots,i_m)}\big)^2.$$

If $m \geq n$, we have

$$\det({}^tA\,A) = \sum_{1 \leq i_1 < \cdots < i_n \leq m} \left(\det A^{(i_1,\ldots,i_n)}\right)^2.$$

Proof. The second formula is obtained from the first by exchanging the matrix A with its transpose. To prove the first, it is enough to use the Cauchy–Binet formula with $B = {}^tA$ and remember that

$$\det(B^{(i_1,\ldots,i_m)}) = \det({}^tA^{(i_1,\ldots,i_m)}) = \det(A_{(i_1,\ldots,i_m)}). \qquad \square$$

Definition 3.2.22. Given a matrix $A = (a_{ij}) \in M_n(K)$, for any $i, j \in \{1, \ldots, n\}$ we denote by A_{ij} the matrix of order $n-1$ obtained from A by deleting its i-th row and j-th column. The determinant of the matrix A_{ij} is called the *minor* of the matrix A corresponding to the indices i and j. The expression

$$a^*_{ij} = (-1)^{i+j}\,|A_{ij}|$$

is called the *cofactor* of the element a_{ij} of A. The transpose of the matrix formed by the cofactors of the elements of A is called the *adjoint matrix* of A, and will be denoted by $\operatorname{adj}(A)$ or A^*:

$$A^* = \operatorname{adj}(A) = {}^t(a^*_{ij}) \in M_n(K).$$

We can now prove the following result, which provides a very useful method for computing the determinant of a matrix.

Proposition 3.2.23 (Laplace formula)**.** *Let $A \in M_n(K)$. For every row index i we have:*

$$\det A = \sum_{h=1}^{n} (-1)^{i+h}\, a_{ih}\, |A_{ih}|. \tag{3.2.1}$$

Similarly for every column index j we have:

$$\det A = \sum_{k=1}^{n} (-1)^{k+j}\, a_{kj}\, |A_{kj}|. \tag{3.2.2}$$

Proof. Since the determinant of a matrix coincides with that of its transpose, exchanging the roles of the rows and columns of the matrix A the formula (3.2.1) reduces to (3.2.2). Therefore it suffices to prove one of the two formulas, for example (3.2.1).

We now want to prove that, in fact, it is enough to prove the formula (3.2.1) for $i = 1$. Let us assume that Laplace formula holds for the first row (i.e., for $i = 1$). Let us choose a row index $i > 1$ and write the matrix A as

$$A = \begin{pmatrix} A^{(1)} \\ \vdots \\ A^{(i-1)} \\ A^{(i)} \\ A^{(i+1)} \\ \vdots \\ A^{(n)} \end{pmatrix}$$

With $i - 1$ exchanges of contiguous rows it is possible to bring the i-th row of A in the place of the first row, obtaining the matrix

$$A' = \begin{pmatrix} A^{(i)} \\ A^{(1)} \\ \vdots \\ A^{(i-1)} \\ A^{(i+1)} \\ \vdots \\ A^{(n)} \end{pmatrix}$$

Since the determinant changes its sign whenever we swap two rows, we have

$$\det A = (-1)^{i-1} \det A'.$$

Now we can compute the determinant of $A' = (a'_{ij})$ along the first row, obtaining

$$\det A' = \sum_{h=1}^{n} (-1)^{1+h} a'_{1h} \, |A'_{1h}|.$$

From the definition of A' it follows that $a'_{1h} = a_{ih}$ and $A'_{1h} = A_{ih}$, hence we have

$$\det A' = \sum_{h=1}^{n} (-1)^{1+h} a_{ih} \, |A_{ih}|.$$

Therefore we deduce that

$$\det A = (-1)^{i-1} \det A' = \sum_{h=1}^{n} (-1)^{i+h} a_{ih} \, |A_{ih}|,$$

that is precisely Laplace formula for the determinant of A computed along the i-th row.

Now all we have to do is prove the formula (3.2.1) for $i = 1$, i.e., we must prove that

$$\det A = \sum_{h=1}^{n} (-1)^{1+h}\, a_{1h}\, |A_{1h}|.$$

Note that A_{1h} is the following square matrix of order $n-1$

$$A_{1h} = \begin{pmatrix} a_{21} & \dots & a_{2,h-1} & a_{2,h+1} & \dots & a_{2n} \\ a_{31} & \dots & a_{3,h-1} & a_{3,h+1} & \dots & a_{3n} \\ \vdots & & \vdots & \vdots & & \vdots \\ a_{n1} & \dots & a_{n,h-1} & a_{n,h+1} & \dots & a_{nn} \end{pmatrix}$$

From the definition of the determinant we have

$$|A_{1h}| = \sum_{\tau \in \mathfrak{S}_{n-1}^{(h)}} \operatorname{sgn}(\tau)\, a_{2,\tau(1)} \cdots a_{h,\tau(h-1)} a_{h+1,\tau(h+1)} \cdots a_{n,\tau(n)},$$

where the summation is over all the permutations τ of the set of $n-1$ elements

$$\{1, 2, \dots, h-1, h+1, \dots, n\}$$

(the symbol $\mathfrak{S}_{n-1}^{(h)}$ denotes the set of such permutations).

We therefore obtain the following expression:

$$\begin{aligned} |A| &= \sum_{h=1}^{n} (-1)^{1+h} a_{1h} \sum_{\tau \in \mathfrak{S}_{n-1}^{(h)}} \operatorname{sgn}(\tau)\, a_{2,\tau(1)} \cdots a_{h,\tau(h-1)} a_{h+1,\tau(h+1)} \cdots a_{n,\tau(n)} \\ &= \sum_{h=1}^{n} \sum_{\tau \in \mathfrak{S}_{n-1}^{(h)}} (-1)^{1+h} \operatorname{sgn}(\tau)\, a_{1h} a_{2,\tau(1)} \cdots a_{h,\tau(h-1)} a_{h+1,\tau(h+1)} \cdots a_{n,\tau(n)}. \end{aligned}$$

Now to the permutation $\tau \in \mathfrak{S}_{n-1}^{(h)}$

$$\tau = \begin{pmatrix} 1 & 2 & \dots & h-1 & h+1 & \dots & n \\ \tau(1) & \tau(2) & \dots & \tau(h-1) & \tau(h+1) & \dots & \tau(n) \end{pmatrix}$$

we associate the permutation $\sigma \in \mathfrak{S}_n$ defined by

$$\sigma = \begin{pmatrix} 1 & 2 & \dots & h-1 & h & h+1 & \dots & n \\ h & \tau(1) & \dots & \tau(h-2) & \tau(h-1) & \tau(h+1) & \dots & \tau(n) \end{pmatrix}$$

More precisely, σ is defined as follows:

$$\sigma(1) = h \quad \text{and} \quad \sigma(k) = \begin{cases} \tau(k-1) & \text{if } 2 \leq k \leq h, \\ \tau(k) & \text{if } k > h. \end{cases}$$

Note that as h varies from 1 to n and τ varies in the set $\mathfrak{S}_{n-1}^{(h)}$, the corresponding permutation σ describes the whole set $\mathfrak{S}_n$ of the permutations of the n elements $\{1, 2, \ldots, n\}$.

The last thing that remains to be understood is the relationship between the sign of τ and that of the corresponding permutation σ.

We note that σ is the composition of the permutation

$$\xi = \begin{pmatrix} 1 & 2 & 3 & \ldots & h & h+1 & \ldots & n \\ h & 1 & 2 & \ldots & h-1 & h+1 & \ldots & n \end{pmatrix}$$

with the permutation τ' defined by

$$\tau' = \begin{pmatrix} 1 & 2 & \ldots & h-1 & h & h+1 & \ldots & n \\ \tau(1) & \tau(2) & \ldots & \tau(h-1) & h & \tau(h+1) & \ldots & \tau(n) \end{pmatrix}$$

In fact we have $\sigma = \tau' \circ \xi$, as can be easily verified. It follows that $\operatorname{sgn}(\sigma) = \operatorname{sgn}(\tau') \operatorname{sgn}(\xi)$. Now we note that the two permutations τ and τ' have the same representation as a product of disjoint cycles (in the representation of τ' there is the cycle of length one (h) whose presence is irrelevant because it corresponds to the identity permutation), hence $\operatorname{sgn}(\tau) = \operatorname{sgn}(\tau')$. Then we have $\operatorname{sgn}(\xi) = (-1)^{h-1} = (-1)^{h+1}$, as we have seen in the Example 3.1.10. It follows that $\operatorname{sgn}(\sigma) = (-1)^{h+1} \operatorname{sgn}(\tau)$ and the Laplace formula can be written as:

$$\begin{aligned} \det A &= \sum_{h=1}^{n} \sum_{\tau \in \mathfrak{S}_{n-1}^{(h)}} (-1)^{1+h} \operatorname{sgn}(\tau)\, a_{1h} a_{2,\tau(1)} \cdots a_{h,\tau(h-1)} a_{h+1,\tau(h+1)} \cdots a_{n,\tau(n)} \\ &= \sum_{\sigma \in \mathfrak{S}_n} \operatorname{sgn}(\sigma)\, a_{1\sigma(1)} a_{2\sigma(2)} \cdots a_{h\sigma(h)} a_{h+1,\sigma(h+1)} \cdots a_{n\sigma(n)}. \end{aligned}$$

Note that this is precisely the definition of the determinant of A. □

Example 3.2.24. We use Laplace formula to compute the determinant of the following matrix:

$$A = \begin{pmatrix} 2 & 0 & 1 & 1 \\ -1 & 3 & 0 & 2 \\ 0 & 2 & -3 & 0 \\ 1 & 4 & -1 & 0 \end{pmatrix}$$

To apply Laplace formula we can choose any row or columns, but it is advisable to choose a row (or a column) among those containing the greatest number of zeros. Choosing, for example, the third row, we get:

$$\det A = -2 \begin{vmatrix} 2 & 1 & 1 \\ -1 & 0 & 2 \\ 1 & -1 & 0 \end{vmatrix} - 3 \begin{vmatrix} 2 & 0 & 1 \\ -1 & 3 & 2 \\ 1 & 4 & 0 \end{vmatrix}$$

Let us consider the first of these two determinants of order three and compute it along the second column:

$$\begin{vmatrix} 2 & 1 & 1 \\ -1 & 0 & 2 \\ 1 & -1 & 0 \end{vmatrix} = -\begin{vmatrix} -1 & 2 \\ 1 & 0 \end{vmatrix} + \begin{vmatrix} 2 & 1 \\ -1 & 2 \end{vmatrix} = -(-2) + 5 = 7.$$

Now we consider the second determinant of order three and compute it along the first row:

$$\begin{vmatrix} 2 & 0 & 1 \\ -1 & 3 & 2 \\ 1 & 4 & 0 \end{vmatrix} = 2 \begin{vmatrix} 3 & 2 \\ 4 & 0 \end{vmatrix} + \begin{vmatrix} -1 & 3 \\ 1 & 4 \end{vmatrix} = -16 - 7 = -23.$$

Finally, we have $\det A = -14 + 69 = 55$.

As a corollary of Laplace theorem we have the following result which provides an explicit formula for the inverse of a square matrix.

Corollary 3.2.25. *Let $A \in M_n(K)$ and denote by A^* its adjoint matrix. The following equality holds:*

$$A\,A^* = \det(A)\,\mathbf{1}_n.$$

Therefore, if $\det(A)$ is invertible, we have

$$A^{-1} = (\det A)^{-1}\,A^*.$$

Proof. The equality $A\,A^* = \det(A)\,\mathbf{1}_n$ is equivalent to the following:

$$\sum_{h=1}^{n} (-1)^{j+h}\,a_{ih}\,|A_{jh}| = \begin{cases} \det A & \text{if } i = j, \\ 0 & \text{if } i \neq j. \end{cases}$$

If $i = j$ the above expression reduces to

$$\sum_{h=1}^{n} (-1)^{i+h}\,a_{ih}\,|A_{ih}|,$$

which is the computation of $\det A$ along the i-th row of A.

Let's consider now the case $i \neq j$. The expression

$$\sum_{h=1}^{n} (-1)^{j+h}\, a_{ih}\, |A_{jh}|$$

is the computation, along the j-th row, of the determinant of the matrix B obtained from A by replacing its j-th row with a copy of the i-th row:

$$B = \begin{pmatrix} a_{11} & a_{12} & \dots & a_{1n} \\ \vdots & \vdots & & \vdots \\ a_{i1} & a_{i2} & \dots & a_{in} \\ \vdots & \vdots & & \vdots \\ a_{j-1,1} & a_{j-1,2} & \dots & a_{j-1,n} \\ a_{i1} & a_{i2} & \dots & a_{in} \\ a_{j+1,1} & a_{j+1,2} & \dots & a_{j+1,n} \\ \vdots & \vdots & & \vdots \\ a_{n1} & a_{n2} & \dots & a_{nn} \end{pmatrix}$$

Since this matrix has two equal rows, its determinant is zero, hence we have

$$\sum_{h=1}^{n} (-1)^{j+h}\, a_{ih}\, |A_{jh}| = 0,$$

for every $i \neq j$. □

Remark 3.2.26 As we have seen in Corollary 3.2.17, from Binet theorem it follows that the invertibility of the determinant of a matrix $A \in M_n(K)$ is a necessary condition for the invertibility of A. The previous result proves that this condition is also sufficient. Therefore we conclude that a matrix $A \in M_n(K)$ is invertible if and only if its determinant is invertible. We can note that this result holds also, more generally, for matrices with coefficients in a unitary commutative ring. For example, a matrix $A \in M_n(\mathbb{Z})$ is invertible if and only if $\det A = \pm 1$ (the condition $\det A \neq 0$ guarantees the existence of A^{-1} in the ring $M_n(\mathbb{Q})$ and not in $M_n(\mathbb{Z})$).

Remark 3.2.27. We recall that a matrix A is invertible if there exists a matrix B such that $AB = \mathbf{1}$ and $BA = \mathbf{1}$. As already observed in the case of linear functions it is possible that a matrix A is *not* invertible but that there still exists a matrix B such that $BA = \mathbf{1}$ (in this case we cannot also have $AB = \mathbf{1}$, otherwise A would be invertible); or it may be the case that $AB = \mathbf{1}$ but not $BA = \mathbf{1}$. Such a matrix B is *not* the inverse of A, but it behaves as such only from the left or only from the right. The same reasoning reported after Definition 2.1.18 allows us to conclude that if a matrix A has both a left and a right inverse, these two must coincide and A is actually invertible.

The following result is analogous to Proposition 2.1.19 and clarifies under which conditions a matrix admits a left or right inverse.

Proposition 3.2.28. *Let A be a matrix with m rows and n columns, with entries in a field K.*

(*i*) *There exists a left inverse of A if and only if the nullity of A is 0.*

(*ii*) *There exists a right inverse of A if and only if the rank of A is m.*

Proof. A is the matrix of the linear function $F_A \colon K^n \to K^m$, $v \mapsto F_A(v) = Av$, with respect to the canonical bases of K^n and K^m. Therefore A has a left (resp., right) inverse if and only if the linear function F_A has a left (resp., right) inverse. The result we want to prove is therefore a consequence of Proposition 2.1.19: it suffices to observe that the function F_A is injective if and only if the nullity of A is 0, and it is surjective if and only if the rank of A is m. □

We now describe an application of these results to the theory of linear systems; it is a formula, known as *Cramer's rule*, which allows to express the solution of a system of n linear equations with n unknowns.

Theorem 3.2.29 (Cramer's rule). *Let $AX = B$ be a system of n linear equations with n unknowns, where*

$$A = \begin{pmatrix} a_{11} & a_{12} & \dots & a_{1n} \\ a_{21} & a_{22} & \dots & a_{2n} \\ \vdots & \vdots & \ddots & \vdots \\ a_{n1} & a_{n2} & \dots & a_{nn} \end{pmatrix} \qquad X = \begin{pmatrix} x_1 \\ x_2 \\ \vdots \\ x_n \end{pmatrix} \qquad B = \begin{pmatrix} b_1 \\ b_2 \\ \vdots \\ b_n \end{pmatrix}$$

If $\det A$ is invertible this system has a unique solution given by

$$x_i = \Delta^{-1}\Delta_i,$$

for every $i = 1, \dots, n$, where $\Delta = \det A$ and Δ_i is the determinant of the matrix obtained from A by replacing its i-th column with the column B.

Proof. If $\det A$ is invertible then the matrix A is invertible, hence the system $AX = B$ has the unique solution given by $X = A^{-1}B$. Recalling that the inverse of A is given by the formula $A^{-1} = (\det A)^{-1} A^*$, by computing the product of A^{-1} and B we obtain

$$x_i = \Delta^{-1} \sum_{h=1}^{n} (-1)^{i+h} |A_{hi}| b_h.$$

Let Δ_i be the determinant of the matrix obtained from A by replacing its i-th column with the column B of the constant terms. By computing Δ_i using Laplace formula along the i-th column, we get

$$\Delta_i = \sum_{h=1}^{n} (-1)^{i+h} b_h |A_{hi}|,$$

from which it follows that $x_i = \Delta^{-1}\Delta_i$. □

3.2.1 Computation of the determinant of a matrix using Gaussian elimination

All the methods for computing the determinant of a square matrix of order n that we have seen so far involve, in general, a sum of $n!$ addends, each of which is, in turn, a product of n elements of A, taken with the appropriate sign. The use of Laplace formula simplifies the computation of the determinant in case the matrix contains many zeros; if instead all the entries of the matrix are different from zero, the number of terms to be added remains $n!$.

We will now see how, by exploiting the properties of the determinant, it is possible to find a much more efficient algorithm.

In the section 2.3.2 of Chapter 2 we described a method for the computation of the rank of a matrix based on Gaussian elimination. This is done by reducing a matrix to an echelon form through a succession of suitable elementary operations on the rows. For a square matrix, the final echelon form is nothing more than an upper triangular matrix. This means that if $A = (a_{ij}) \in M_n(K)$ is a square matrix of order n, by means of elementary operations on the rows it is possible to transform it into an upper triangular matrix

$$A' = \begin{pmatrix} a'_{11} & a'_{12} & a'_{13} & \dots & a'_{1n} \\ 0 & a'_{22} & a'_{23} & \dots & a'_{2n} \\ 0 & 0 & a'_{33} & \dots & a'_{3n} \\ \vdots & \vdots & \vdots & \ddots & \vdots \\ 0 & 0 & 0 & \dots & a'_{nn} \end{pmatrix}$$

The computation of the determinant of A' is straightforward:

$$\det A' = a'_{11}a'_{22}\cdots a'_{nn}.$$

Since the effect on the determinant of a matrix of each elementary operation on the rows is known, it is possible to take it into account to find the relationship between the determinant of the matrix A and that of A'. It is worth mentioning the three types of elementary operations on the rows and their effect on the determinant of a matrix:

(1) swap two rows: in this case the determinant changes its sign;

(2) multiply a row by a scalar λ: in this case the determinant gets multiplied by λ;

(3) add to a row a multiple of another row: in this case the determinant does not change.

This method thus allows us to compute the determinant of a matrix A by using only elementary operations on its rows, in order to reduce it to an upper triangular form (of course, one could also use elementary operations on the columns, or one could reduce the matrix to a lower triangular form).

We can make a rough estimate of the maximum number of operations necessary to compute the determinant of a square matrix of order n using the strategy just described. The number of elements below the main diagonal is

$$1 + 2 + \cdots + (n-1) = \frac{(n-1)\,n}{2}$$

and all these must be transformed into zeros. One of these zeros is produced by adding to a row a multiple of another row, and the sum of two row vectors consists of n sums of numbers (actually, the further one advances in the algorithm, the fewer sums are necessary, because many elements of the rows to be added are zeros). Therefore the number of necessary "operations" (neglecting possible row swaps) is approximately equal to

$$n\,\frac{(n-1)\,n}{2} = \frac{n^3 - n^2}{2}.$$

Since, for large n, the number $\frac{n^3-n^2}{2}$ is much smaller than $n!$, this algorithm for the computation of the determinant is much more efficient than those previously described.

Example 3.2.30. We describe the computation of the determinant of the following matrix using the Gaussian elimination method.

$$A = \begin{pmatrix} 2 & 1 & -1 & 3 \\ -1 & 3 & 1 & 2 \\ 2 & -1 & 4 & 1 \\ 3 & -2 & 1 & 4 \end{pmatrix}$$

By exchanging the first with the second row, we have:

$$\det A = - \begin{vmatrix} -1 & 3 & 1 & 2 \\ 2 & 1 & -1 & 3 \\ 2 & -1 & 4 & 1 \\ 3 & -2 & 1 & 4 \end{vmatrix}$$

Now to the second row we add the first row multiplied by 2, to the third row we add the first multiplied by 2, and to the fourth row we add the first multiplied by 3:

$$\det A = - \begin{vmatrix} -1 & 3 & 1 & 2 \\ 0 & 7 & 1 & 7 \\ 0 & 5 & 6 & 5 \\ 0 & 7 & 4 & 10 \end{vmatrix}$$

Now we exchange the second with the third column:

$$\det A = \begin{vmatrix} -1 & 1 & 3 & 2 \\ 0 & 1 & 7 & 7 \\ 0 & 6 & 5 & 5 \\ 0 & 4 & 7 & 10 \end{vmatrix}$$

Now to the third row we add the second row multiplied by -6, and to the fourth row we add the second multiplied by -4:

$$\det A = \begin{vmatrix} -1 & 1 & 3 & 2 \\ 0 & 1 & 7 & 7 \\ 0 & 0 & -37 & -37 \\ 0 & 0 & -21 & -18 \end{vmatrix}$$

Now we collect -37 from the third row:

$$\det A = -37 \begin{vmatrix} -1 & 1 & 3 & 2 \\ 0 & 1 & 7 & 7 \\ 0 & 0 & 1 & 1 \\ 0 & 0 & -21 & -18 \end{vmatrix}$$

Finally, to the fourth row we add the third row multiplied by 21:

$$\det A = -37 \begin{vmatrix} -1 & 1 & 3 & 2 \\ 0 & 1 & 7 & 7 \\ 0 & 0 & 1 & 1 \\ 0 & 0 & 0 & 3 \end{vmatrix}$$

It follows that

$$\det A = -37 \cdot (-3) = 111.$$

Example 3.2.31 (THE VANDERMONDE DETERMINANT). Let $x_1, \ldots, x_n \in K$, with $n \geq 2$, and consider the following matrix, called the *Vandermonde matrix*:

$$\begin{pmatrix} 1 & 1 & \ldots & 1 \\ x_1 & x_2 & \ldots & x_n \\ x_1^2 & x_2^2 & \ldots & x_n^2 \\ \vdots & \vdots & & \vdots \\ x_1^{n-1} & x_2^{n-1} & \ldots & x_n^{n-1} \end{pmatrix}$$

The determinant of this matrix is called the *Vandermonde determinant*, denoted by $V(x_1, x_2, \ldots, x_n)$.

We will prove that

$$V(x_1, x_2, \ldots, x_n) = \prod_{1 \leq i < j \leq n} (x_j - x_i). \tag{3.2.3}$$

The proof is done by induction on n.

For $n = 2$ we have

$$V(x_1, x_2) = \begin{vmatrix} 1 & 1 \\ x_1 & x_2 \end{vmatrix} = x_2 - x_1.$$

Now we assume that $n > 2$ and that the formula (3.2.3) holds for $n - 1$.

To compute $V(x_1, x_2, \ldots, x_n)$ we perform the following elementary operations on the rows of the Vandermonde matrix: to the n-th row we subtract the $(n-1)$-th row multiplied by x_1, to the $(n-1)$-th row we subtract the $(n-2)$-th row multiplied by x_1, ..., to the second row we subtract the first multiplied by x_1. We obtain:

$$V(x_1, \dots, x_n) = \begin{vmatrix} 1 & 1 & 1 & \dots & 1 \\ 0 & x_2 - x_1 & x_3 - x_1 & \dots & x_n - x_1 \\ 0 & x_2(x_2 - x_1) & x_3(x_3 - x_1) & \dots & x_n(x_n - x_1) \\ \vdots & \vdots & \vdots & & \vdots \\ 0 & x_2^{n-2}(x_2 - x_1) & x_3^{n-2}(x_3 - x_1) & \dots & x_n^{n-2}(x_n - x_1) \end{vmatrix}$$

$$= \begin{vmatrix} x_2 - x_1 & x_3 - x_1 & \dots & x_n - x_1 \\ x_2(x_2 - x_1) & x_3(x_3 - x_1) & \dots & x_n(x_n - x_1) \\ \vdots & \vdots & & \vdots \\ x_2^{n-2}(x_2 - x_1) & x_3^{n-2}(x_3 - x_1) & \dots & x_n^{n-2}(x_n - x_1) \end{vmatrix}$$

Now from the first column we can collect the factor $x_2 - x_1$, from the second column the factor $x_3 - x_1$, ..., from the last column we can collect the factor $x_n - x_1$, and we obtain

$$V(x_1, \dots, x_n) = (x_2 - x_1)(x_3 - x_1) \cdots (x_n - x_1) \begin{vmatrix} 1 & 1 & \dots & 1 \\ x_2 & x_3 & \dots & x_n \\ \vdots & \vdots & & \vdots \\ x_2^{n-2} & x_3^{n-2} & \dots & x_n^{n-2} \end{vmatrix}$$

This last determinant is the Vandermonde determinant $V(x_2, \dots, x_n)$ which, by the inductive hypothesis, is equal to $\prod_{2 \leq i < j \leq n}(x_j - x_i)$. Substituting in the previous formula we get

$$V(x_1, x_2, \dots, x_n) = \prod_{1 \leq i < j \leq n} (x_j - x_i).$$

3.2.2 The determinant of an endomorphism

Let f be an endomorphism of a finite dimensional vector space V over K. Choosing a basis of V we can associate to f a matrix $A \in M_n(K)$, where $n = \dim V$. To different bases of V there correspond different matrices of f, however all these matrices are similar to each other.

Proposition 3.2.32. *Similar matrices have the same determinant.*

Proof. Two matrices $A, A' \in M_n(K)$ are similar if and only if there exists an invertible matrix P such that

$$A' = PAP^{-1}.$$

From Binet theorem it follows that

$$\det A' = (\det P)(\det A)(\det P^{-1}) = (\det P)(\det A)(\det P)^{-1} = \det A. \qquad \square$$

This result allows us to define the determinant of an endomorphism f as the determinant of a matrix A associated with f by choosing a basis of V, since this determinant does not depend on the chosen basis.

Definition 3.2.33. Let $f \in \operatorname{End} V$. The *determinant* of f is the determinant of the matrix of f with respect to any choice of a basis of V.

An immediate consequence of Binet theorem is the following:

Proposition 3.2.34. *Let f and g be two endomorphisms of V. Then*

$$\det(g \circ f) = (\det g)(\det f).$$

If f is invertible then

$$\det(f^{-1}) = (\det f)^{-1}.$$

Furthermore, the following result holds:

Proposition 3.2.35. *An endomorphism f of a finite dimensional vector space V over K is an isomorphism if and only if* $\det f \neq 0$.

Proof. Let A be the matrix of f with respect to any basis of V. Then f is an isomorphism if and only if A is invertible, and A is invertible if and only if $\det A = \det f \neq 0$ (cf. Corollary 3.2.25). $\square$

3.3 Rank and determinant

In Chapter 2 the (column) rank of a matrix was defined as the maximum number of linearly independent columns. One can also define the row rank as the maximum number of linearly independent rows. In this section we will prove that the two definitions of rank coincide and we will analyze the relationships between the rank of a matrix A and the determinants of the square submatrices that can be extracted from A.

Theorem 3.3.1. *The row rank and the column rank of $A \in M_{m,n}(K)$ coincide.*

Proof. Let us denote by r the row rank and by c the column rank of A. If $r = 0$ then all the elements of A are zero, hence $c = 0$. So let's assume that $r > 0$. A linear dependence relation among the columns of A

$$A_{(1)}x_1 + A_{(2)}x_2 + \cdots + A_{(n)}x_n = \mathbf{0}$$

is a nonzero solution of the homogeneous system

$$AX = \mathbf{0}, \tag{3.3.1}$$

where $X = {}^t(x_1, \ldots, x_n)$. Hence the column rank of A is determined[1] by the set of solutions of the system (3.3.1).

Since a permutation of the rows of A does not change the set of solutions of the system (3.3.1), nor does it change the row rank of A, it is not restrictive to assume that the r linearly independent rows of A are just the first r. It follows that the rows $A^{(r+1)}, A^{(r+2)}, \ldots, A^{(m)}$ are linear combinations of the first r rows of A. This means that the last $m - r$ equations of the system $AX = \mathbf{0}$ are linear combinations of the first r, hence the system $AX = \mathbf{0}$ is equivalent to the system $A'X = \mathbf{0}$, where A' is the matrix formed by the first r rows of A:

$$A' = \begin{pmatrix} A^{(1)} \\ \vdots \\ A^{(r)} \end{pmatrix}$$

Since A and A' determine linear systems with the same set of solutions, and since the set of solutions determine the column rank of the matrix, we deduce that A and A' must have the same column rank, hence the column rank of A' is c. Since the columns of A' are vectors of K^r, it follows that $c \leq r$.

Applying the same reasoning to the matrix ${}^tA$, we deduce that $r \leq c$, hence $r = c$. □

From this result it follows that the distinction between row rank and column rank is useless. We can therefore speak of the *rank* of a matrix to indicate the common value of the two types of rank.

Corollary 3.3.2. *Elementary operations on the rows or columns of a matrix do not change its rank.*

---

[1] We recall that $\mathrm{rk}(A) = n - \mathrm{null}(A)$, where $\mathrm{null}(A)$ is the dimension of the space of solutions of the homogeneous system $AX = \mathbf{0}$ (see Chapter 2, Definition 2.2.15 and Remark 2.1.12).

Proof. It is obvious that the elementary operations on the rows of a matrix do not change its row rank, and that the elementary operations on the columns of a matrix do not change its column rank. The conclusion follows because the row and column ranks coincide. □

Remark 3.3.3. From this last result it follows that in the Gauss' elimination method described in Chapter 2, Section 2.3.2, to compute the rank of a matrix, both elementary operations on rows and columns can be used.

Proposition 3.3.4. *Let $A \in M_{m,n}(K)$ and $B \in M_{n,r}(K)$.*

(*i*) *We have:*

$$\operatorname{rk}(AB) \leq \min\big(\operatorname{rk}(A), \operatorname{rk}(B)\big). \tag{3.3.2}$$

(*ii*) *If $m = n$ and A is invertible, then*

$$\operatorname{rk}(AB) = \operatorname{rk}(B).$$

(*iii*) *If $n = r$ and B is invertible, then*

$$\operatorname{rk}(AB) = \operatorname{rk}(A).$$

Proof. (*i*) Let $F_A \colon K^n \to K^m$ and $F_B \colon K^r \to K^n$ be the linear functions defined by the matrices A and B, respectively. Then AB is the matrix of the composition $F_A \circ F_B \colon K^r \to K^m$. Recalling that the rank of a matrix is equal to the dimension of the image of the corresponding linear function, the inequality (3.3.2) is equivalent to the following:

$$\dim \operatorname{Im}(F_A \circ F_B) \leq \min\big(\dim \operatorname{Im}(F_A), \dim \operatorname{Im}(F_B)\big),$$

whose proof is an easy exercise.

(*ii*) If $m = n$ the fact that A is invertible is equivalent to the fact that $F_A \colon K^n \to K^n$ is an isomorphism. It follows that $\operatorname{Im}(F_B) \cong F_A(\operatorname{Im} F_B)$, hence $\dim \operatorname{Im}(F_A \circ F_B) = \dim \operatorname{Im}(F_B)$, i.e., $\operatorname{rk}(AB) = \operatorname{rk}(B)$.

(*iii*) Finally, if $n = r$ the invertibility of B is equivalent to the invertibility of F_B. It follows that $\operatorname{Im}(F_A \circ F_B) \cong \operatorname{Im}(F_A)$, hence $\operatorname{rk}(AB) = \operatorname{rk}(A)$. □

For square matrices we have:

Theorem 3.3.5. *A square matrix of order n is invertible if and only if its rank is n.*

Proof. If A is invertible, from the previous result we have

$$\operatorname{rk}(A) = \operatorname{rk}(A^{-1}A) = \operatorname{rk}(\mathbf{1}_n) = n.$$

Conversely, let us suppose that A has rank n. Then its columns $A_{(1)}, \ldots, A_{(n)}$ are a basis of K^n, hence for every $j = 1, \ldots, n$ there exist $b_{1j}, b_{2j}, \ldots, b_{nj} \in K$ such that

$$A_{(1)}b_{1j} + A_{(2)}b_{2j} + \cdots + A_{(n)}b_{nj} = e_j, \tag{3.3.3}$$

where e_j is the j-th vector of the canonical basis of K^n. If we set $B = (b_{ij})$, the equations (3.3.3) can be written as follows:

$$AB = \mathbf{1}_n,$$

from which it follows that A is invertible. □

Let $A \in M_{m,n}(K)$ be a matrix with m rows and n columns. Let us choose row indices $1 \leq i_1 < i_2 < \cdots < i_p \leq m$ and column indices $1 \leq j_1 < j_2 < \cdots < j_q \leq n$ and set $I = \{i_1, i_2, \ldots, i_p\}$ and $J = \{j_1, j_2, \ldots, j_q\}$. We denote by A_{IJ} the $p \times q$ submatrix of A formed by the elements located at the intersections of the p rows and the q columns determined by the indices of the sets I and J, respectively.

Proposition 3.3.6. *If B is a submatrix of A then $\operatorname{rk}(B) \leq \operatorname{rk}(A)$.*

Proof. Let $A \in M_{m,n}(K)$, $I = \{i_1, i_2, \ldots, i_p\}$ and $J = \{j_1, j_2, \ldots, j_q\}$ be two sets of indices, and let $B = A_{IJ}$. Let us consider the submatrix $C = A_{IN}$, where $N = \{1, 2, \ldots, n\}$. If we consider the row rank, the inequality

$$\operatorname{rk}(C) \leq \operatorname{rk}(A)$$

is obvious. Note that B is also a submatrix of C, and if we now consider the column rank, the inequality

$$\operatorname{rk}(B) \leq \operatorname{rk}(C)$$

is obvious. From these two inequalities it follows that $\operatorname{rk}(B) \leq \operatorname{rk}(A)$. □

We are now able to prove the following theorem, which relates the notion of rank to that of determinant.

Theorem 3.3.7. *The rank of a matrix A with entries in a field is equal to the maximum order of its invertible square submatrices, i.e., to the maximum order of the nonzero minors of A.*

Proof. Let ρ be the maximum order of the invertible square submatrices of A. From Theorem 3.3.5 and Proposition 3.3.6 it follows that $\rho \leq \operatorname{rk}(A)$. On the other hand, let $r = \operatorname{rk}(A)$ and let $A^{(i_1)}$, $A^{(i_2)}$, ..., $A^{(i_r)}$ be r linearly independent rows of A. Then the submatrix A_{IN} of A (where $I = \{i_1, i_2, \ldots, i_r\}$ and $N = \{1, 2, \ldots, n\}$) has rank r, hence it has r columns, with indices j_1, j_2, ..., j_r, that are linearly independent. If we set $J = \{j_1, j_2, \ldots, j_r\}$, this means that the square submatrix A_{IJ} (of order r) has rank r, and hence it is invertible. From this it follows that $\rho \geq r = \operatorname{rk}(A)$, hence we necessarily have $\rho = \operatorname{rk}(A)$. □

Remark 3.3.8. Based on this result, the rank of a matrix A can also be defined as the maximum order of nonzero minors of A. Using the latter as the definition of the rank of a matrix, and remembering that the determinant of a matrix coincides with that of its transpose, the fact that the rank of a matrix coincides with the rank of its transpose is completely obvious.

From this last result we immediately deduce the following criterion to establish whether n vectors of K^n form a basis:

Corollary 3.3.9. *The vectors $v_1, v_2, \ldots, v_n \in K^n$ form a basis if and only if the determinant of the matrix whose rows (or columns) are the given vectors in nonzero.*

Bordered minors

Let $A \in M_{m,n}(K)$ and suppose we want to compute its rank by finding the maximum order of the nonzero minors of A.

We suppose that A is not the zero matrix, hence its rank is ≥ 1. We will therefore have to compute the minors of increasing order, starting from those of order 2. When for a certain r a nonzero minor of order r has been found, while all minors of order $r+1$ are zero (or do not exist, if $r = \min(m,n)$), we will deduce that $\operatorname{rk}(A) = r$. In fact, from the vanishing of all minors of order $r+1$ follows the vanishing of each minor of higher order: this is easily proved by induction on s, using Laplace formula to compute a minor of order $s > r+1$ along one of its rows or columns.

Actually, in the situation just described it is not necessary to check that all minors of order $r+1$ are zero; it suffices to limit ourselves to those minors of order $r+1$ which contain the square submatrix of order r with nonzero determinant that we have considered. In fact, the following result holds:

Theorem 3.3.10 (Bordered minors theorem). *Let $A \in M_{m,n}(K)$ and let $B = A_{IJ}$ be a square submatrix of order r such that* $\det B \neq 0$. *Suppose that the determinant of every square submatrix of order $r+1$ of A constructed by adding to B one row and one column of A (these are the so-called* bordered minors *of B) is zero. Then A has rank r.*

Proof. Let $B = A_{IJ}$ with $I = \{i_1, i_2, \ldots, i_r\}$ and $J = \{j_1, j_2, \ldots, j_r\}$. Since $\det B \neq 0$ we deduce that the columns of A with indices $j_1, \ldots, j_r$ are linearly independent. The hypothesis of the vanishing of all bordered minors of B implies that every other column of A is a linear combination of the columns with indices $j_1, \ldots, j_r$, hence A has rank r. □

Exercises

Exercise 3.3.1. Compute

$$\det \begin{pmatrix} 2 & 0 & 3 & -1 \\ 4 & 1 & 2 & 0 \\ 0 & 3 & 0 & 1 \\ 2 & 0 & 2 & -1 \end{pmatrix}$$

Exercise 3.3.2. Compute

$$\det \begin{pmatrix} 2 & 3 & 4 & -1 & 2 \\ 2 & 3 & 4 & -1 & 1 \\ 2 & 3 & 5 & 1 & 1 \\ 2 & 1 & 1 & 1 & 1 \\ 1 & 2 & -1 & 2 & -1 \end{pmatrix}$$

Exercise 3.3.3. Compute the inverse of the matrix

$$A = \begin{pmatrix} 2 & -3 & -1 \\ 2 & -1 & -3 \\ 1 & -3 & -1 \end{pmatrix}$$

Exercise 3.3.4. Compute the inverse of the matrix

$$A = \begin{pmatrix} 1 & 2 & 1 \\ 1 & 3 & 0 \\ 1 & 2 & 2 \end{pmatrix}$$

Exercise 3.3.5. Compute the rank of the following matrix by computing its minors:

$$A = \begin{pmatrix} 1 & 2 & 3 & 1 \\ 4 & 5 & 6 & 1 \\ 7 & 8 & 9 & 1 \end{pmatrix}$$

Exercise 3.3.6. Compute the determinant of the following $n \times n$ matrix by using Gauss' elimination method:

$$\begin{pmatrix} 1 & 1 & 1 & \cdots & 1 \\ 1 & 2 & 1 & \cdots & 1 \\ 1 & 1 & 3 & \cdots & 1 \\ \vdots & \vdots & \vdots & & \vdots \\ 1 & 1 & 1 & \cdots & n \end{pmatrix}$$

Exercise 3.3.7. Let A_n denote the following $n \times n$ matrix:

$$A_n = \begin{pmatrix} 2 & 1 & 0 & \cdots & 0 \\ 3 & 2 & 1 & \cdots & 0 \\ 0 & 3 & 2 & \ddots & \vdots \\ \vdots & \vdots & \ddots & \ddots & 1 \\ 0 & 0 & \cdots & 3 & 2 \end{pmatrix}$$

Find a recursive formula for computing $\det A_n$, for every integer $n \geq 1$.

Exercise 3.3.8. Let us denote by D_n (with $n \geq 1$) the determinant of the following matrix of order n:

$$D_n = \det \begin{pmatrix} 1 & -1 & 0 & \cdots & 0 \\ 1 & 1 & -1 & \cdots & 0 \\ 0 & 1 & 1 & \ddots & \vdots \\ \vdots & \vdots & \ddots & \ddots & -1 \\ 0 & 0 & \cdots & 1 & 1 \end{pmatrix}$$

Prove that, for $n \geq 3$, we have $D_n = D_{n-1} + D_{n-2}$. Note that the succession of D_n coincides with the succession of *Fibonacci's numbers*: 1, 2, 3, 5, 8, 13, ...

Exercise 3.3.9. Let A_n denote the following $n \times n$ matrix:

$$A_n = \begin{pmatrix} 1+a_1 & 1 & 1 & \cdots & 1 \\ 1 & 1+a_2 & 1 & \cdots & 1 \\ 1 & 1 & 1+a_3 & \cdots & 1 \\ \vdots & \vdots & \vdots & \ddots & \vdots \\ 1 & 1 & 1 & \cdots & 1+a_n \end{pmatrix}$$

where $a_1, \ldots, a_n \in K$. Prove that, for every integer $n \geq 2$, we have

$$\det A_n = a_1 a_2 \cdots a_n + \sum_{i=1}^{n} a_1 a_2 \cdots \hat{a}_i \cdots a_n,$$

where $\hat{a}_i$ means that the element a_i is not present in the product.

Exercise 3.3.10. Let A be an $n \times n$ matrix with $\det A \neq 0$ and A^* be its adjoint matrix. Prove the following formula, due to Cauchy:

$$|A^*| = |A|^{n-1}.$$

Exercise 3.3.11. Let $K = \mathbb{R}$ be the field of real numbers (actually, it is sufficient to assume that K is a field of characteristic $\neq 2$) and let $A \in M_n(K)$ be a skew-symmetric matrix, i.e., a matrix such that ${}^tA = -A$. Prove that if n is odd, then $\det A = 0$.

Exercise 3.3.12. Given a matrix $A \in M_{m,n}(\mathbb{R})$, prove that $\det({}^tAA) \geq 0$. Prove also that $\det({}^tAA) > 0$ if and only if $\operatorname{rk}(A) = n$.

4

Diagonalization of Endomorphisms

Given an endomorphism f of a finite dimensional vector space V, is it possible to find a basis of V with respect to which the matrix of f takes a particularly simple form (for example, it is a diagonal matrix)?

To answer this question we will introduce the notions of eigenvalues and eigenvectors of an endomorphism f of a vector space V and we will determine the necessary and sufficient conditions for f to be diagonalizable. Finally, we will describe the Jordan canonical form of an endomorphism of V.

4.1 Eigenvalues and eigenvectors

Let V be a vector space of dimension n over the field K and $f\colon V \to V$ be a linear function. In Chapter 2 we have seen how, by choosing a basis of V, it is possible to associate f with a square matrix of order n, with coefficients in K. Of course, different bases of V correspond to different matrices of f, all of which are similar to each other. We can therefore ask whether it is possible to find a basis of V in such a way that the corresponding matrix of f takes a (particularly simple) canonical form.

Suppose, for example, that there exists a basis of V formed by vectors $v_1, v_2, \ldots, v_n$ such that $f(v_i) = \lambda_i v_i$, for every $i = 1, \ldots, n$, and suitable scalars $\lambda_i \in K$. With respect to this basis, the matrix of f would take the following diagonal form

$$\begin{pmatrix} \lambda_1 & 0 & \cdots & 0 \\ 0 & \lambda_2 & \cdots & 0 \\ \vdots & \vdots & \ddots & \vdots \\ 0 & 0 & \cdots & \lambda_n \end{pmatrix}$$

We then give the following definition:

Definition 4.1.1. An endomorphism f of V is *diagonalizable* if there exists a basis of V such that the matrix of f with respect to such basis is diagonal.

Recalling that two square matrices are said to be similar when they represent the same endomorphism with respect to different bases, we can also give the following definition:

Definition 4.1.2. A square matrix A is *diagonalizable* if it is similar to a diagonal matrix, i.e., if there exist an invertible matrix S and a diagonal matrix D such that $A = SDS^{-1}$.

Our goal will therefore be to try to determine under which conditions a square matrix (or an endomorphism of a vector space) is diagonalizable.

We start by giving the following definition:

Definition 4.1.3. An *eigenvalue* of an endomorphism f of V is a scalar $\lambda \in K$ such that there exists at least one nonzero vector $v \in V$ such that $f(v) = \lambda v$. Such a vector v is called an *eigenvector* of f associated to the eigenvalue λ.

Given $\lambda \in K$ we denote by $\lambda \colon V \to V$ the linear function $\lambda\,\mathrm{id}_V$, i.e., the function that sends a vector v to λv. Then the equation $f(v) = \lambda v$ can be written in the following form: $(f - \lambda)(v) = 0$. From this it follows that the set of eigenvectors associated to the eigenvalue λ (together with the zero vector) is the kernel of the linear function

$$f - \lambda \colon V \to V, \quad v \mapsto f(v) - \lambda v.$$

We set

$$V_\lambda = \mathrm{Ker}(f - \lambda) = \{v \in V \mid f(v) = \lambda v\}.$$

We have:

Proposition 4.1.4. *For every eigenvalue λ of f, the set*

$$V_\lambda = \{v \in V \mid f(v) = \lambda v\}$$

is a vector subspace of V. This is called the eigenspace *of f associated to the eigenvalue λ.*

Before we begin the study of the main properties of eigenvalues and eigenvectors, let's see how it is possible to determine them.

Let f be an endomorphism of a vector space V of dimension n over the field K. From the definition it follows that $\lambda \in K$ is an eigenvalue of f if and only if $\operatorname{Ker}(f-\lambda) \neq \{\mathbf{0}\}$, which is equivalent to saying that the linear function $f - \lambda \colon V \to V$ is not injective. We now recall that requiring that $f - \lambda$ is not injective is equivalent to requiring that $\det(f - \lambda) = 0$ (see Chapter 3, Proposition 3.2.35). Therefore we can deduce that λ is an eigenvalue of f if and only if $\det(f - \lambda) = 0$.

To compute this determinant we can arbitrarily choose a basis of V and consider the corresponding matrix A associated with f. If we denote by

$$\lambda \cdot \mathbf{1} = \begin{pmatrix} \lambda & 0 & \cdots & 0 \\ 0 & \lambda & \cdots & 0 \\ \vdots & \vdots & \ddots & \vdots \\ 0 & 0 & \cdots & \lambda \end{pmatrix}$$

the matrix associated with the linear function $\lambda \operatorname{id}_V$, we have

$$\det(f - \lambda) = \det(A - \lambda \cdot \mathbf{1}).$$

We now give the following definition:

Definition 4.1.5. Let A be a square matrix of order n with entries in K and x be an indeterminate. The *characteristic polynomial* of A is

$$P_A(x) = \det(A - x \cdot \mathbf{1}).$$

We observe that, if $A \in M_n(K)$, $P_A(x)$ is a polynomial of degree n with coefficients in K, whose highest degree monomial is $(-1)^n x^n$.

We point out that some authors define the characteristic polynomial of a matrix A by setting

$$P_A(x) = \det(x \cdot \mathbf{1} - A) = (-1)^n \det(A - x \cdot \mathbf{1}),$$

to make $P_A(x)$ a monic polynomial.

The next result shows that the characteristic polynomial of a square matrix depends, in fact, only on its similarity class.

Proposition 4.1.6. *Let f be an endomorphism of V and A and A' be two matrices of f, corresponding to two different bases of V. Then we have $P_A(x) = P_{A'}(x)$.*

Proof. Recall that two matrices A and A' are associated to the same endomorphism f of V if and only if they are similar, i.e., if and only if there exists an invertible matrix S such that $A' = SAS^{-1}$. In that case we have:

$$\begin{aligned}\det(A' - x\cdot\mathbf{1}) &= \det(SAS^{-1} - x\cdot\mathbf{1})\\ &= \det\left(S(A - x\cdot\mathbf{1})S^{-1}\right)\\ &= \det(S)\det(A - x\cdot\mathbf{1})\det(S^{-1})\\ &= \det(A - x\cdot\mathbf{1}),\end{aligned}$$

since $\det(S^{-1}) = (\det S)^{-1}$. □

Based on this result, we can give the following definition:

Definition 4.1.7. The *characteristic polynomial* $P_f(x)$ of an endomorphism f of a finite dimensional vector space V is the polynomial $P_f(x) = \det(f - x)$. It coincides with the characteristic polynomial $P_A(x)$ of any matrix associated to f.

From the above, we conclude that $\lambda \in K$ is an eigenvalue of f if and only if λ is a root of the characteristic polynomial of f, i.e., if and only if $P_f(\lambda) = 0$.

Remark 4.1.8. The equation $\det(A - x\cdot\mathbf{1}) = 0$ is the *characteristic equation* of the matrix A.

Remark 4.1.9. We note that, since the eigenvalues of a square matrix of order n are the solutions of its characteristic equation, which has degree n, a square matrix with coefficients in a field K does not necessarily have eigenvalues in K. Already in the case where $K = \mathbb{R}$ and $n = 2$, it is known that there are quadratic equations which have no real solutions. If instead K is an algebraically closed field, such as for example the field $\mathbb{C}$ of complex numbers, then every polynomial of degree $n \geq 1$ with coefficients in K has n zeros in K (counted with the appropriate multiplicities), hence every square matrix with entries in an algebraically closed field has eigenvalues.

Remark 4.1.10. We have seen that if two square matrices are similar, they have the same characteristic polynomial. However, the reverse implication does not hold. For example, the following matrices of order n

$$\begin{pmatrix}\alpha & 0 & \cdots & 0 & 0\\ 0 & \alpha & \cdots & 0 & 0\\ \vdots & \vdots & \ddots & \vdots & \vdots\\ 0 & 0 & \cdots & \alpha & 0\\ 0 & 0 & \cdots & 0 & \alpha\end{pmatrix} \quad \text{and} \quad \begin{pmatrix}\alpha & 1 & 0 & \cdots & 0\\ 0 & \alpha & 1 & \cdots & 0\\ \vdots & \vdots & \ddots & \ddots & \vdots\\ 0 & 0 & \cdots & \alpha & 1\\ 0 & 0 & \cdots & 0 & \alpha\end{pmatrix}$$

have the same characteristic polynomial $(\alpha - x)^n$, but, obviously, they are not similar, since a scalar matrix is similar only to itself.

Once the eigenvalues are known, determining the eigenvectors does not present any difficulty. If λ is an eigenvalue of f (or of a matrix A) the corresponding eigenvectors are the non-zero elements of the vector subspace $\operatorname{Ker}(f - \lambda)$. We therefore have to determine the non-zero solutions of the following system of linear equations:

$$(A - \lambda \cdot \mathbf{1}) \begin{pmatrix} x_1 \\ \vdots \\ x_n \end{pmatrix} = \mathbf{0}.$$

Now let's see some examples.

Example 4.1.11. Let us consider the matrix with real coefficients

$$A = \begin{pmatrix} 0 & -1 \\ 1 & 0 \end{pmatrix}.$$

It corresponds to the following linear function:

$$f\colon \mathbb{R}^2 \to \mathbb{R}^2 \qquad \begin{pmatrix} a \\ b \end{pmatrix} \mapsto \begin{pmatrix} -b \\ a \end{pmatrix}$$

The characteristic polynomial of the matrix A is

$$\det(A - x \cdot \mathbf{1}) = \det \begin{pmatrix} -x & -1 \\ 1 & -x \end{pmatrix} = x^2 + 1$$

which has no real zeros. The matrix A therefore has no real eigenvalues (it does however have two complex eigenvalues, given by $x_1 = \sqrt{-1}$ and $x_2 = -\sqrt{-1}$).

Example 4.1.12. Let us consider the matrix with real coefficients

$$A = \begin{pmatrix} 0 & 4 \\ -1 & 4 \end{pmatrix}.$$

The characteristic polynomial of A is

$$\det(A - x \cdot \mathbf{1}) = \det \begin{pmatrix} -x & 4 \\ -1 & 4 - x \end{pmatrix} = x^2 - 4x + 4 = (x - 2)^2.$$

This polynomial has a unique[1] real zero $x = 2$. The eigenvectors are therefore the non-zero solutions of the following system:

$$(A - 2 \cdot \mathbf{1}) \begin{pmatrix} x_1 \\ x_2 \end{pmatrix} = \mathbf{0},$$

that is

$$\begin{pmatrix} -2 & 4 \\ -1 & 2 \end{pmatrix} \begin{pmatrix} x_1 \\ x_2 \end{pmatrix} = \begin{pmatrix} 0 \\ 0 \end{pmatrix}$$

This system can be rewritten as follows

$$\begin{cases} -2x_1 + 4x_2 = 0 \\ -x_1 + 2x_2 = 0 \end{cases}$$

and its solutions are therefore the solutions of the single equation $x_1 = 2x_2$. The solution space is

$$V_2 = \left\{ \begin{pmatrix} 2\alpha \\ \alpha \end{pmatrix} \middle| \alpha \in \mathbb{R} \right\}$$

whose dimension is 1. We have seen that the eigenspace relative to the eigenvalue $\lambda = 2$ has dimension 1. Since there are no other eigenvalues, this means that there does not exist a basis of $\mathbb{R}^2$ formed by eigenvectors of A, hence A is not diagonalizable.

Example 4.1.13. Let us consider the matrix with real coefficients

$$A = \begin{pmatrix} -5 & 8 \\ -4 & 7 \end{pmatrix}.$$

The characteristic polynomial of A is

$$\det(A - x \cdot \mathbf{1}) = \det \begin{pmatrix} -5 - x & 8 \\ -4 & 7 - x \end{pmatrix} = x^2 - 2x - 3.$$

This polynomial has two real roots $x_1 = -1$ and $x_2 = 3$; these are the two eigenvalues of A.

Consider the eigenvalue $\lambda_1 = -1$; the corresponding eigenvectors are the non-zero solutions of the following system:

$$(A - (-1) \cdot \mathbf{1}) \begin{pmatrix} x_1 \\ x_2 \end{pmatrix} = \mathbf{0},$$

[1]In this case, the unique solution of the equation $(x - 2)^2 = 0$ must be counted with *multiplicity* 2, i.e., considered as *two* coinciding solutions.

that is

$$\begin{pmatrix} -4 & 8 \\ -4 & 8 \end{pmatrix} \begin{pmatrix} x_1 \\ x_2 \end{pmatrix} = \begin{pmatrix} 0 \\ 0 \end{pmatrix}.$$

This system can be rewritten as follows

$$\begin{cases} -4x_1 + 8x_2 = 0 \\ -4x_1 + 8x_2 = 0 \end{cases}$$

and its solutions are therefore the solutions of the single equation $x_1 = 2x_2$. The solution space is

$$V_{-1} = \left\{ \begin{pmatrix} 2\alpha \\ \alpha \end{pmatrix} \middle| \alpha \in \mathbb{R} \right\}$$

whose dimension is 1. A basis of this subspace is given, for example, by the vector $v_1 = (2, 1)$.

Consider now the eigenvalue $\lambda_2 = 3$; the corresponding eigenvectors are the non-zero solutions of the following system:

$$(A - 3 \cdot \mathbf{1}) \begin{pmatrix} x_1 \\ x_2 \end{pmatrix} = \mathbf{0},$$

that is

$$\begin{pmatrix} -8 & 8 \\ -4 & 4 \end{pmatrix} \begin{pmatrix} x_1 \\ x_2 \end{pmatrix} = \begin{pmatrix} 0 \\ 0 \end{pmatrix}.$$

This system can be rewritten as follows

$$\begin{cases} -8x_1 + 8x_2 = 0 \\ -4x_1 + 4x_2 = 0 \end{cases}$$

and its solutions are therefore the solutions of the single equation $x_1 = x_2$. The solution space is

$$V_3 = \left\{ \begin{pmatrix} \beta \\ \beta \end{pmatrix} \middle| \beta \in \mathbb{R} \right\}$$

whose dimension is 1. A basis of this subspace is given, for example, by the vector $v_2 = (1, 1)$.

It is easy to verify that the vectors $v_1 = (2, 1)$ and $v_2 = (1, 1)$ are linearly independent, hence they are a basis of $\mathbb{R}^2$.

In conclusion, there exists a basis of $\mathbb{R}^2$ formed by eigenvectors of A, therefore A is diagonalizable.

Let us now study in more detail some properties of the eigenvalues and eigenvectors of an endomorphism f of V.

Definition 4.1.14. Let f be an endomorphism of a finite dimensional vector space V and let $P_f(x)$ be its characteristic polynomial. Let $\lambda \in K$ be an eigenvalue of f. The *(algebraic) multiplicity* of λ is the largest integer m such that $(x-\lambda)^m$ divides $P_f(x)$. The dimension of the eigenspace $V_\lambda = \operatorname{Ker}(f-\lambda)$ is called the *geometric multiplicity* of λ.

Proposition 4.1.15. *Eigenvectors related to distinct eigenvalues are linearly independent, i.e.: let $\lambda_1, \dots, \lambda_r$ be eigenvalues of an endomorphism f, with $\lambda_i \neq \lambda_j$ if $i \neq j$, and let v_i be an eigenvector for the eigenvalue λ_i, for $i = 1, \dots, r$. Then the vectors $v_1, \dots, v_r$ are linearly independent.*

Proof. We prove the assertion by induction on r. If $r = 1$ we have one eigenvector v_1 which, being non-zero, is linearly independent. Suppose therefore that the statement is true for $r-1$ eigenvectors. Let $v_1, \dots, v_r$ be r eigenvectors and consider a linear combination

$$\alpha_1 v_1 + \cdots + \alpha_{r-1} v_{r-1} + \alpha_r v_r = \mathbf{0}. \tag{4.1.1}$$

By applying the endomorphism f, and recalling that $f(v_i) = \lambda_i v_i$, we obtain

$$\alpha_1 \lambda_1 v_1 + \cdots + \alpha_{r-1} \lambda_{r-1} v_{r-1} + \alpha_r \lambda_r v_r = \mathbf{0}. \tag{4.1.2}$$

Multiplying (4.1.1) by λ_r and subtracting (4.1.2), we get

$$\alpha_1(\lambda_r - \lambda_1)v_1 + \cdots + \alpha_{r-1}(\lambda_r - \lambda_{r-1})v_{r-1} = \mathbf{0}.$$

By the inductive hypothesis, the vectors $v_1, \dots, v_{r-1}$ are linearly independent, hence

$$\alpha_1(\lambda_r - \lambda_1) = \alpha_2(\lambda_r - \lambda_2) = \cdots = \alpha_{r-1}(\lambda_r - \lambda_{r-1}) = 0.$$

Since the eigenvalues λ_i are pairwise distinct, it follows that

$$\alpha_1 = \alpha_2 = \cdots = \alpha_{r-1} = 0.$$

Substituting these values into equation (4.1.1), it reduces to $\alpha_r v_r = \mathbf{0}$, which implies $\alpha_r = 0$. We have thus proved that the vectors $v_1, \dots, v_r$ are linearly independent. □

The two multiplicities, algebraic and geometric, of an eigenvalue of an endomorphism satisfy the following property:

Proposition 4.1.16. *Let V be a finite dimensional vector space over K and f be an endomorphism of V. Let $\lambda \in K$ be an eigenvalue of f with algebraic multiplicity m. Then we have*

$$\dim V_\lambda \leq m,$$

i.e., the geometric multiplicity of an eigenvalue is always less than or equal to its algebraic multiplicity.

Proof. Let $r = \dim V_\lambda$ and let $v_1, \dots, v_r$ be a basis of V_λ. Let us complete these vectors to a basis $v_1, \dots, v_r, v_{r+1}, \dots, v_n$ of V. With respect to this basis the matrix A of f has the following form

$$A = \begin{pmatrix} \lambda \cdot \mathbf{1}_r & B \\ \mathbf{0} & C \end{pmatrix}$$

where $\mathbf{1}_r$ is the identity matrix of order r, B is a matrix with r rows and $n-r$ columns and C is a square matrix of order $n-r$.

The matrix A can be used to compute the characteristic polynomial of f:

$$P_f(x) = \det(A - x\cdot\mathbf{1}) = \det\big((\lambda - x)\cdot\mathbf{1}_r\big)\det(C - x\cdot\mathbf{1}) = (\lambda - x)^r \det(C - x\cdot\mathbf{1}).$$

From this we deduce that λ is a root of the polynomial $P_f(x)$ of multiplicity $\geq r$, hence we have $m \geq r = \dim V_\lambda$. □

We are now able to prove the following result:

Theorem 4.1.17. *Let $f\colon V \to V$ be an endomorphism of a vector space V of dimension n over the field K. Let us denote by $\lambda_1, \lambda_2, \dots, \lambda_r$ the eigenvalues of f in K and by $m_1, m_2, \dots, m_r$ their respective algebraic multiplicities. Then f is diagonalizable if and only if $m_1 + m_2 + \dots + m_r = n$ and for every eigenvalue λ_i, its geometric multiplicity is equal to its algebraic multiplicity.*

Proof. If f is diagonalizable there exists a basis of V with respect to which the matrix of f is a block matrix of the form

$$\begin{pmatrix} \lambda_1 \cdot \mathbf{1}_{m_1} & \mathbf{0} & \cdots & \mathbf{0} \\ \mathbf{0} & \lambda_2 \cdot \mathbf{1}_{m_2} & & \mathbf{0} \\ \vdots & & \ddots & \vdots \\ \mathbf{0} & \cdots & \mathbf{0} & \lambda_r \cdot \mathbf{1}_{m_r} \end{pmatrix}$$

Using this matrix to compute the characteristic polynomial of f, we find

$$P_f(x) = (\lambda_1 - x)^{m_1}(\lambda_2 - x)^{m_2} \cdots (\lambda_r - x)^{m_r},$$

from which it follows that $m_1 + m_2 + \cdots + m_r = \deg P_f(x) = n$. Moreover, since f is diagonalizable, there exists a basis of V consisting of eigenvectors of f, from which it follows that

$$\dim V_{\lambda_1} + \dim V_{\lambda_2} + \cdots + \dim V_{\lambda_r} = n = \dim V.$$

Conversely, suppose there are r eigenvalues of f in K, $\lambda_1, \lambda_2, \dots, \lambda_r$, with algebraic multiplicities $m_1, m_2, \dots, m_r$, respectively, such that $m_1 + m_2 + \cdots + m_r = n$ and that for every eigenvalue λ_i of f, its geometric multiplicity coincides with its algebraic multiplicity. We have to show that there is a basis of V consisting of eigenvectors of f.

Since by hypothesis $\dim V_{\lambda_i} = m_i$ for every eigenvalue λ_i, it follows that there exists a basis $v_1^{(i)}, v_2^{(i)}, \dots, v_{m_i}^{(i)}$ of the eigenspace V_{λ_i}, for every $i = 1, \dots, r$. Since we assumed that $m_1 + \cdots + m_r = n$, the set

$$\{v_1^{(1)}, \dots, v_{m_1}^{(1)}, v_1^{(2)}, \dots, v_{m_2}^{(2)}, \dots, v_1^{(r)}, \dots, v_{m_r}^{(r)}\}$$

contains n vectors which are linearly independent, as is easily verified by recalling that eigenvectors related to distinct eigenvalues are linearly independent (Proposition 4.1.15). Therefore these n vectors are a basis of V. Thus we have constructed a basis of V formed by eigenvectors of f, hence f is diagonalizable. □

Remark 4.1.18. Based on what we have seen, we can state that an endomorphism f of a finite dimensional vector space V over K is diagonalizable if and only if f has eigenvalues $\lambda_1, \dots, \lambda_r \in K$ and we have

$$V = V_{\lambda_1} \oplus V_{\lambda_2} \oplus \cdots \oplus V_{\lambda_r}.$$

4.2 Jordan canonical form

Let V be a vector space of dimension n over the field $\mathbb{C}$ of complex numbers, and $f\colon V \to V$ be a linear function.

For every $\lambda \in \mathbb{C}$ we will denote by $\lambda\colon V \to V$ the linear function given by $v \mapsto \lambda v$.

Recall that $\lambda \in \mathbb{C}$ is an eigenvalue of f if $\mathrm{Ker}(f - \lambda) \neq \{\mathbf{0}\}$ and a nonzero vector $v \in V$ is an eigenvector for the eigenvalue λ if $v \in \mathrm{Ker}(f - \lambda)$, i.e., if $f(v) = \lambda v$. If λ is an eigenvalue of f, the vector subspace $\mathrm{Ker}(f - \lambda)$ of V is the eigenspace corresponding to the eigenvalue λ.

Now we will generalize the notion of eigenvector. For any integer $m > 0$ we denote by $(f - \lambda)^m$ the composition

$$\underbrace{(f - \lambda) \circ (f - \lambda) \circ \cdots \circ (f - \lambda)}_{m}$$

Definition 4.2.1. Let $\lambda \in \mathbb{C}$ be an eigenvalue of $f \colon V \to V$. A nonzero vector $v \in V$ is called a *generalized eigenvector* of f for the eigenvalue λ, if $v \in \operatorname{Ker}(f - \lambda)^m$, for some $m > 0$.

The minimum m for which $v \in \operatorname{Ker}(f - \lambda)^m$ is called the *period* of v (if $m = 1$, v is an eigenvector).

Note that the following inclusions hold

$$\operatorname{Ker}(f - \lambda) \subseteq \operatorname{Ker}(f - \lambda)^2 \subseteq \cdots \subseteq \operatorname{Ker}(f - \lambda)^m \subseteq \cdots \subseteq V. \qquad (4.2.1)$$

We denote by

$$V_\lambda = \bigcup_{m>0} \operatorname{Ker}(f - \lambda)^m$$

the vector subspace of V of all generalized eigenvectors corresponding to the eigenvalue λ. Obviously, it is $V_\lambda = \operatorname{Ker}(f - \lambda)^r$, when r is sufficiently large.

Lemma 4.2.2. *Let v be a generalized eigenvector for f corresponding to the eigenvalue λ, and m be the period of v (this means that $v \in \operatorname{Ker}(f - \lambda)^m$ but $v \notin \operatorname{Ker}(f - \lambda)^{m-1}$). Then the m vectors*

$$v,\ (f - \lambda)(v),\ (f - \lambda)^2(v),\ \ldots,\ (f - \lambda)^{m-1}(v)$$

are linearly independent.

Proof. We consider a linear combination

$$\alpha_0 v + \alpha_1 (f - \lambda)(v) + \alpha_2 (f - \lambda)^2(v) + \cdots + \alpha_{m-1}(f - \lambda)^{m-1}(v) = \mathbf{0}.$$

By applying $(f - \lambda)^{m-1}$, and recalling that $(f - \lambda)^m(v) = \mathbf{0}$, we get

$$\alpha_0 (f - \lambda)^{m-1}(v) + \mathbf{0} = \mathbf{0},$$

hence $\alpha_0 = 0$. Therefore the previous linear combination reduces to

$$\alpha_1 (f - \lambda)(v) + \alpha_2 (f - \lambda)^2(v) + \cdots + \alpha_{m-1}(f - \lambda)^{m-1}(v) = \mathbf{0}.$$

Now applying $(f - \lambda)^{m-2}$ we obtain

$$\alpha_1 (f - \lambda)^{m-1}(v) + \mathbf{0} = \mathbf{0},$$

hence $\alpha_1 = 0$. Continuing in this way we prove that all the coefficients α_i are zero. $\square$

From this result it follows that the period of every generalized eigenvector of f is $\leq n$. In fact, if a generalized eigenvector v has period $m > n$, the m vectors v, $(f-\lambda)(v)$, ..., $(f-\lambda)^{m-1}(v)$ would be linearly independent, which is impossible since $m > n = \dim V$. Therefore we have

$$V_\lambda = \operatorname{Ker}(f-\lambda)^n.$$

The following result generalizes the fact that eigenvectors related to distinct eigenvalues are linearly independent.

Proposition 4.2.3. *Let V and f be as above and $\lambda_1, \dots, \lambda_r$ be eigenvalues of f, with $\lambda_i \neq \lambda_j$ if $i \neq j$. For every $i = 1, \dots, r$ let $v_i \in V_{\lambda_i}$ be a generalized eigenvector for the eigenvalue λ_i. Then the vectors $v_1, \dots, v_r$ are linearly independent.*

Proof. For every $i = 1, \dots, r$ we denote by m_i the period of v_i and we consider the endomorphism $(f-\lambda_i)^{m_i-1}$.

First we observe that, for each i and j, the endomorphisms $(f-\lambda_i)^s$ and $(f-\lambda_j)^t$ commute, for every $s, t \geq 1$. This follows from the fact that powers of f commute with each other and f commutes with multiplication by a scalar λ. Moreover, for every $i = 1, \dots, r$, the vector

$$w_i = (f-\lambda_i)^{m_i-1}(v_i)$$

is an eigenvector of f for the eigenvalue λ_i, in fact we have

$$(f-\lambda_i)(w_i) = (f-\lambda_i)(f-\lambda_i)^{m_i-1}(v_i) = (f-\lambda_i)^{m_i}(v_i) = \mathbf{0}.$$

It follows that

$$(f-\lambda_j)^m(w_i) = (\lambda_i-\lambda_j)^m w_i,$$

for every $i, j = 1, \dots, r$ and every $m \geq 1$.

Now we consider a linear combination

$$\alpha_1 v_1 + \alpha_2 v_2 + \cdots + \alpha_r v_r = \mathbf{0}$$

and we apply the endomorphism

$$(f-\lambda_1)^{m_1-1}(f-\lambda_2)^{m_2}(f-\lambda_3)^{m_3}\cdots(f-\lambda_r)^{m_r}.$$

From what we have seen above, we get:

$$\alpha_1(\lambda_1-\lambda_2)^{m_2}(\lambda_1-\lambda_3)^{m_3}\cdots(\lambda_1-\lambda_r)^{m_r}(f-\lambda_1)^{m_1-1}(v_1) = \mathbf{0}$$

and recalling that the eigenvalues $\lambda_1, \dots, \lambda_r$ are pairwise distinct, it follows that $\alpha_1 = 0$.

The above linear combination then reduces to

$$\alpha_2 v_2 + \cdots + \alpha_r v_r = \mathbf{0}.$$

Now applying the endomorphism

$$(f - \lambda_2)^{m_2 - 1}(f - \lambda_3)^{m_3} \cdots (f - \lambda_r)^{m_r}$$

we obtain:

$$\alpha_2(\lambda_2 - \lambda_3)^{m_3} \cdots (\lambda_2 - \lambda_r)^{m_r}(f - \lambda_2)^{m_2 - 1}(v_2) = \mathbf{0}$$

hence $\alpha_2 = 0$.

Continuing in this way we prove that all α_i, $i = 1, \dots, r$, are zero. □

Recall that, even if a linear function $f\colon V \to V$ has all its eigenvalues in the field K (which always happens, if K is algebraically closed), the eigenvectors of f may fail to generate the vector space V. In this case there is no basis of V consisting of eigenvectors of f, so f is not diagonalizable. We will now see that with generalized eigenvectors such a problem does not arise.

Proposition 4.2.4. *Let V be a finite dimensional vector space over $\mathbb{C}$. For every endomorphism $f\colon V \to V$ there exists a basis of V formed by generalized eigenvectors of f.*

Proof. We will prove the statement by induction on the dimension of V. If $\dim V = 1$ every linear function $f\colon V \to V$ is given by the multiplication by a scalar λ, hence every nonzero vector $v \in V$ is an eigenvector of f.

Suppose therefore that the result is true for all vector spaces of dimension strictly smaller than $n = \dim V$. Let $\lambda \in \mathbb{C}$ be an eigenvalue of f and $V_\lambda = \operatorname{Ker}(f - \lambda)^n$ be the subspace of V formed by the generalized eigenvectors of f corresponding to the eigenvalue λ. If $V = V_\lambda$ the proof is finished (see Lemma 4.2.2). Otherwise we set $W_\lambda = \operatorname{Im}(f - \lambda)^n$. We want to prove that $V = V_\lambda \oplus W_\lambda$.

Since $\dim V_\lambda + \dim W_\lambda = \dim V$, it suffices to prove that $V_\lambda \cap W_\lambda = \{\mathbf{0}\}$. Let $v \in V_\lambda \cap W_\lambda$. Since $v \in W_\lambda = \operatorname{Im}(f - \lambda)^n$, we have $v = (f - \lambda)^n(u)$, for some $u \in V$. But since $v \in V_\lambda = \operatorname{Ker}(f - \lambda)^n$, we have $(f - \lambda)^n(v) = (f - \lambda)^{2n}(u) = \mathbf{0}$. This means that u is a generalized eigenvector of f for the eigenvalue λ, and since the period of any generalized eigenvector is $< n$, we must have $(f - \lambda)^n(u) = \mathbf{0}$, i.e., $v = \mathbf{0}$.

Now we prove that W_λ is stable for f, i.e., that $f(W_\lambda) \subseteq W_\lambda$. Let $v \in W_\lambda$, then $v = (f-\lambda)^n(u)$, for some $u \in V$. We have

$$f(v) = f((f-\lambda)^n(u)) = (f-\lambda)^n(f(u)),$$

because f commutes with $(f-\lambda)^n$. This means that $f(v) \in W_\lambda$, which is what we wanted to prove.

Therefore we can consider the vector space W_λ with the restriction of the linear function f. Since $\dim W_\lambda < \dim V$, by the inductive hypothesis it follows that W_λ is generated by generalized eigenvectors of f. This concludes the proof. □

We can summarize what we have seen so far in the following theorem:

Theorem 4.2.5. *Let V and f be as above, $\lambda_1, \dots, \lambda_r$ be all the eigenvalues of f, with $\lambda_i \neq \lambda_j$ if $i \neq j$, and let*

$$P(x) = (x-\lambda_1)^{e_1}(x-\lambda_2)^{e_2} \cdots (x-\lambda_r)^{e_r}$$

be the characteristic polynomial of f. Then we have:

(*i*) $V = V_{\lambda_1} \oplus V_{\lambda_2} \oplus \cdots \oplus V_{\lambda_r}$, *where V_{λ_i} is the vector subspace of the generalized eigenvectors of f corresponding to the eigenvalue λ_i.*

(*ii*) *Each V_{λ_i} is stable for f, i.e., we have $f(V_{\lambda_i}) \subseteq V_{\lambda_i}$, for every $i = 1, \dots, r$.*

(*iii*) $\dim V_{\lambda_i} = e_i$, *hence we have*

$$V_{\lambda_i} = \operatorname{Ker}(f-\lambda_i)^n = \operatorname{Ker}(f-\lambda_i)^{e_i},$$

for every $i = 1, \dots, r$.

Proof. (*i*) By the previous proposition V is generated by the generalized eigenvectors of f, i.e., $V = V_{\lambda_1} + V_{\lambda_2} + \cdots + V_{\lambda_r}$. Since we have already proved that generalized eigenvectors corresponding to distinct eigenvalues are linearly independent, this sum is direct.

(*ii*) We recall that f commutes with $(f-\lambda)^m$, for every $\lambda \in K$ and every $m \geq 1$. If $v \in V_{\lambda_i}$ we have $(f-\lambda_i)^n(v) = \mathbf{0}$, hence

$$\mathbf{0} = f((f-\lambda_i)^n(v)) = (f-\lambda_i)^n(f(v)),$$

i.e., $f(v) \in V_{\lambda_i}$.

(*iii*) From what we have seen in (*ii*) it follows that f induces an endomorphism of V_{λ_i}, for every $i = 1, \ldots, r$. The restriction of f to V_{λ_i} has the unique eigenvalue λ_i, hence its characteristic polynomial is $(x - \lambda_i)^{d_i}$, where $d_i = \dim V_{\lambda_i}$. Since the characteristic polynomial of f is the product of the characteristic polynomials of the restriction of f to the subspaces V_{λ_i}, we necessarily have

$$(x - \lambda_1)^{e_1} \cdots (x - \lambda_r)^{e_r} = (x - \lambda_1)^{d_1} \cdots (x - \lambda_r)^{d_r},$$

from which it follows that $d_i = e_i$, for every $i = 1, \ldots, r$. From Lemma 4.2.2 it follows that

$$V_{\lambda_i} = \mathrm{Ker}(f - \lambda_i)^n = \mathrm{Ker}(f - \lambda_i)^{e_i}. \qquad \square$$

Remark 4.2.6. Point (*iii*) of the previous theorem states that the period of a generalized eigenvector of f relative to the eigenvalue λ_i is less than or equal to the exponent e_i with which the factor $(x-\lambda_i)$ appears in the characteristic polynomial of f, i.e., the period of each generalized eigenvector is less than or equal to the algebraic multiplicity of the corresponding eigenvalue.

From (*i*) and (*ii*) of the previous theorem it follows that, if we choose a basis $\{w_1^{(i)}, w_2^{(i)}, \ldots, w_{d_i}^{(i)}\}$ of V_{λ_i}, for each $i = 1, \ldots, r$ (where $d_i = e_i = \dim V_{\lambda_i}$), then the set

$$\{w_1^{(1)}, \ldots, w_{d_1}^{(1)}, w_1^{(2)}, \ldots, w_{d_2}^{(2)}, \ldots, w_1^{(r)}, \ldots, w_{d_r}^{(r)}\}$$

is a basis of V with respect to which the matrix of f is a block matrix

$$A = \left(\begin{array}{c|c|c|c} A_1 & \mathbf{0} & \cdots & \mathbf{0} \\ \hline \mathbf{0} & A_2 & \cdots & \mathbf{0} \\ \hline \vdots & \vdots & \ddots & \vdots \\ \hline \mathbf{0} & \mathbf{0} & \cdots & A_r \end{array}\right)$$

where each matrix A_i (for each $i = 1, \ldots, r$) is the matrix of the restriction of f to the subspace V_{λ_i}, with respect to the basis $\{w_1^{(i)}, w_2^{(i)}, \ldots, w_{d_i}^{(i)}\}$.

We now focus our attention on the restriction of f to a single subspace V_{λ_i}. Let m_i be the maximum of the periods of the elements of V_{λ_i}; it is $m_i \leq d_i = \dim V_{\lambda_i}$ and we have the following chain of inclusions

$$\mathrm{Ker}(f - \lambda_i) \subset \mathrm{Ker}(f - \lambda_i)^2 \subset \cdots \subset \mathrm{Ker}(f - \lambda_i)^{m_i} = V_{\lambda_i},$$

where all inclusions are strict. Therefore we can choose the basis of V_{λ_i} as follows: we start by choosing a basis of $\text{Ker}(f - \lambda_i)$, then complete it to a basis of $\text{Ker}(f - \lambda_i)^2$, which we will complete in turn to a basis of $\text{Ker}(f - \lambda_i)^3$ and so on, until we obtain a basis for V_{λ_i}.

Now we note that, if $v \in \text{Ker}(f - \lambda_i)^k$, we have

$$f(v) = \lambda_i v + (f - \lambda_i)(v),$$

and $(f - \lambda_i)(v) \in \text{Ker}(f - \lambda_i)^{k-1}$. It follows that, if $\{w_1^{(i)}, w_2^{(i)}, \dots, w_{d_i}^{(i)}\}$ is the basis of V_{λ_i} constructed in the way just described, we have

$$f(w_j^{(i)}) = \lambda_i w_j^{(i)} + \sum_{h=1}^{j-1} \alpha_h w_h^{(i)},$$

for every $j = 1, \dots, d_i$. This means that the matrix A_i of f with respect to this basis has the following form

$$A_i = \begin{pmatrix} \lambda_i & * & * & \cdots & * \\ 0 & \lambda_i & * & \cdots & * \\ 0 & 0 & \lambda_i & \cdots & * \\ \vdots & \vdots & \ddots & \ddots & * \\ 0 & 0 & \cdots & 0 & \lambda_i \end{pmatrix}$$

i.e., is an upper triangular matrix with all diagonal entries equal to the eigenvalue λ_i.

This is a good result: it states that, on the field $\mathbb{C}$ of complex numbers, every square matrix A is similar to a block diagonal matrix, where the diagonal blocks A_i are upper triangular matrices in which the elements on the main diagonal are the eigenvalues λ_i of the matrix A.

But if we try a little harder, we can get a better result. Before continuing, however, we want to introduce the notion of *minimal polynomial* of an endomorphism of V (or, equivalently, of a square matrix).

Let V and f be as above and let

$$P(x) = (x - \lambda_1)^{e_1}(x - \lambda_2)^{e_2} \cdots (x - \lambda_r)^{e_r}$$

be the characteristic polynomial of f, where we assume that the eigenvalues $\lambda_1, \dots, \lambda_r$ are pairwise distinct. We have already observed that the period of each generalized eigenvector of f, corresponding to the eigenvalue λ_i, is less than or equal to e_i. So let m_i be the maximum of the periods of the elements of V_{λ_i}; obviously it is $m_i \leq e_i$, for every $i = 1, \dots, r$.

We now give the following definition:

Definition 4.2.7. With the previous notations, the polynomial

$$Q(x) = (x - \lambda_1)^{m_1}(x - \lambda_2)^{m_2} \cdots (x - \lambda_r)^{m_r}$$

is called the *minimal polynomial*[2] of the endomorphism f of V.

We note immediately that, since $m_i \leq e_i$, for each $i = 1, \ldots, r$, the minimal polynomial of f divides its characteristic polynomial. Furthermore, the minimal polynomial and the characteristic polynomial of an endomorphism f have the same zeros, which are precisely the eigenvalues of f.

We can now provide a diagonalizability criterion of an endomorphism in terms of its minimal polynomial.

Proposition 4.2.8. *Let V and f be as above. The endomorphism f is diagonalizable if and only if its minimal polynomial is a product of distinct linear factors.*

Proof. To say that the minimal polynomial of f is a product of distinct linear factors is equivalent to saying that the periods $m_1, \ldots, m_r$ of the generalized eigenvectors are all equal to 1. But this is equivalent to saying that the generalized eigenvectors are, in fact, true eigenvectors. From Theorem 4.2.5 (i) we therefore deduce the existence of a basis of V consisting of eigenvectors of f, which is equivalent to saying that f is diagonalizable. □

We can also prove the following result:

Theorem 4.2.9 (Hamilton–Cayley). *Let V and f be as above, and let*

$$Q(x) = (x - \lambda_1)^{m_1}(x - \lambda_2)^{m_2} \cdots (x - \lambda_r)^{m_r}$$

be the minimal polynomial of f. Then we have $Q(f) = 0$.

Proof. $Q(f)$ is the endomorphism of V given by

$$(f - \lambda_1)^{m_1}(f - \lambda_2)^{m_2} \cdots (f - \lambda_r)^{m_r}.$$

If $v \in V_{\lambda_i}$, for some $i = 1, \ldots, r$, we have $(f - \lambda_i)^{m_i}(v) = \mathbf{0}$, hence $Q(f)(v) = \mathbf{0}$. Since every vector of V can be written as a sum of vectors belonging to the subspaces V_{λ_i} (Theorem 4.2.5 (i)), it follows that $Q(f)(v) = \mathbf{0}$ for every $v \in V$, which means that $Q(f)$ is the zero endomorphism. □

[2] The name *minimal polynomial* derives from the fact that it is the monic polynomial of minimum degree vanishing on f, i.e., such that $Q(f)$ is the zero endomorphism.

Corollary 4.2.10. *Let A be a square matrix of order n and $Q(x)$ be its minimal polynomial. Then $Q(A) = \mathbf{0}$. In particular, if $P(x)$ is the characteristic polynomial of A, we also have $P(A) = \mathbf{0}$ (it suffices to recall that $Q(x)$ divides $P(x)$).*

We now return to the problem of improving the choice of the basis of each subspace V_{λ_i}, in order to further simplify the matrices A_i which appear as diagonal blocks of the matrix A of f.

We focus our attention on a single eigenvalue, which we will denote by λ, and on the corresponding subspace of generalized eigenvectors V_λ. We denote by m the maximum of the periods of the elements of V_λ and consider the chain of strict inclusions

$$\operatorname{Ker}(f-\lambda) \subset \operatorname{Ker}(f-\lambda)^2 \subset \cdots \subset \operatorname{Ker}(f-\lambda)^m = V_\lambda.$$

For each $j = 1, \dots, m$, we set $d_j = \dim \operatorname{Ker}(f-\lambda)^j$.

If we choose a nonzero vector $v \in \operatorname{Ker}(f-\lambda)^m \setminus \operatorname{Ker}(f-\lambda)^{m-1}$, the m vectors

$$w_m = v,\ w_{m-1} = (f-\lambda)(v),\ w_{m-2} = (f-\lambda)^2(v), \dots,\ w_1 = (f-\lambda)^{m-1}(v),$$

are linearly independent (cf. Lemma 4.2.2). Note that $(f-\lambda)(w_1) = (f-\lambda)^m(v) = \mathbf{0}$, i.e.,

$$f(w_1) = \lambda w_1,$$

and

$$(f-\lambda)(w_j) = (f-\lambda)(f-\lambda)^{m-j}(v) = (f-\lambda)^{m-j+1}(v) = w_{j-1},$$

that is

$$f(w_j) = \lambda w_j + w_{j-1},$$

for each $j = 2, \dots, m$.

This means that f induces an endomorphism of the subspace of V_λ generated by the vectors $w_1, w_2, \dots, w_m$, whose matrix, with respect to the basis $\{w_1, w_2, \dots, w_m\}$ of this subspace is the following square matrix of order m:

$$J_\lambda = \begin{pmatrix} \lambda & 1 & 0 & 0 & \cdots & 0 \\ 0 & \lambda & 1 & 0 & \cdots & 0 \\ 0 & 0 & \lambda & 1 & \cdots & 0 \\ \vdots & \vdots & \ddots & \ddots & \ddots & \vdots \\ 0 & 0 & \cdots & 0 & \lambda & 1 \\ 0 & 0 & \cdots & 0 & 0 & \lambda \end{pmatrix}$$

Such a matrix is called a *Jordan block* of order m corresponding to the eigenvalue λ.

Let $s = d_m - d_{m-1} = \dim \operatorname{Ker}(f-\lambda)^m - \dim \operatorname{Ker}(f-\lambda)^{m-1}$. Recall that $s \geq 1$. We want to prove that the matrix of the restriction of f to the subspace V_λ contains, along the main diagonal, s Jordan blocks J_λ of order m. We consider s vectors

$$v_1, v_2, \ldots, v_s \in \operatorname{Ker}(f-\lambda)^m \setminus \operatorname{Ker}(f-\lambda)^{m-1}$$

such that

$$\operatorname{Ker}(f-\lambda)^m = \operatorname{Ker}(f-\lambda)^{m-1} \oplus \langle v_1, \ldots, v_s \rangle. \tag{4.2.2}$$

By applying to each of the vectors $v_1, \ldots, v_s$ the procedure described above, we obtain the following set of vectors:

$$\begin{array}{l} v_1,\ (f-\lambda)(v_1),\ (f-\lambda)^2(v_1),\ \ldots,\ (f-\lambda)^{m-1}(v_1), \\ v_2,\ (f-\lambda)(v_2),\ (f-\lambda)^2(v_2),\ \ldots,\ (f-\lambda)^{m-1}(v_2), \\ \cdot\ \cdot\ \cdot\ \cdot\ \cdot\ \cdot\ \cdot\ \cdot\ \cdot\ \cdot\ \cdot\ \cdot\ \cdot\ \cdot\ \cdot \\ v_s,\ (f-\lambda)(v_s),\ (f-\lambda)^2(v_s),\ \ldots,\ (f-\lambda)^{m-1}(v_s) \end{array} \tag{4.2.3}$$

(each of these rows corresponds to a Jordan block of order m).

Now we have to prove that all these vectors are linearly independent. We consider a linear combination

$$\begin{aligned} \alpha_1^{(0)} v_1 + \cdots + \alpha_s^{(0)} v_s + \alpha_1^{(1)}(f-\lambda)(v_1) + \cdots + \alpha_s^{(1)}(f-\lambda)(v_s) + \\ + \alpha_1^{(2)}(f-\lambda)^2(v_1) + \cdots + \alpha_s^{(2)}(f-\lambda)^2(v_s) + \cdots \\ \cdots + \alpha_1^{(m-1)}(f-\lambda)^{m-1}(v_1) + \cdots + \alpha_s^{(m-1)}(f-\lambda)^{m-1}(v_s) = \mathbf{0}. \end{aligned}$$

Leaving only $\alpha_1^{(0)} v_1 + \cdots + \alpha_s^{(0)} v_s$ on the left and bringing all the other addends to the right of the equal sign, we deduce that

$$\alpha_1^{(0)} v_1 + \cdots + \alpha_s^{(0)} v_s \in \operatorname{Ker}(f-\lambda)^{m-1},$$

from which it follows that $\alpha_1^{(0)} = \alpha_2^{(0)} = \cdots = \alpha_s^{(0)} = 0$ (recall that, according to (4.2.2), we have $\operatorname{Ker}(f-\lambda)^{m-1} \cap \langle v_1, \ldots, v_s \rangle = \{\mathbf{0}\}$).

Therefore the previous linear combination reduces to

$$\begin{aligned} \alpha_1^{(1)}(f-\lambda)(v_1) + \cdots + \alpha_s^{(1)}(f-\lambda)(v_s) + \\ + \alpha_1^{(2)}(f-\lambda)^2(v_1) + \cdots + \alpha_s^{(2)}(f-\lambda)^2(v_s) + \cdots \\ \cdots + \alpha_1^{(m-1)}(f-\lambda)^{m-1}(v_1) + \cdots + \alpha_s^{(m-1)}(f-\lambda)^{m-1}(v_s) = \mathbf{0}. \end{aligned}$$

Proceeding as before, we have

$$\alpha_1^{(1)}(f-\lambda)(v_1)+\cdots+\alpha_s^{(1)}(f-\lambda)(v_s)\in \operatorname{Ker}(f-\lambda)^{m-2},$$

hence

$$\alpha_1^{(1)}v_1+\cdots+\alpha_s^{(1)}v_s\in \operatorname{Ker}(f-\lambda)^{m-1},$$

from which it follows that $\alpha_1^{(1)}=\alpha_2^{(1)}=\cdots=\alpha_s^{(1)}=0$.

Continuing in this way we prove that all the coefficients $\alpha_j^{(i)}$ are zero, which concludes the proof of linear independence.

Now we observe that if $d_m-d_{m-1}=s$, then $d_i-d_{i-1}\geq s$, for every $i=1,\dots,m$ (where we set $d_0=\dim\operatorname{Ker}(f-\lambda)^0=0$). In fact, since $v_1,\dots,v_s$ have period m, the vectors

$$(f-\lambda)^{m-i}(v_1),(f-\lambda)^{m-i}(v_2),\dots,(f-\lambda)^{m-i}(v_s)$$

have period i and, as we have seen, are linearly independent.

If it happens that $d_i-d_{i-1}=s$, for every $i=1,\dots,m$, we have

$$\begin{aligned}\dim\operatorname{Ker}(f-\lambda)&=s\\ \dim\operatorname{Ker}(f-\lambda)^2&=2s\\ \cdot\quad\cdot\quad\cdot&\quad\cdot\quad\cdot\quad\cdot\quad\cdot\\ \dim\operatorname{Ker}(f-\lambda)^m&=m\,s\end{aligned}$$

and, since $V_\lambda=\operatorname{Ker}(f-\lambda)^m$, the $m\,s$ vectors of (4.2.3) are a basis of V_λ. Recalling what we have previously seen, it is now immediate to verify that the matrix of the restriction of f to V_λ consists of s Jordan blocks of order m:

$$\left(\begin{array}{c|c|c|c} J_\lambda & \mathbf{0} & \cdots & \mathbf{0}\\ \hline \mathbf{0} & J_\lambda & \cdots & \mathbf{0}\\ \hline \vdots & \vdots & \ddots & \vdots\\ \hline \mathbf{0} & \mathbf{0} & \cdots & J_\lambda \end{array}\right)$$

If instead we have $d_i-d_{i-1}>s$, for some i, we denote by j the maximum index $<m$ such that $d_j-d_{j-1}=t>s$ and choose vectors $w_1,w_2,\dots,w_{t-s}$ such that we have

$$\begin{aligned}\operatorname{Ker}(f-\lambda)^j=\operatorname{Ker}(f-\lambda)^{j-1}\oplus\langle(f-\lambda)^{m-j}(v_1),\dots,(f-\lambda)^{m-j}(v_s)\rangle\\ \oplus\langle w_1,\dots,w_{t-s}\rangle.\end{aligned}$$

Reasoning in a similar way to what was done previously, it can be shown that the vectors

$$\begin{array}{l} w_1, (f-\lambda)(w_1), (f-\lambda)^2(w_1), \ldots, (f-\lambda)^{j-1}(w_1), \\ w_2, (f-\lambda)(w_2), (f-\lambda)^2(w_2), \ldots, (f-\lambda)^{j-1}(w_2), \\ \cdot\ \cdot\ \cdot\ \cdot\ \cdot\ \cdot\ \cdot\ \cdot\ \cdot\ \cdot\ \cdot\ \cdot\ \cdot\ \cdot\ \cdot\ \cdot \\ w_{t-s}, (f-\lambda)(w_{t-s}), (f-\lambda)^2(w_{t-s}), \ldots, (f-\lambda)^{j-1}(w_{t-s}) \end{array}$$

are linearly independent. From this it follows that, in this case, the matrix of the restriction of f to V_λ also contains, along the main diagonal, $t-s$ Jordan blocks of order j associated to the eigenvalue λ, in addition to the s blocks J_λ of order m already mentioned.

Reasoning in a similar way with the generalized eigenvectors of increasingly smaller period, we arrive at the conclusion that, in general, the matrix of the restriction of f to V_λ is a block diagonal matrix, in which the diagonal blocks are Jordan blocks of order $\leq m$ for the eigenvalue λ, where m is the maximum of the periods of the elements of V_λ (furthermore there exists at least one Jordan block of order m). This integer m is precisely the exponent with which the factor $(x-\lambda)$ appears in the minimum polynomial of f.

We have thus proved the following theorem:

Theorem 4.2.11 (Jordan). *Let V be a vector space of dimension n over the field*[3] $\mathbb{C}$ *and $f\colon V \to V$ be a linear function. Let*

$$P(x) = (x-\lambda_1)^{e_1}(x-\lambda_2)^{e_2}\cdots(x-\lambda_r)^{e_r}$$

be its characteristic polynomial (we assume that the eigenvalues $\lambda_1, \ldots, \lambda_r$ are pairwise distinct). Then there exists a basis of V with respect to which the matrix of f is a block matrix of the form

$$J = \left(\begin{array}{c|c|c|c} J_1 & \mathbf{0} & \cdots & \mathbf{0} \\ \hline \mathbf{0} & J_2 & \cdots & \mathbf{0} \\ \hline \vdots & \vdots & \ddots & \vdots \\ \hline \mathbf{0} & \mathbf{0} & \cdots & J_r \end{array}\right)$$

[3] An analogous result holds for any algebraically closed field.

where each J_i is, in turn, a block matrix of the form

$$J_i = \left(\begin{array}{c|c|c|c} J^1_{\lambda_i} & \mathbf{0} & \cdots & \mathbf{0} \\ \hline \mathbf{0} & J^2_{\lambda_i} & \cdots & \mathbf{0} \\ \hline \vdots & \vdots & \ddots & \vdots \\ \hline \mathbf{0} & \mathbf{0} & \cdots & J^{s_i}_{\lambda_i} \end{array}\right)$$

where $J^1_{\lambda_i}, \ldots, J^{s_i}_{\lambda_i}$ are Jordan blocks corresponding to the eigenvalue λ_i.

A matrix J of this form is called a *canonical Jordan form* and a basis of V with respect to which f has such a matrix is called a *Jordan basis.*

Let us now try to clarify the above by applying the results obtained so far in a concrete example.

Exercise. Let V be a vector space of dimension 5 over the complex field $\mathbb{C}$ and let $\{v_1, \ldots, v_5\}$ be a basis of V. Let $f\colon V \to V$ be the linear function given by the matrix

$$A = \begin{pmatrix} 1 & 0 & -2 & 0 & 0 \\ -2 & 3 & -2 & 0 & 0 \\ 2 & 0 & 5 & 0 & 0 \\ 0 & 2 & 2 & 2 & -1 \\ 0 & 0 & 0 & 1 & 4 \end{pmatrix}$$

with respect to the chosen basis. Determine the characteristic polynomial and the minimal polynomial of f, a Jordan canonical form of f, and a Jordan basis of V.

The characteristic polynomial of f is

$$P(x) = \det(x \cdot \mathbf{1} - A) = \begin{vmatrix} x-1 & 0 & 2 & 0 & 0 \\ 2 & x-3 & 2 & 0 & 0 \\ -2 & 0 & x-5 & 0 & 0 \\ 0 & -2 & -2 & x-2 & 1 \\ 0 & 0 & 0 & -1 & x-4 \end{vmatrix}$$

$$= \begin{vmatrix} x-1 & 0 & 2 \\ 2 & x-3 & 2 \\ -2 & 0 & x-5 \end{vmatrix} \cdot \begin{vmatrix} x-2 & 1 \\ -1 & x-4 \end{vmatrix}$$

$$= (x-3)[(x-1)(x-5)+4][(x-2)(x-4)+1]$$

$$= (x-3)(x-3)^2(x-3)^2 = (x-3)^5,$$

hence f has the unique eigenvalue $\lambda = 3$, with multiplicity 5. Since the minimal polynomial $Q(x)$ must divide the characteristic polynomial, it must be

$$Q(x) = (x-3)^m, \qquad \text{with } 1 \leq m \leq 5.$$

Note that

$$A - 3 = \begin{pmatrix} -2 & 0 & -2 & 0 & 0 \\ -2 & 0 & -2 & 0 & 0 \\ 2 & 0 & 2 & 0 & 0 \\ 0 & 2 & 2 & -1 & -1 \\ 0 & 0 & 0 & 1 & 1 \end{pmatrix}$$

$$(A-3)^2 = \begin{pmatrix} 0 & 0 & 0 & 0 & 0 \\ 0 & 0 & 0 & 0 & 0 \\ 0 & 0 & 0 & 0 & 0 \\ 0 & -2 & -2 & 0 & 0 \\ 0 & 2 & 2 & 0 & 0 \end{pmatrix}$$

and finally $(A-3)^3 = \mathbf{0}$, therefore the minimal polynomial is $Q(x) = (x-3)^3$. This means that the maximum period of the generalized eigenvectors of f (for the unique eigenvalue 3) is $m = 3$, hence we have the chain of strict inclusions

$$\operatorname{Ker}(f-3) \subset \operatorname{Ker}(f-3)^2 \subset \operatorname{Ker}(f-3)^3 = V.$$

We have:

$$\begin{aligned} \dim \operatorname{Ker}(f-3) &= 5 - \operatorname{rk}(A - 3 \cdot \mathbf{1}) = 5 - 3 = 2, \\ \dim \operatorname{Ker}(f-3)^2 &= 5 - \operatorname{rk}(A - 3 \cdot \mathbf{1})^2 = 5 - 1 = 4, \\ \dim \operatorname{Ker}(f-3)^3 &= \dim V = 5. \end{aligned}$$

Setting $d_i = \dim \operatorname{Ker}(f-3)^i$, the integer $s = d_m - d_{m-1} = d_3 - d_2$ of the proof of Jordan theorem is $s = 5-4 = 1$, hence the Jordan matrix J contains a Jordan block of order $m = 3$ for the eigenvalue $\lambda = 3$,

$$\begin{pmatrix} 3 & 1 & 0 \\ 0 & 3 & 1 \\ 0 & 0 & 3 \end{pmatrix}$$

Since we have $\dim \operatorname{Ker}(f-3)^2 - \dim \operatorname{Ker}(f-3) = 4 - 2 = 2$, using the same notations used in the proof of Jordan theorem, we have $j = 2$ and $d_2 - d_1 = t = 2 > s = 1$, hence $t - s = 1$ and the Jordan matrix J also contains a Jordan block of order $j = 2$

$$\begin{pmatrix} 3 & 1 \\ 0 & 3 \end{pmatrix}$$

In conclusion, the Jordan matrix of f is:

$$J = \left(\begin{array}{ccc|cc} 3 & 1 & 0 & 0 & 0 \\ 0 & 3 & 1 & 0 & 0 \\ 0 & 0 & 3 & 0 & 0 \\ \hline 0 & 0 & 0 & 3 & 1 \\ 0 & 0 & 0 & 0 & 3 \end{array}\right)$$

Now we must find a basis $\{w_1, \dots, w_5\}$ of V with respect to which the matrix of f is J. We choose a vector $v \in \operatorname{Ker}(f-3)^3$ such that $v \notin \operatorname{Ker}(f-3)^2$: for instance, the vector v_2 satisfies such requests. Then we set

$$w_3 = v_2, \quad w_2 = (f-3)(v_2) = 2v_4, \quad w_1 = (f-3)^2(v_2) = -2v_4 + 2v_5.$$

The vectors $\{w_1, w_2, w_3\}$ are linearly independent and form the part of the basis of V which is responsible for the presence of the Jordan block of order 3.

To continue we must now choose a vector u such that we have

$$\operatorname{Ker}(f-3)^2 = \operatorname{Ker}(f-3) \oplus \langle (f-3)^{3-2}(v_2) \rangle \oplus \langle u \rangle,$$

i.e., u must be a vector in $\operatorname{Ker}(f-3)^2$ that does not belong to the subspace generated by $\operatorname{Ker}(f-3)$ and the vector $(f-3)(v_2) = w_2 = 2v_4$.

We need to determine the kernel of $f-3$: it is given by the vectors $x_1v_1 + x_2v_2 + \cdots + x_5v_5$ such that the x_i are solutions of the following linear system:

$$\begin{cases} -2x_1 - 2x_3 = 0 \\ 2x_2 + 2x_3 - x_4 - x_5 = 0 \\ x_4 + x_5 = 0. \end{cases}$$

By solving this system, we find

$$\begin{cases} x_1 = -x_3 \\ x_2 = -x_3 \\ x_4 = -x_5 \end{cases}$$

hence a basis of $\operatorname{Ker}(f-3)$ is given by the vectors $v_1 + v_2 - v_3$ and $v_4 - v_5$. Therefore we have

$$\begin{aligned} \operatorname{Ker}(f-3) \oplus \langle w_2 \rangle &= \langle v_1 + v_2 - v_3, v_4 - v_5, 2v_4 \rangle \\ &= \langle v_1 + v_2 - v_3, v_4, v_5 \rangle. \end{aligned}$$

It is now easy to see that the vector v_1 does not belong to this subspace, but belongs to the kernel of $(f-3)^2$: this is the vector u we are looking for. The two vectors missing to complete the Jordan basis are therefore

$$w_5 = v_1, \quad w_4 = (f-3)(v_1) = -2v_1 - 2v_2 + 2v_3.$$

In conclusion, the vectors

$$\begin{aligned} w_1 &= -2v_4 + 2v_5 \\ w_2 &= 2v_4 \\ w_3 &= v_2 \\ w_4 &= -2v_1 - 2v_2 + 2v_3 \\ w_5 &= v_1 \end{aligned}$$

are a basis of V with respect to which the matrix of f is

$$J = \begin{pmatrix} 3 & 1 & 0 & 0 & 0 \\ 0 & 3 & 1 & 0 & 0 \\ 0 & 0 & 3 & 0 & 0 \\ 0 & 0 & 0 & 3 & 1 \\ 0 & 0 & 0 & 0 & 3 \end{pmatrix}$$

Finally, we observe that the basis change matrix (i.e., the matrix whose columns contain the coordinates of the vectors of the new basis $\{w_1, \ldots, w_5\}$ computed with respect to the original basis $\{v_1, \ldots, v_5\}$) is

$$P = \begin{pmatrix} 0 & 0 & 0 & -2 & 1 \\ 0 & 0 & 1 & -2 & 0 \\ 0 & 0 & 0 & 2 & 0 \\ -2 & 2 & 0 & 0 & 0 \\ 2 & 0 & 0 & 0 & 0 \end{pmatrix}$$

Therefore we have

$$J = P^{-1}AP$$

or, equivalently,

$$PJ = AP,$$

as can be easily verified by a direct computation.

4.3 Symmetric and skew-symmetric matrices

In this section we will study some properties of the eigenvalues of symmetric or skew-symmetric matrices with real entries. We start by giving the following definition:

Definition 4.3.1. Let $A \in M_n(K)$ be a square matrix of order n with entries in a field K. A is *symmetric* if ${}^tA = A$, it is *skew-symmetric* (or *anti-symmetric*) if ${}^tA = -A$.

Remark 4.3.2. Every square matrix A can always[4] be written as the sum of a symmetric matrix A_s and a skew-symmetric matrix A_a. In fact, by setting

$$A_s = \frac{A + {}^tA}{2}, \qquad A_a = \frac{A - {}^tA}{2},$$

we have $A = A_s + A_a$.

Now we consider symmetric or skew-symmetric matrices with real entries. We will prove that a symmetric matrix with real entries has all its eigenvalues in $\mathbb{R}$. We need the following result:

Lemma 4.3.3. *Given a vector $v = {}^t(a_1, a_2, \ldots, a_n) \in \mathbb{C}^n$ we will denote by $\bar{v} = {}^t(\bar{a}_1, \bar{a}_2, \ldots, \bar{a}_n)$ the complex conjugate of v. Then, for every $v \in \mathbb{C}^n$, we have ${}^tv\bar{v} \geq 0$, and ${}^tv\bar{v} = 0$ if and only if $v = \mathbf{0}$.*

Proof. The proof consists of a simple computation:

$$\begin{aligned} {}^tv\bar{v} &= (a_1, \ldots, a_n) \begin{pmatrix} \bar{a}_1 \\ \vdots \\ \bar{a}_n \end{pmatrix} \\ &= a_1\bar{a}_1 + a_2\bar{a}_2 + \cdots + a_n\bar{a}_n \\ &= |a_1|^2 + |a_2|^2 + \cdots + |a_n|^2, \end{aligned}$$

where we have used the fact that, for any complex number z, we have $z\bar{z} = |z|^2$. We therefore conclude that ${}^tv\bar{v}$ is a sum of squares of real numbers, from which the statement of the lemma immediately follows. □

We are now able to prove the following result:

[4]If the characteristic of the field K is not 2.

Theorem 4.3.4. *Let $A \in M_n(\mathbb{R})$ be a symmetric matrix of order n. Then A has n real eigenvalues (not necessarily distinct).*

Proof. Let $P_A(x) = \det(A - x \cdot \mathbf{1})$ be the characteristic polynomial of A: it is a polynomial of degree n with real coefficients. According to the Fundamental Theorem of Algebra (Theorem A.3.3), the equation $P_A(x) = 0$ has n solutions in the complex field, hence A has n eigenvalues (not necessarily distinct) $\lambda_1, \lambda_2, \ldots, \lambda_n \in \mathbb{C}$. We will now prove that, in fact, all eigenvalues of A are real.

Let $\lambda \in \mathbb{C}$ be one of the eigenvalues of A and $v \in \mathbb{C}^n$ be an eigenvector for the eigenvalue λ, hence we have $Av = \lambda v$. By taking the complex conjugate and recalling that $\overline{A} = A$ because A has real entries, we get $A\bar{v} = \bar{\lambda}\bar{v}$. By taking the transpose of both sides of the equality $Av = \lambda v$, we obtain ${}^t v {}^t A = \lambda {}^t v$, which is equivalent to ${}^t v A = \lambda {}^t v$, since ${}^t A = A$ (because A is symmetric). Now we multiply both sides of this last equality, on the right, by $\bar{v}$:

$$ {}^t v A \bar{v} = \lambda {}^t v \bar{v}. $$

Recalling that $A\bar{v} = \bar{\lambda}\bar{v}$, we get $\bar{\lambda} {}^t v \bar{v} = \lambda {}^t v \bar{v}$, i.e., $(\bar{\lambda} - \lambda) {}^t v \bar{v} = 0$. Since $v \neq \mathbf{0}$, by the previous lemma we have ${}^t v \bar{v} > 0$, from which it follows that $\bar{\lambda} - \lambda = 0$, i.e., $\bar{\lambda} = \lambda$, which means that $\lambda \in \mathbb{R}$. □

Remark 4.3.5. We will prove later that every symmetric matrix of order n with real entries is diagonalizable; that is, it not only has n real (not necessarily distinct) eigenvalues, but it also has n linearly independent eigenvectors.

Finally, let us consider the case of skew-symmetric matrices with real coefficients. A proof completely analogous to that of the previous theorem allows us to obtain the following result:

Theorem 4.3.6. *Let $A \in M_n(\mathbb{R})$ be a skew-symmetric matrix of order n. Then A has n purely imaginary eigenvalues (not necessarily distinct).*

Proof. The proof is identical to that of the previous theorem, up to the point where, by applying the transposition to the equality $Av = \lambda v$, we get ${}^t v {}^t A = \lambda {}^t v$. In this case the matrix A is skew-symmetric, hence ${}^t A = -A$. The above equality thus becomes ${}^t v A = -\lambda {}^t v$, and multiplying on the right by $\bar{v}$ and recalling that $A\bar{v} = \bar{\lambda}\bar{v}$, we obtain $\bar{\lambda} {}^t v \bar{v} = -\lambda {}^t v \bar{v}$, i.e., $(\bar{\lambda} + \lambda) {}^t v \bar{v} = 0$. As in the proof of the previous theorem, from this equality we get $\bar{\lambda} + \lambda = 0$, i.e., $\bar{\lambda} = -\lambda$, which means that λ is purely imaginary, i.e, it is $\lambda = i\alpha$, with $\alpha \in \mathbb{R}$. □

Exercises

Exercise 4.1. Let V be a vector space with a basis $\{v_1, v_2, v_3, v_4\}$, and f be the endomorphism of V defined by $f(v_1) = 3v_1 + 2v_3$, $f(v_2) = 2v_2 + 3v_4$, $f(v_3) = 2v_1 + 3v_3$, $f(v_4) = 2v_2 + 3v_4$. Find a basis of V with respect to which the matrix of f is diagonal.

Exercise 4.2. Consider the vectors $v_1 = (1,1,1)$, $v_2 = (2,0,1)$, $v_3 = (1,1,3)$, and let $f\colon \mathbb{R}^3 \to \mathbb{R}^3$ be a linear function such that $f(v_1) = 3v_1$, $f(v_2) = 2v_2$, $f(v_3) = 2v_3 + 2v_2$.

(1) Write the matrix of f with respect to the canonical bases.

(2) Find the eigenvalues and the eigenvectors of f.

(3) Check that the sum of the eigenspaces of f is direct.

Exercise 4.3. Let $f\colon \mathbb{R}^3 \to \mathbb{R}^3$ be the linear function defined by setting

$$f(e_1 + e_2) = 6e_1 - 6e_2 + 4e_3, \quad f(e_1 - e_2) = 2e_1 - 2e_2, \quad f(e_3) = e_1 - 2e_2 + 3e_3,$$

where e_1, e_2, e_3 is the canonical basis of $\mathbb{R}^3$.

(1) Write the matrix of f with respect to the canonical basis.

(2) Find the eigenvalues and the eigenvectors of f and determine if f is diagonalizable.

(3) Find for which values of t the vector $v = (-2, t, -2)$ is an eigenvector of f.

Exercise 4.4. Given the matrix

$$A = \begin{pmatrix} -t & 3+t & -t \\ 3-t & t & 3-t \\ 6 & -6 & 6 \end{pmatrix}$$

determine for which value of t A is diagonalizable. For this value of t find a diagonal matrix D and an invertible matrix P such that $A = PDP^{-1}$.

Exercise 4.5. Find the Jordan canonical form of the endomorphism $\phi\colon \mathbb{R}^3 \to \mathbb{R}^3$ whose matrix (with respect to the canonical basis) is

$$A = \begin{pmatrix} 3 & 4 & 4 \\ -3 & -6 & -8 \\ 3 & 7 & 9 \end{pmatrix}.$$

Write a basis of $\mathbb{R}^3$ with respect to which the matrix of ϕ is the canonical form found.

Exercise 4.6. Find the Jordan canonical form of the endomorphism $\phi\colon \mathbb{Q}^4 \to \mathbb{Q}^4$ whose matrix (with respect to the canonical basis) is

$$A = \begin{pmatrix} -1 & 0 & 0 & 0 \\ 0 & 1 & 0 & 0 \\ 1 & -3 & -1 & -1 \\ 0 & -2 & 0 & 1 \end{pmatrix}.$$

Find a basis of $\mathbb{Q}^4$ with respect to which the matrix of ϕ is the canonical form found.

Exercise 4.7. Find the Jordan canonical form of the endomorphism $\phi\colon \mathbb{R}^4 \to \mathbb{R}^4$ whose matrix (with respect to the canonical basis) is

$$A = \begin{pmatrix} 2 & 0 & 0 & 0 \\ -4 & 2 & -4 & -3 \\ 4 & 0 & 5 & 3 \\ -3 & 0 & -2 & 0 \end{pmatrix}.$$

Write a basis of $\mathbb{R}^4$ with respect to which the matrix of ϕ is the canonical form found.

Exercise 4.8. Find the Jordan canonical form of the endomorphism $\phi\colon \mathbb{R}^4 \to \mathbb{R}^4$ whose matrix (with respect to the canonical basis) is

$$A = \begin{pmatrix} 0 & 0 & 2 & -4 \\ 1 & 2 & -1 & 2 \\ 0 & 0 & 2 & 0 \\ 1 & 0 & -1 & 4 \end{pmatrix}.$$

Find a basis of $\mathbb{R}^4$ with respect to which the matrix of ϕ is the canonical form found.

Exercise 4.9. Let $A \in M_n(\mathbb{R})$ be such that $A^2 = A$. Prove that A is diagonalizable and that its eigenvalues are only 0 or 1. Prove that if $B \in M_n(\mathbb{R})$ is another matrix such that $B^2 = B$, then A and B are similar if and only if they have the same rank.

Exercise 4.10. Give an example of two square matrices A and B of the same order, having the same characteristic polynomial and the same minimal polynomial, but such that A is not similar to B.

Exercise 4.11. Let V be a complex vector space of dimension 9. Determine all the endomorphisms ϕ of V that satisfy the following conditions:

$$\begin{aligned} &\dim \operatorname{Ker}(\phi - 2) = 1, \quad \dim \operatorname{Ker}(\phi - 2)^3 = 3, \\ &\dim \operatorname{Ker}(\phi - 3) = 2, \quad \dim \operatorname{Ker}(\phi - 3)^2 = 4, \\ &\dim \operatorname{Im}(\phi^2) = 7. \end{aligned}$$

Exercise 4.12. Let V be a vector space of dimension 10 over $\mathbb{Q}$. Determine the characteristic polynomial, the minimal polynomial and the Jordan matrix of all the endomorphisms ϕ of V that satisfy the following conditions:

$$\begin{aligned}
&\dim \operatorname{Ker}(\phi-5)=2, \quad \dim \operatorname{Ker}(\phi-5)^2=3, \quad \dim \operatorname{Ker}(\phi-5)^3=4,\\
&\dim \operatorname{Ker}(\phi+2)=2, \quad \dim \operatorname{Ker}(\phi+2)^2=4,\\
&\dim \operatorname{Im}(\phi)=8.
\end{aligned}$$

Exercise 4.13. Let V be a complex vector space of dimension 6. Determine all the endomorphisms ϕ of V of rank ≥ 4 whose minimal polynomial is $x^4-6x^3+9x^2$.

5

Euclidean Vector Spaces

In this chapter we will introduce the notion of *scalar product* (also called *dot product*) of two vectors (it is a particular kind of product of two vectors whose result is a scalar). We will then see how this product allows us to define the length of a vector and the angle between two vectors. These notions will then allow us, in turn, to compute areas, volumes, etc.

5.1 Lengths and angles

In Euclidean geometry, the notions of length of a segment (or distance between two points) and of angle between two lines are defined in a completely natural way. One of the most important results concerning the lengths is the Pythagorean Theorem, which allows to derive the length of the hypotenuse of a right triangle when the lengths of the two sides are known. We will take a cue from the validity of the Pythagorean Theorem in $\mathbb{R}^2$ and $\mathbb{R}^3$ to define the length of a vector in $\mathbb{R}^n$.

5.1.1 Length of a vector

Let's start by considering a vector $v = (a_1, a_2) \in \mathbb{R}^2$ in the Euclidean plane, that we can represent as in the figure on the side.

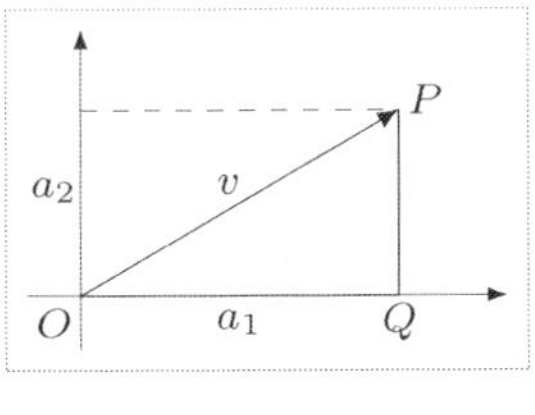

The vector v is represented by the oriented segment $\overrightarrow{OP}$, while the lengths of the segments OQ and PQ are respectively $|a_1|$ and $|a_2|$, that is, the two components of the vector v taken in absolute value to avoid problems if they are negative. Since the triangle $\triangle OPQ$ is a right triangle, it follows from the Pythagorean theorem that

$$\|OP\|^2 = \|OQ\|^2 + \|PQ\|^2,$$

where with the symbol $\|AB\|$ we denote the length of a segment AB. By recalling that $\|OQ\| = |a_1|$ and $\|PQ\| = |a_2|$, we obtain the following formula expressing the length of the segment representing the vector v:

$$\|OP\| = \sqrt{a_1^2 + a_2^2}.$$

Let's consider now a vector $v = (a_1, a_2, a_3) \in \mathbb{R}^3$, represented by the oriented segment $\overrightarrow{OP}$ as in the following figure:

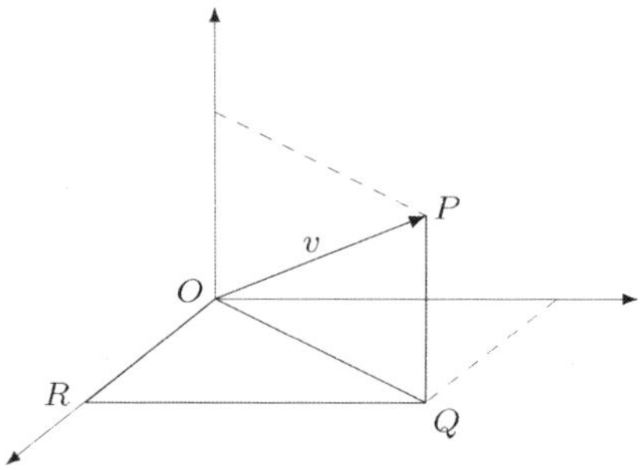

Since $\triangle OPQ$ is a right triangle, we have

$$\|OP\|^2 = \|OQ\|^2 + \|PQ\|^2$$

and, noting that also $\triangle OQR$ is a right triangle, we have

$$\|OQ\|^2 = \|OR\|^2 + \|QR\|^2,$$

hence it follows that

$$\|OP\|^2 = \|OR\|^2 + \|QR\|^2 + \|PQ\|^2.$$

Noting that $\|OR\| = |a_1|$, $\|QR\| = |a_2|$ and $\|PQ\| = |a_3|$, we conclude that the length of the (segment representing the) vector $v = (a_1, a_2, a_3)$ is given by

$$\|OP\| = \sqrt{a_1^2 + a_2^2 + a_3^2}.$$

These two results motivate the following definition:

Definition 5.1.1. Let $v = (a_1, a_2, \ldots, a_n) \in \mathbb{R}^n$. The *length* of v (also called the *norm* of v) is given by

$$\|v\| = \sqrt{a_1^2 + a_2^2 + \cdots + a_n^2}.$$

It follows that the *norm* defines a function

$$\|\cdot\|\colon \mathbb{R}^n \to \mathbb{R}, \qquad v \mapsto \|v\|$$

that satisfies the following properties:

(*i*) $\|v\| \geq 0$, for any $v \in \mathbb{R}^n$, and $\|v\| = 0$ if and only if $v = \mathbf{0}$;

(*ii*) $\|\lambda v\| = |\lambda|\|v\|$, for any $\lambda \in \mathbb{R}$ and any $v \in \mathbb{R}^n$;

(*iii*) $\|v + w\| \leq \|v\| + \|w\|$, for any $v, w \in \mathbb{R}^n$.

The property (*i*) is completely obvious (a sum of squares of real numbers is always ≥ 0 and is zero if and only if all the numbers are zero), while the property (*ii*) can be verified with an easy calculation.

The property (*iii*), called *triangular inequality*, has an obvious geometric meaning: it is equivalent to the well-known result of Euclidean geometry which states that, in each triangle, the length of one side is less than the sum of the lengths of the other two (see figure on the side). The algebraic proof of this property will be given later (see Proposition 5.1.7).

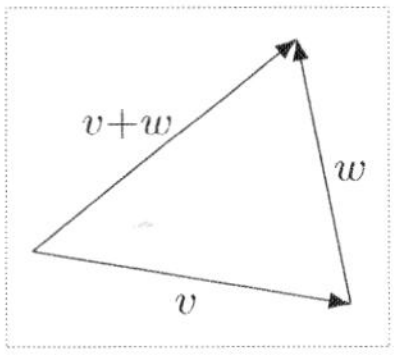

Example. As an example, we will now see how to compute the length of the diagonal of a (hyper)cube in $\mathbb{R}^n$, for any $n \geq 2$.

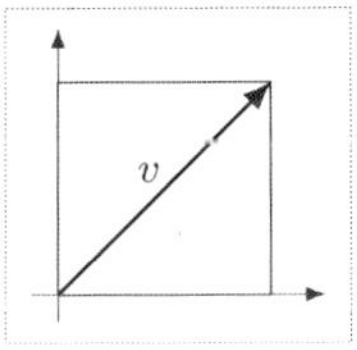

In the case of the plane ($n = 2$), we consider a square with a side of length 1, which we can always assume to have a vertex at the origin and the sides parallel to the coordinate axes. The vertices of this square are therefore the points of coordinates $(0, 0)$, $(0, 1)$, $(1, 0)$ and $(1, 1)$ and as a diagonal we can consider the segment whose end points are the vertices $(0, 0)$ and $(1, 1)$.
This segment represents the vector $v = (1, 1)$ whose length is given by $\|v\| = \sqrt{2}$.

In the case of three-dimensional space ($n = 3$), we consider a cube with a side of length 1 having a vertex at the origin and sides parallel to the coordinate axes. The vertices of the cube are therefore the points of coordinates $(0, 0, 0)$, $(0, 0, 1)$, $(0, 1, 0)$, $(0, 1, 1)$, $(1, 0, 0)$, $(1, 0, 1)$, $(1, 1, 0)$ and $(1, 1, 1)$, and the diagonal is the segment whose end points are the vertices $(0, 0, 0)$ and $(1, 1, 1)$. This segment represents the vector $v = (1, 1, 1)$ whose length is $\|v\| = \sqrt{3}$.

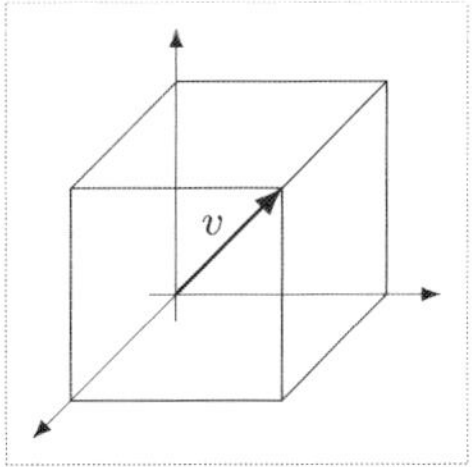

In a completely analogous way we can define the *unitary hypercube* in $\mathbb{R}^n$: it is the "solid" whose vertices are the 2^n points with coordinates $(0, 0, \ldots, 0)$, $(0, 0, \ldots, 1)$, $(0, \ldots, 1, 0)$, $\ldots$, $(1, 1, \ldots, 1)$. The diagonal is represented by the vector $v = (1, 1, \ldots, 1)$, whose length is $\|v\| = \sqrt{n}$. It can be observed that, as n increases, the diagonal of the unitary hypercube increases tending to become infinitely long, while the length of the side of the hypercube itself remains, of course, always equal to 1.

5.1.2 Angles

Two non-zero vectors v and w in $\mathbb{R}^2$ (or in $\mathbb{R}^3$) determine two angles whose sum is a round angle. With the expression "angle between two vectors" we will always mean the *convex* (undirected) angle, i.e., the one whose value is in the interval $[0, \pi]$.

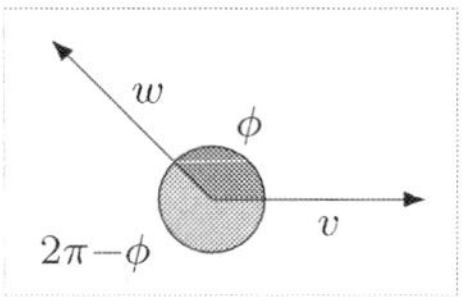

Let us consider the vector space $V = \mathbb{R}^2$, or $V = \mathbb{R}^3$. In physics, the *scalar product* of two vectors v and $w \in V$ is defined by setting

$$v \cdot w = \|v\| \|w\| \cos \phi,$$

where ϕ is the angle between the two vectors (note that $\cos \phi = \cos(2\pi - \phi)$ hence, in the definition of the scalar product, it is not important which of the two angles determined by v and w is considered). It follows that the scalar product defines a function

$$\cdot : V \times V \to \mathbb{R}, \qquad (v, w) \mapsto v \cdot w.$$

We want to show that this function is *bilinear* (i.e., linear with respect to each of its two arguments) and *symmetric*, i.e., that it has the following properties:

(i) $(\lambda v) \cdot w = \lambda(v \cdot w)$,

(ii) $(v_1 + v_2) \cdot w = v_1 \cdot w + v_2 \cdot w$,

(iii) $v \cdot (\mu w) = \mu(v \cdot w)$,

(iv) $v \cdot (w_1 + w_2) = v \cdot w_1 + v \cdot w_2$,

(v) $v \cdot w = w \cdot v$,

for every $\lambda, \mu \in \mathbb{R}$ and every $v, v_1, v_2, w, w_1, w_2 \in V$.

The symmetry property (v) is obvious. Thanks to this it is therefore sufficient to prove the linearity of the scalar product with respect to only one of its two arguments, for example with respect to the second (properties (iii) and (iv)).

To prove (iii) we will consider the following cases: $\mu = 0$, $\mu > 0$ and $\mu < 0$. If $\mu = 0$ we have $v \cdot (0w) = v \cdot \mathbf{0} = 0 = 0(v \cdot w)$. If $\mu > 0$ we have $\|\mu w\| = |\mu|\|w\| = \mu\|w\|$ and the vector μw is parallel and has the same orientation of the vector w. It follows that the angle between the vectors v and μw is equal to the angle ϕ between v and w. Hence we have:

$$v \cdot (\mu w) = \|v\|\|\mu w\| \cos \phi = \mu\|v\|\|w\| \cos \phi = \mu(v \cdot w).$$

If $\mu < 0$ we have $\|\mu w\| = |\mu|\|w\| = -\mu\|w\|$ and, in this case, the vector μw is parallel to w but has the opposite orientation. It follows that the angle between the vectors v and μw is $\pi - \phi$, where ϕ is the angle between v and w. Hence we have:

$$v \cdot (\mu w) = \|v\|\|\mu w\| \cos(\pi - \phi) = -\mu\|v\|\|w\|(-\cos \phi) = \mu(v \cdot w).$$

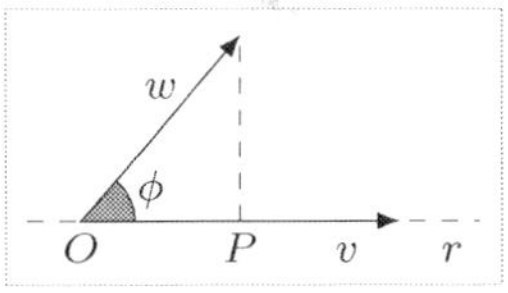

Now we will prove the additivity of the scalar product with respect to the second argument (property (iv)). Let's start with the following observation. Consider two vectors $v, w \in V$ (with $v \neq \mathbf{0}$) and let us denote by $\overrightarrow{OP}$ the orthogonal projection of the vector w on the line r spanned by the vector v, as shown in the figure on the side. The product $\|w\| \cos \phi$ is the length of the segment OP (considered with the minus sign if the vector $\overrightarrow{OP}$ has opposite orientation of the vector v, i.e., if $\frac{\pi}{2} < \phi \leq \pi$). We can therefore state that the dot product of v and w is the product of the norm of v times the norm of the orthogonal projection of w on the line generated by the vector v (always with the convention that this norm must be considered with the negative sign if $\frac{\pi}{2} < \phi \leq \pi$).

Let w_1 and w_2 be two vectors of V and let $w = w_1 + w_2$. We must prove that $v \cdot w = v \cdot w_1 + v \cdot w_2$. We therefore consider the orthogonal projections of the vectors w_1, w_2 and w on the line r generated by the vector v, as illustrated in the following figure

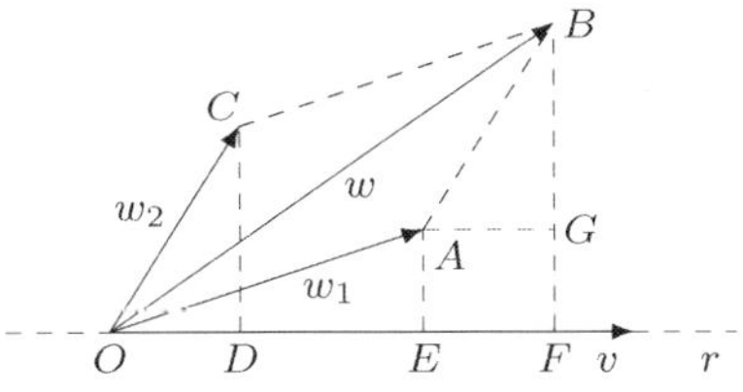

where, for simplicity, we have considered only the case in which the angles between v and w_1, between v and w_2 and between v and w are all smaller than a right angle. By denoting with $\overrightarrow{OE}$ the orthogonal projection of the vector w_1 on the line r, with $\overrightarrow{OD}$ that of w_2 and by $\overrightarrow{OF}$ that of w, we have:

$$v \cdot w_1 = \|v\|\|OE\|, \qquad v \cdot w_2 = \|v\|\|OD\|, \qquad v \cdot w = \|v\|\|OF\|.$$

From the equality of the triangles $\triangle OCD$ and $\triangle ABG$ we obtain the equality

$$\|OD\| = \|AG\| = \|EF\|,$$

from which it follows

$$\|OF\| = \|OE\| + \|EF\| = \|OE\| + \|OD\|.$$

Hence we have:

$$v \cdot w = \|v\|\|OF\| = \|v\|\|OE\| + \|v\|\|OD\| = v \cdot w_1 + v \cdot w_2,$$

as stated.

Now we can prove the following result:

Proposition 5.1.2. *Let* $v = (a_1, a_2)$, $w = (b_1, b_2) \in \mathbb{R}^2$ *(or,* $v = (a_1, a_2, a_3)$, $w = (b_1, b_2, b_3) \in \mathbb{R}^3$*). Then we have:*

$$v \cdot w = a_1 b_1 + a_2 b_2 \qquad (\textit{resp.}, \; v \cdot w = a_1 b_1 + a_2 b_2 + a_3 b_3).$$

Proof. We carry out the proof for $\mathbb{R}^3$ (the case of $v, w \in \mathbb{R}^2$ is analogous). Let $v = (a_1, a_2, a_3)$, $w = (b_1, b_2, b_3) \in \mathbb{R}^3$ and let us denote by e_1, e_2, e_3 the vectors of the canonical basis of $\mathbb{R}^3$. We can write $v = a_1 e_1 + a_2 e_2 + a_3 e_3$ and $w = b_1 e_1 + b_2 e_2 + b_3 e_3$. By using the bilinearity of the scalar product, we have:

$$v \cdot w = (a_1 e_1 + a_2 e_2 + a_3 e_3) \cdot (b_1 e_1 + b_2 e_2 + b_3 e_3) = \sum_{i,j=1}^{3} a_i b_j \, (e_i \cdot e_j).$$

Note that $e_i \cdot e_i = \|e_i\|^2 = 1$ for $i = 1, 2, 3$, while $e_i \cdot e_j = 0$ for every $i \neq j$, because the vectors of the canonical basis are pairwise orthogonal. Substituting in the previous expression we obtain

$$v \cdot w = \sum_{i=1}^{3} a_i b_i. \qquad \square$$

Remark 5.1.3. In the case of two vectors of $\mathbb{R}^2$, a direct proof of the previous result can be provided, without resorting to the bilinearity of the scalar product (but using some trigonometric results).

Let $v = (a_1, a_2)$ and $w = (b_1, b_2)$ be two vectors of $\mathbb{R}^2$, as illustrated in the following figure:

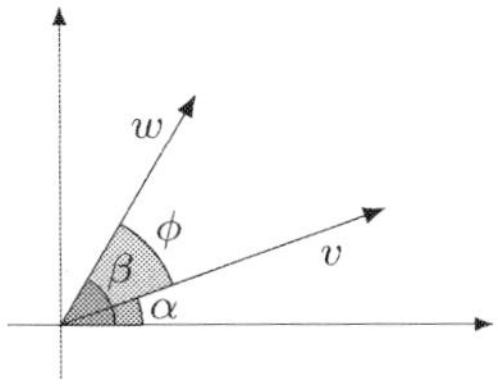

The angle ϕ between v and w is the difference $\phi = \beta - \alpha$ between the angle β that the vector w forms with the X axis and the angle α between the X axis and the vector v. Hence the scalar product of v and w is:

$$\begin{aligned} v \cdot w &= \|v\|\|w\| \cos\phi \\ &= \|v\|\|w\| \cos(\beta - \alpha) \\ &= \|v\|\|w\| \cos\beta\cos\alpha + \|v\|\|w\| \sin\beta\sin\alpha \\ &= a_1 b_1 + a_2 b_2, \end{aligned}$$

since $a_1 = \|v\| \cos\alpha$, $a_2 = \|v\| \sin\alpha$, $b_1 = \|w\| \cos\beta$ and $b_2 = \|w\| \sin\beta$.

The previous results allow us to find the angle ϕ between two non-zero vectors v and w of $\mathbb{R}^2$ or $\mathbb{R}^3$. In fact, we have:

$$\cos\phi = \frac{v \cdot w}{\|v\|\|w\|}. \tag{5.1.1}$$

We can therefore conclude that, in the case of vectors of $\mathbb{R}^2$ or $\mathbb{R}^3$, the scalar product allows us to compute both the norm of a vector and the (undirected) angle formed by two non-zero vectors. In the next section we will see how these results can be extended to vectors of $\mathbb{R}^n$, for every $n \geq 2$.

5.1.3 The scalar product in $\mathbb{R}^n$

The results obtained in the previous paragraphs justify the following definition:

Definition 5.1.4. Given two vectors $v = (a_1, a_2, \ldots, a_n)$, $w = (b_1, b_2, \ldots, b_n)$ of $\mathbb{R}^n$, their *scalar product* is

$$v \cdot w = a_1 b_1 + a_2 b_2 + \cdots + a_n b_n = \sum_{i=1}^{n} a_i b_i.$$

According to this definition, for any $v \in \mathbb{R}^n$, we have

$$v \cdot v = \sum_{i=1}^{n} a_i^2 = \|v\|^2,$$

hence

$$\|v\| = \sqrt{v \cdot v}.$$

Since we know that in $\mathbb{R}^2$ and $\mathbb{R}^3$ the formula (5.1.1) holds, we would like to use a similar formula to define the angle ϕ between two non-zero vectors of $\mathbb{R}^n$, for every $n \geq 2$. However, since the cosine of an angle takes only values in the range $[-1, 1]$, we must first verify that, for each pair of non-zero vectors $v, w \in \mathbb{R}^n$, we have

$$-1 \leq \frac{v \cdot w}{\|v\|\|w\|} \leq 1.$$

Proposition 5.1.5 (Cauchy–Schwarz inequality). *For every pair of vectors $v, w \in \mathbb{R}^n$, we have*

$$|v \cdot w| \leq \|v\|\|w\|. \tag{5.1.2}$$

Furthermore, equality holds if and only if the two vectors are linearly dependent.

Proof. Let's start by proving the first statement. We remark that if one of the two vectors is zero, the result is trivially verified. So we assume that v and w are non-zero vectors. If we set $u_\alpha = v + \alpha w$, with $\alpha \in \mathbb{R}$, we have:

$$\|u_\alpha\|^2 = u_\alpha \cdot u_\alpha = (v + \alpha w) \cdot (v + \alpha w) = \|v\|^2 + 2\alpha\,(v \cdot w) + \alpha^2\|w\|^2 \geq 0,$$

for any $\alpha \in \mathbb{R}$. This shows that the degree 2 polynomial in the variable α

$$\alpha^2\|w\|^2 + 2\alpha\,(v \cdot w) + \|v\|^2$$

takes only non-negative values hence its discriminant must be ≤ 0:

$$\Delta = 4(v \cdot w)^2 - 4\|v\|^2\|w\|^2 \leq 0.$$

It follows that $|v \cdot w| \leq \|v\|\|w\|$.

Now we prove the second statement. If the two vectors are linearly dependent, one of them must be a multiple of the other. We can assume, for example, that $w = \lambda v$, for some $\lambda \in \mathbb{R}$. In this case $\|w\| = |\lambda|\|v\|$ and we have

$$|v \cdot w| = |v \cdot \lambda v| = |\lambda|\|v\|^2 = \|v\|\|w\|.$$

Conversely, suppose that in the expression (5.1.2) the equality holds. We have

$$v \cdot w = \pm\|v\|\|w\|.$$

If $w = \mathbf{0}$ the two vectors are (trivially) linearly dependent; otherwise we set

$$\bar{\alpha} = \begin{cases} -\frac{\|v\|}{\|w\|} & \text{if } v \cdot w = \|v\|\|w\| \\ \frac{\|v\|}{\|w\|} & \text{if } v \cdot w = -\|v\|\|w\| \end{cases}$$

and consider the vector $u = v + \bar{\alpha}w$. We have:

$$\begin{aligned} \|u\|^2 &= (v + \bar{\alpha}w) \cdot (v + \bar{\alpha}w) \\ &= \|v\|^2 + 2\bar{\alpha}\,(v \cdot w) + \bar{\alpha}^2\|w\|^2 \\ &= \|v\|^2 \pm 2\bar{\alpha}\,\|v\|\|w\| + \bar{\alpha}^2\|w\|^2 \\ &= \left(\|v\| \pm \bar{\alpha}\,\|w\|\right)^2 \\ &= \left(\|v\| - \tfrac{\|v\|}{\|w\|}\|w\|\right)^2 = 0. \end{aligned}$$

Since the only vector whose norm is zero is the zero vector, it follows that $u = v + \bar{\alpha}w = \mathbf{0}$, which shows that v and w are linearly dependent. □

From the Cauchy–Schwarz inequality it follows that, for every pair of non-zero vectors $v, w \in \mathbb{R}^n$, we have

$$-1 \leq \frac{v \cdot w}{\|v\|\|w\|} \leq 1.$$

Moreover, from the proof of Proposition 5.1.5, it follows easily that we have $\frac{v \cdot w}{\|v\|\|w\|} = 1$ if and only if the vectors v and w are parallel and have the same orientation, while $\frac{v \cdot w}{\|v\|\|w\|} = -1$ if and only if v and w are parallel but have the opposite orientation. Note that, for every real number $t \in [-1, 1]$ there exists a unique angle $\phi \in [0, \pi]$ such that $t = \cos\phi$; hence we can give the following definition:

Definition 5.1.6. Given two non-zero vectors $v, w \in \mathbb{R}^n$, the (non-oriented) *angle* between them is $\phi \in [0, \pi]$ such that

$$\cos\phi = \frac{v \cdot w}{\|v\|\|w\|}. \tag{5.1.3}$$

Note that, based on the previous observations, the angle between two vectors is zero if and only if the two vectors are parallel and have the same orientation, while it is equal to π if and only if they are parallel but have opposite orientations.

Another consequence of the Cauchy–Schwarz inequality is the so-called *triangular inequality*, that we will now prove.

Proposition 5.1.7 (TRIANGULAR INEQUALITY). *For every $v, w \in \mathbb{R}^n$, we have*

$$\|v + w\| \leq \|v\| + \|w\|.$$

Proof. We have:

$$\begin{aligned} \|v + w\|^2 &= (v + w) \cdot (v + w) \\ &= \|v\|^2 + 2\,(v \cdot w) + \|w\|^2 \\ &\leq \|v\|^2 + 2\,|v \cdot w| + \|w\|^2 \\ &\leq \|v\|^2 + 2\,\|v\|\|w\| + \|w\|^2 \\ &= (\|v\| + \|w\|)^2. \end{aligned}$$

By taking square roots we get $\|v + w\| \leq \|v\| + \|w\|$. □

Remark 5.1.8. From the definition of angle between two vectors we deduce the following condition of orthogonality: two non-zero vectors are orthogonal if and only if their scalar product is zero. In fact, we have $v \cdot w = 0$ if and only if $\|v\| = 0$ (hence $v = \mathbf{0}$) or $\|w\| = 0$ (hence $w = \mathbf{0}$) or $\cos\phi = 0$ (hence $\phi = \frac{\pi}{2}$).

5.2 Areas and volumes

In this section we will see how, using the scalar product, it is possible to compute areas and volumes in $\mathbb{R}^n$.

Let us consider two vectors $v, w \in \mathbb{R}^n$. Since any two vectors are always contained in a plane (i.e., in a subspace of dimension 2 of $\mathbb{R}^n$) we can focus our attention on that plane. Let us denote by

$$\mathscr{P}(v, w) = \{\lambda v + \mu w \,|\, 0 \leq \lambda, \mu \leq 1\}$$

the parallelogram determined by the two vectors v and w, as in the following figure:

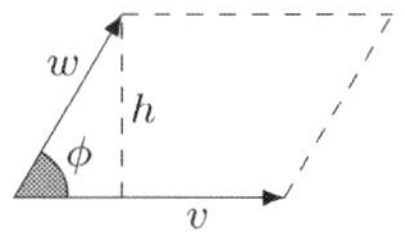

The area of this parallelogram is the product of the length of the vector v (the base) by the height h relative to this base. If we denote by ϕ the angle between the vectors v and w, we have $h = \|w\| \sin\phi$. It follows that:

$$\text{Area}\, \mathscr{P}(v, w) = \|v\| \|w\| \sin\phi.$$

By recalling that $\sin^2\phi = 1 - \cos^2\phi$ and using the formula (5.1.3), we have

$$\sin^2\phi = 1 - \frac{(v \cdot w)^2}{\|v\|^2\|w\|^2} = \frac{\|v\|^2\|w\|^2 - (v \cdot w)^2}{\|v\|^2\|w\|^2}.$$

Hence we have

$$\big(\text{Area}\, \mathscr{P}(v, w)\big)^2 = \|v\|^2\|w\|^2 - (v \cdot w)^2 = (v \cdot v)(w \cdot w) - (v \cdot w)^2.$$

This formula may be rewritten as follows:

$$\text{Area}\, \mathscr{P}(v, w) = \sqrt{\det \begin{pmatrix} v \cdot v & v \cdot w \\ w \cdot v & w \cdot w \end{pmatrix}}. \tag{5.2.1}$$

Now we will see that if $n = 2$, i.e., for two vectors $v, w \in \mathbb{R}^2$, it is possible to compute the area of the parallelogram $\mathscr{P}(v, w)$ without using the scalar product.

Let $v = (a_1, a_2)$, $w = (b_1, b_2) \in \mathbb{R}^2$ and consider the situation described in the figure on the side (where, for simplicity, we have assumed that $a_1, a_2, b_1, b_2 \geq 0$). The area of the parallelogram $\mathscr{P}(v, w)$ can be obtained, by difference of areas, from the area of the rectangle $OABC$ by subtracting the areas of the two rectangles $ADEF$ and $CGHI$ and of the four triangles ODE, BEF, BHI and OGH. Noting that $\|OD\| = \|BI\| = a_1$, $\|DE\| = \|AF\| = \|CG\| = \|HI\| = a_2$, $\|CI\| = \|GH\| = \|DA\| = \|EF\| = b_1$, $\|OG\| = \|BF\| = b_2$, we have:

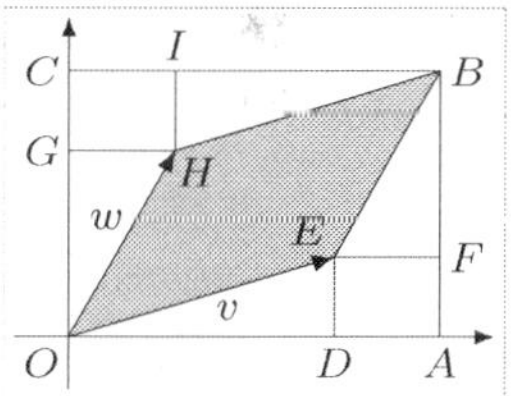

$$\begin{aligned}
\text{Area}(OABC) &= (a_1 + b_1)(a_2 + b_2), \\
\text{Area}(ODE) &= \text{Area}(BHI) = \tfrac{1}{2}\, a_1 a_2, \\
\text{Area}(OGH) &= \text{Area}(BEF) = \tfrac{1}{2}\, b_1 b_2, \\
\text{Area}(ADEF) &= \text{Area}(CGHI) = b_1 a_2,
\end{aligned}$$

from which it follows:

$$\begin{aligned}\text{Area}\,\mathscr{P}(v,w) &= (a_1+b_1)(a_2+b_2) - a_1a_2 - b_1b_2 - 2b_1a_2 \\ &= a_1b_2 - b_1a_2 = \det\begin{pmatrix} a_1 & b_1 \\ a_2 & b_2 \end{pmatrix}.\end{aligned}$$

The determinant of the square matrix consisting of the components of the two vectors v and w (written either in column or in row) therefore provides the value of the area of the parallelogram $\mathscr{P}(v,w)$. However, remembering that

$$\det\begin{pmatrix} a_1 & b_1 \\ a_2 & b_2 \end{pmatrix} = -\det\begin{pmatrix} b_1 & a_1 \\ b_2 & a_2 \end{pmatrix}$$

it is convenient to take this determinant as an absolute value, to avoid the risk of finding negative values for the area of a parallelogram. Hence we obtain the following formula:

$$\text{Area}\,\mathscr{P}(v,w) = \left|\det\begin{pmatrix} a_1 & b_1 \\ a_2 & b_2 \end{pmatrix}\right|, \tag{5.2.2}$$

for every $v=(a_1,a_2)$, $w=(b_1,b_2)\in\mathbb{R}^2$.

Let us now turn to the case of volumes. We consider three vectors $u,v,w\in\mathbb{R}^n$ and denote with

$$\mathscr{P}(u,v,w) = \{\lambda u + \mu v + \nu w \,|\, 0\le\lambda,\mu,\nu\le 1\}$$

the parallelepiped determined by them. Since any three vectors of $\mathbb{R}^n$ are always contained in a subspace of dimension 3, we can focus our attention on that 3-dimensional subspace. Let us therefore consider the situation schematized in the following figure:

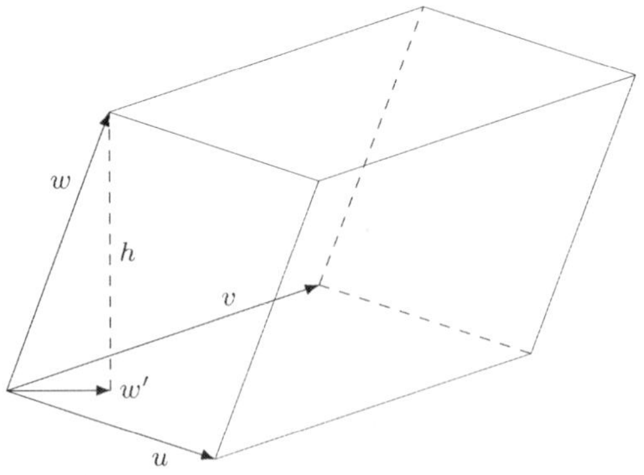

The volume of such parallelepiped is the product of the area of the parallelogram $\mathscr{P}(u,v)$ (the area of the base) by the height h relative to that base. To determine the height h we can proceed as follows: we consider a generic vector $\alpha u+\beta v$ belonging to the plane spanned by u and v. We must determine the values of α and β that minimize[1] the norm of the vector $w-\alpha u-\beta v$. The minimum value of the norm of this vector is the height h we are looking for. Developing the calculations, we find:

$$\begin{aligned}\|w-\alpha u-\beta v\|^2 &= (w-\alpha u-\beta v)\cdot(w-\alpha u-\beta v)\\ &= w\cdot w-2\alpha\,(u\cdot w)-2\beta\,(v\cdot w)+2\alpha\beta\,(u\cdot v)\\ &\quad+\alpha^2(u\cdot u)+\beta^2(v\cdot v).\end{aligned}$$

To find the minimum value of this expression it is sufficient to calculate its two partial derivatives, with respect to α and β, and impose that these are zero. The following system is thus obtained:

$$\begin{cases}2\alpha\,(u\cdot u)+2\beta\,(u\cdot v)-2\,(u\cdot w)=0\\ 2\alpha\,(u\cdot v)+2\beta\,(v\cdot v)-2\,(v\cdot w)=0\end{cases}\tag{5.2.3}$$

whose solution is

$$\begin{cases}\bar\alpha=\dfrac{(u\cdot w)(v\cdot v)-(u\cdot v)(v\cdot w)}{(u\cdot u)(v\cdot v)-(u\cdot v)^2}\\[2ex] \bar\beta=\dfrac{(u\cdot u)(v\cdot w)-(u\cdot v)(u\cdot w)}{(u\cdot u)(v\cdot v)-(u\cdot v)^2}.\end{cases}$$

By setting $w'=\bar\alpha u+\bar\beta v$, the vector $w-w'$ represents the height of the parallelepiped $\mathscr{P}(u,v,w)$, relative to the base $\mathscr{P}(u,v)$. Remembering that the area of the parallelogram $\mathscr{P}(u,v)$ is given by

$$\operatorname{Area}\mathscr{P}(u,v)=\sqrt{\det\begin{pmatrix}u\cdot u & u\cdot v\\ v\cdot u & v\cdot v\end{pmatrix}},$$

[1] Alternatively, it could be required that the vector $w-\alpha u-\beta v$ be orthogonal to the plane spanned by u and v (the height of a parallelepiped must be perpendicular to its base). This is tantamount to requiring that

$$\begin{cases}(w-\alpha u-\beta v)\cdot u=0\\ (w-\alpha u-\beta v)\cdot v=0.\end{cases}$$

Thus we find a system equivalent to (5.2.3).

we are now able to compute the volume of the parallelepiped $\mathscr{P}(u, v, w)$ by multiplying this expression by $h = \|w - w'\|$. Developing the computations we find the following formula for the volume:

$$\operatorname{Vol} \mathscr{P}(u, v, w) = \sqrt{\det \begin{pmatrix} u \cdot u & u \cdot v & u \cdot w \\ v \cdot u & v \cdot v & v \cdot w \\ w \cdot u & w \cdot v & w \cdot w \end{pmatrix}}.$$

If $n = 3$, i.e., for three vectors $u, v, w \in \mathbb{R}^3$, it is also possible to compute the volume of the parallelepiped $\mathscr{P}(u, v, w)$ as the absolute value of the determinant of the matrix whose columns (or rows) are formed by the components of the given vectors. In fact, it can be proved that, given $u = (a_1, a_2, a_3)$, $v = (b_1, b_2, b_3)$, $w = (c_1, c_2, c_3) \in \mathbb{R}^3$, we have

$$\operatorname{Vol} \mathscr{P}(u, v, w) = \left| \det \begin{pmatrix} a_1 & b_1 & c_1 \\ a_2 & b_2 & c_2 \\ a_3 & b_3 & c_3 \end{pmatrix} \right|. \tag{5.2.4}$$

The results we obtained for the areas of the parallelograms and the volumes of the parallelepipeds justify the following definition:

Definition 5.2.1. Given r vectors $v_1, v_2, \ldots, v_r \in \mathbb{R}^n$ (with $r \leq n$), the *parallelotope* spanned by them is the subset of $\mathbb{R}^n$ given by

$$\mathscr{P}(v_1, v_2, \ldots, v_r) = \left\{ \sum_{i=1}^{r} \lambda_i v_i \;\middle|\; 0 \leq \lambda_i \leq 1, i = 1, \ldots, r \right\}.$$

Its r-dimensional *measure* (or volume) is defined by setting

$$\operatorname{Meas} \mathscr{P}(v_1, \ldots, v_r) = \sqrt{\det \left(v_i \cdot v_j \right)_{i,j=1,\ldots,r}}.$$

Note that, in the cases $r = 2$ or $r = 3$, this definition reduces respectively to the definition of the area of a parallelogram or volume of a parallelepiped, while, for $r = 1$, it is the definition of the norm of a vector.

Remark 5.2.2. If the r vectors $v_1, v_2, \ldots, v_r$ are linearly dependent it is easy to prove that the r-dimensional measure of $\mathscr{P}(v_1, v_2, \ldots, v_r)$ is zero. In fact, in this case one of the r vectors can be expressed as a linear combination of the others. Up to a rearrangement, we can then assume it is

$$v_r = \lambda_1 v_1 + \lambda_2 v_2 + \cdots + \lambda_{r-1} v_{r-1}.$$

It is now immediate to verify that, in the matrix $\left(v_i \cdot v_j\right)_{i,j=1,\dots,r}$, the last column is a linear combination of the previous columns, therefore the determinant of this matrix is zero. This is the reason why, in Definition 5.2.1, we have assumed that $r \le n$; in fact, if $r > n$ the r vectors $v_1, \dots, v_r$ will always be linearly dependent.

Writing the components of the r vectors $v_1 = {}^t(a_{11}, a_{21}, \dots, a_{n1})$, $v_2 = {}^t(a_{12}, a_{22}, \dots, a_{n2})$, $\dots$, $v_r = {}^t(a_{1r}, a_{2r}, \dots, a_{nr})$ as columns, we obtain the following matrix with n rows and r columns (we are assuming $r \le n$)

$$A = \begin{pmatrix} a_{11} & a_{12} & \cdots & a_{1r} \\ a_{21} & a_{22} & \cdots & a_{2r} \\ \vdots & \vdots & & \vdots \\ a_{n1} & a_{n2} & \cdots & a_{nr} \end{pmatrix}.$$

The matrix of scalar products $v_i \cdot v_j$ can be written as follows:

$$\left(v_i \cdot v_j\right)_{i,j=1,\dots,r} = {}^tAA.$$

By the Cauchy–Binet formula (Chapter 3, Theorem 3.2.18), we have:

$$\begin{aligned} \det\left(v_i \cdot v_j\right)_{i,j=1,\dots,r} &= \det\left({}^tAA\right) \\ &= \sum_{1 \le i_1 < i_2 < \dots < i_r \le n} \det({}^tA_{(i_1,i_2,\dots,i_r)}) \det(A^{(i_1,i_2,\dots,i_r)}) \\ &= \sum_{1 \le i_1 < i_2 < \dots < i_r \le n} \det(A^{(i_1,i_2,\dots,i_r)})^2. \end{aligned} \tag{5.2.5}$$

In the particular case $r = n$, A is a square matrix and from Binet's theorem it follows that

$$\det\left(v_i \cdot v_j\right)_{i,j=1,\dots,n} = (\det A)^2,$$

hence

$$\operatorname{Meas} \mathscr{P}(v_1, \dots, v_n) = |\det A|, \tag{5.2.6}$$

where A is the matrix whose columns are the n vectors $v_1, \dots, v_n \in \mathbb{R}^n$. For $n = 2$ and $n = 3$ we thus find the two formulas (5.2.2) and (5.2.4) mentioned above. We note that, from this last formula, it is evident that $\operatorname{Meas} \mathscr{P}(v_1, \dots, v_n) = 0$ if and only if $\det A = 0$, that is, if and only if the vectors v_1, v_2, $\dots$, v_n are linearly dependent.

As we will now see, the formula (5.2.5) can be considered a generalization, to the case of r-dimensional volumes, of the classic Pythagorean theorem.

Let us try to explain this statement by taking as an example the case of two vectors in $\mathbb{R}^3$. Let $v_1 = {}^t(a_{11}, a_{21}, a_{31})$, $v_2 = {}^t(a_{12}, a_{22}, a_{32})$ be two vectors in $\mathbb{R}^3$ and let us consider the parallelogram $\mathscr{P}(v_1, v_2)$ determined by these vectors. The formula (5.2.5), in this case, reduces to

$$\big(\operatorname{Area}\mathscr{P}(v_1, v_2)\big)^2 = \det(A^{(1,2)})^2 + \det(A^{(1,3)})^2 + \det(A^{(2,3)})^2, \qquad (5.2.7)$$

where

$$A^{(1,2)} = \begin{pmatrix} a_{11} & a_{12} \\ a_{21} & a_{22} \end{pmatrix}, \qquad A^{(1,3)} = \begin{pmatrix} a_{11} & a_{12} \\ a_{31} & a_{32} \end{pmatrix}, \qquad A^{(2,3)} = \begin{pmatrix} a_{21} & a_{22} \\ a_{31} & a_{32} \end{pmatrix}.$$

We can note that the columns of $A^{(1,2)}$ are the orthogonal projections of the vectors v_1 and v_2 on the plane OXY, therefore the square of the determinant of the matrix $A^{(1,2)}$ represents the square of the area of the parallelogram obtained by orthogonally projecting the parallelogram $\mathscr{P}(v_1, v_2)$ on the OXY plane, as shown in the following figure:

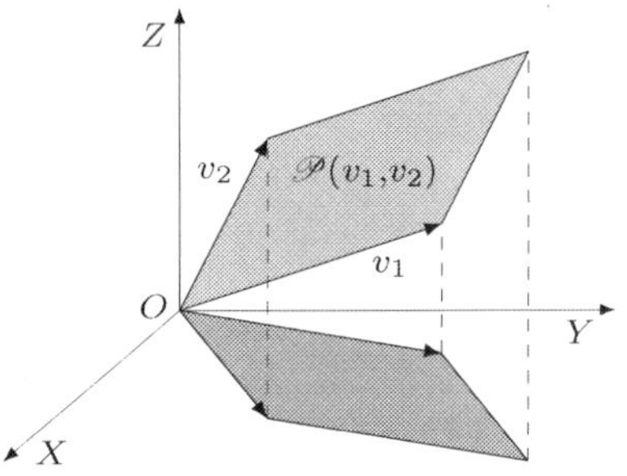

Similarly, the columns of $A^{(1,3)}$ are the orthogonal projections of the vectors v_1 and v_2 on the plane OXZ, hence the square of the determinant of the matrix $A^{(1,3)}$ represents the square of the area of the parallelogram obtained by orthogonally projecting the parallelogram $\mathscr{P}(v_1, v_2)$ on the OXZ plane.

Finally, the columns of $A^{(2,3)}$ are the orthogonal projections of the vectors v_1 and v_2 on the plane OYZ, hence the square of the determinant of the matrix $A^{(2,3)}$ is the square of the area of the parallelogram obtained by orthogonally projecting $\mathscr{P}(v_1, v_2)$ onto the OYZ plane.

The formula (5.2.7) therefore states that the square of the area of a parallelogram in $\mathbb{R}^3$ is the sum of the squares of the areas of the orthogonal projections of that parallelogram on the coordinate planes.

Recall that the Pythagorean theorem states that the square of the length of a segment is the sum of the squares of the lengths of its orthogonal projections on the coordinate axes. The formula (5.2.7) therefore represents an analogue of the Pythagorean theorem for areas.

In the general case, the geometric meaning of the formula (5.2.5) can be expressed as follows: given r vectors $v_1, v_2, \dots, v_r$ in $\mathbb{R}^n$, the square of the r-dimensional measure of the parallelotope $\mathscr{P}(v_1, \dots, v_r)$ is the sum of the squares of the r-dimensional measures of the orthogonal projections of $\mathscr{P}(v_1, \dots, v_r)$ on the r-coordinate subspaces (i.e., on all subspaces of dimension r generated by the vectors $e_{i_1}, e_{i_2}, \dots, e_{i_r}$, with $1 \leq i_1 < i_2 < \cdots < i_r \leq n$, where $e_1, \dots, e_n$ denote the vectors of the canonical basis of $\mathbb{R}^n$).

5.3 Bilinear forms

Let V be a vector space over the field K.

Definition 5.3.1. A *bilinear form* on V is a function

$$g\colon V \times V \to K, \qquad (v, w) \mapsto g(v, w),$$

that is linear with respect to each of its two arguments, i.e., such that

(*i*) $g(\lambda_1 v_1 + \lambda_2 v_2, w) = \lambda_1 g(v_1, w) + \lambda_2 g(v_2, w)$,

(*ii*) $g(v, \mu_1 w_1 + \mu_2 w_2) = \mu_1 g(v, w_1) + \mu_2 g(v, w_2)$,

for any $\lambda_1, \lambda_2, \mu_1, \mu_2 \in K$ and any $v, v_1, v_2, w, w_1, w_2 \in V$.

Definition 5.3.2. A bilinear form $g\colon V \times V \to K$ is *symmetric* if $g(v, w) = g(w, v)$, for every $v, w \in V$. If the characteristic of the field K is not 2, g is *antisymmetric* if, for every $v, w \in V$, we have $g(v, w) = -g(w, v)$.

Remark 5.3.3. Note that, if K is a field whose characteristic is not 2, every bilinear form g can be decomposed as follows

$$g(v, w) = g_s(v, w) + g_a(v, w),$$

where g_s and g_a are the two bilinear forms defined by setting

$$g_s(v, w) = \frac{g(v, w) + g(w, v)}{2}, \qquad g_a(v, w) = \frac{g(v, w) - g(w, v)}{2}.$$

Since g_s is a symmetric bilinear form while g_a is antisymmetric, this means that any bilinear form on V can be expressed as the sum of a symmetric bilinear form and an antisymmetric one. The study of bilinear forms can therefore be reduced to the study of the symmetric ones and the antisymmetric ones. In the following we will deal exclusively with the study of symmetric bilinear forms.

Let V be a vector space and U, W two vector subspaces of V such that $V = U \oplus W$. Let

$$g_U : U \times U \to K, \qquad g_W : W \times W \to K,$$

be two bilinear forms defined on U and W, respectively. By recalling that every vector $v \in V$ can be written in a unique way as $v = u + w$, with $u \in U$ and $w \in W$, we can define a function

$$g_V : V \times V \to K$$

by setting

$$g_V(u_1 + w_1, u_2 + w_2) = g_U(u_1, u_2) + g_W(w_1, w_2),$$

for every $u_1, u_2 \in U$, $w_1, w_2 \in W$. It is easy to verify that g_V is a bilinear form on V; it is symmetric if and only if both g_U and g_W are symmetric.

Such bilinear form will be denoted by $g_V = g_U \oplus g_W$ and called the *direct sum*[2] of g_U and g_W.

Let's consider now a symmetric bilinear form g defined on a vector space V. We give the following definition:

Definition 5.3.4. The *kernel* of g is the following subset of V:

$$\mathrm{Ker}(g) = \{v \in V \mid g(v, w) = 0, \text{ for every } w \in V\}.$$

Remark 5.3.5. It is easily shown, using the bilinearity of g, that $\mathrm{Ker}(g)$ is a vector subspace of V. Furthermore, from the symmetry of g, it also follows that

$$\mathrm{Ker}(g) = \{w \in V \mid g(v, w) = 0, \text{ for every } v \in V\}.$$

Definition 5.3.6. A bilinear form $g\colon V \times V \to K$ is *non-degenerate* if $\mathrm{Ker}(g) = \{\mathbf{0}\}$. Otherwise it is called *degenerate*.

For any vector space V we will denote by g_0 the *null bilinear form* on V, i.e., the bilinear form defined by setting $g_0(v, w) = 0$, for each $v, w \in V$. Note that, for any symmetric bilinear form g on V, the restriction of g to $\mathrm{Ker}(g) \times \mathrm{Ker}(g)$ is the null bilinear form on the subspace $\mathrm{Ker}(g)$ of V.

We are now able to state and prove the following result:

---

[2] The bilinear form $g_V = g_U \oplus g_W$ has the following property: $g_V(v_1, v_2) = 0$ if $v_1 \in U$ and $v_2 \in W$, or if $v_1 \in W$ and $v_2 \in U$. For this reason it is also called the *orthogonal sum* of g_U and g_W and sometimes denoted by $g_V = g_U \boxplus g_W$.

Proposition 5.3.7. *Let g be a symmetric bilinear form defined on a vector space V. There exists a vector subspace U of V such that:*

(*i*) $V = U \oplus \mathrm{Ker}(g)$;

(*ii*) *the restriction of g to $U \times U$, denoted by g_U, is a non-degenerate symmetric bilinear form on U;*

(*iii*) $g = g_U \oplus g_0$, *where g_0 is the null bilinear form on* $\mathrm{Ker}(g)$.

Proof. Given V and g, there certainly exists a vector subspace U of V such that $V = U \oplus \mathrm{Ker}(g)$. Denoting by g_U the restriction of g to $U \times U$ and by g_0 the null bilinear form defined on $\mathrm{Ker}(g)$, we consider the bilinear form $g_U \oplus g_0$ defined on $U \oplus \mathrm{Ker}(g) = V$. We must prove that $g = g_U \oplus g_0$.

Let us consider two vectors $v_1, v_2 \in V$ and write $v_1 = u_1 + w_1$ and $v_2 = u_2 + w_2$, where $u_1, u_2 \in U$ and $w_1, w_2 \in \mathrm{Ker}(g)$. We have:

$$\begin{aligned}(g_U \oplus g_0)(v_1, v_2) &= (g_U \oplus g_0)(u_1 + w_1, u_2 + w_2)\\ &= g_U(u_1, u_2) + g_0(w_1, w_2)\\ &= g_U(u_1, u_2) = g(u_1, u_2),\end{aligned}$$

and

$$\begin{aligned}g(v_1, v_2) &= g(u_1 + w_1, u_2 + w_2)\\ &= g(u_1, u_2) + g(u_1, w_2) + g(w_1, u_2) + g(w_1, w_2).\end{aligned}$$

Since $w_1, w_2 \in \mathrm{Ker}(g)$, we have $g(u_1, w_2) = g(w_1, u_2) = g(w_1, w_2) = 0$ hence $g(v_1, v_2) = (g_U \oplus g_0)(v_1, v_2)$.

It remains only to prove that the bilinear form $g_U : U \times U \to K$ is non-degenerate. Let's consider a vector $u_0 \in \mathrm{Ker}(g_U)$ and a generic vector $v \in V$. By writing v as $v = u + w$, with $u \in U$ and $w \in \mathrm{Ker}(g)$, we have

$$\begin{aligned}g(u_0, v) &= g(u_0, u + w)\\ &= g(u_0, u) + g(u_0, w)\\ &= g_U(u_0, u) + 0 = 0\end{aligned}$$

because $w \in \mathrm{Ker}(g)$ and $u_0 \in \mathrm{Ker}(g_U)$. It follows that $u_0 \in \mathrm{Ker}(g)$, hence we must have $u_0 = \mathbf{0}$, because $U \cap \mathrm{Ker}(g) = \{\mathbf{0}\}$. We have thus shown that $\mathrm{Ker}(g_U) = \{\mathbf{0}\}$, hence g_U is non-degenerate. □

Remark 5.3.8. The previous proposition states that every symmetric bilinear form is a direct sum of a non-degenerate symmetric bilinear form and a null bilinear form. We can therefore limit ourselves to the study of the non-degenerate symmetric bilinear forms.

Recalling that the usual scalar product in $\mathbb{R}^n$ is a non-degenerate symmetric bilinear form, and also recalling the condition of orthogonality between two vectors of $\mathbb{R}^n$, we can give the following definition:

Definition 5.3.9. Let V be a vector space with a non-degenerate symmetric bilinear form g. Two vectors $v, w \in V$ are *orthogonal* if $g(v, w) = 0$. Two vector subspaces $U_1, U_2 \subseteq V$ are *orthogonal* if $g(u_1, u_2) = 0$, for every $u_1 \in U_1$ and every $u_2 \in U_2$.

Example 5.3.10. In the vector space $V = \mathbb{R}^2$ let us consider the symmetric bilinear form g defined by setting

$$g\big((x_1, x_2), (y_1, y_2)\big) = x_1 y_2 + x_2 y_1.$$

It is easy to check that $\mathrm{Ker}(g) = \{(0,0)\}$, hence g is non-degenerate. Note that $g\big((1,1), (1,-1)\big) = 0$, hence the vectors $v_1 = (1,1)$ e $v_2 = (1,-1)$ are orthogonal. However, we also have $g\big((1,0), (1,0)\big) = 0$, hence the vector $e_1 = (1,0)$ turns out to be orthogonal to itself!

Definition 5.3.11. Let V be a vector space with a symmetric bilinear form g. A vector $v \in V$ such that $g(v, v) = 0$ is called *isotropic*. A subspace $U \subseteq V$ is called *totally isotropic* if the restriction of g to $U \times U$ is the null bilinear form, i.e., if $g(u_1, u_2) = 0$, for every $u_1, u_2 \in U$.

Remark 5.3.12. If $U \subseteq V$ is a totally isotropic subspace, every vector $u \in U$ is isotropic. However, note that a subspace that is spanned by isotropic vectors is not necessarily totally isotropic. If $u \in V$ is an isotropic vector, then the subspace $\langle u \rangle$ spanned by u is totally isotropic, however, if u_1, $u_2 \in V$ are two isotropic vectors, it is not true, in general, that the subspace $U = \langle u_1, u_2 \rangle$ is totally isotropic. In fact, we obviously have $g(u_1, u_1) = 0$ and $g(u_2, u_2) = 0$, but this does not imply that $g(u_1, u_2) = 0$. The bilinear form of Example 5.3.10 shows this fact: the two vectors $e_1 = (1,0)$ and $e_2 = (0,1)$ are isotropic, but we have $g(e_1, e_2) = 1$.

Definition 5.3.13. Let V be a vector space with a symmetric bilinear form g. Given a subset S of V we define its *orthogonal* as follows:

$$S^\perp = \{v \in V \mid g(v, w) = 0, \forall w \in S\}.$$

Proposition 5.3.14. *Let V be a vector space with a symmetric bilinear form g and S be a subset of V. Then:*

(*i*) $S^{\perp}$ *is a vector subspace of* V*;*

(*ii*) *We have* $S \subseteq \langle S \rangle \subseteq (S^{\perp})^{\perp}$*, where* $\langle S \rangle$ *denotes the vector subspace of* V *generated by* S*.*

Proof. To prove the first claim we consider two vectors $v_1, v_2 \in S^{\perp}$ and two scalars $\lambda_1, \lambda_2 \in K$. For any $w \in S$, we have:

$$g(\lambda_1 v_1 + \lambda_2 v_2, w) = \lambda_1 g(v_1, w) + \lambda_2 g(v_2, w) = \lambda_1 0 + \lambda_2 0 = 0,$$

hence $\lambda_1 v_1 + \lambda_2 v_2 \in S^{\perp}$.

As for the second statement, the inclusion $S \subseteq \langle S \rangle$ is obvious. To prove that $\langle S \rangle \subseteq (S^{\perp})^{\perp}$ let us consider a generic vector $w \in \langle S \rangle$; it can be written as

$$w = \lambda_1 w_1 + \lambda_2 w_2 + \cdots + \lambda_r w_r,$$

for some $\lambda_1, \ldots, \lambda_r \in K$ and some $w_1, \ldots, w_r \in S$. For every $v \in S^{\perp}$, we have

$$g(w, v) = g\Big(\sum_{i=1}^{r} \lambda_i w_i, v\Big) = \sum_{i=1}^{r} \lambda_i \, g(w_i, v) = 0,$$

since $w_i \in S$ and $v \in S^{\perp}$. This means that w is orthogonal to every vector of $S^{\perp}$, hence $w \in (S^{\perp})^{\perp}$. □

Proposition 5.3.15. *Let* V *be a vector space with a symmetric bilinear form* g *and* S*,* T *be two subsets of* V*. The following properties hold:*

(*i*) $S \subseteq T \Rightarrow T^{\perp} \subseteq S^{\perp}$*;*

(*ii*) $S^{\perp} = \langle S \rangle^{\perp}$*.*

Proof. (*i*) Let $v \in T^{\perp}$; we have $g(v, w) = 0$, for every $w \in T$. Since $S \subseteq T$, we also have $g(v, w) = 0$, for every $w \in S$, which means that $v \in S^{\perp}$.

(*ii*) From the inclusion $S \subseteq \langle S \rangle$ and the property (*i*) it follows that $\langle S \rangle^{\perp} \subseteq S^{\perp}$. To prove the opposite inclusion we consider a vector $v \in S^{\perp}$. Every vector $w \in \langle S \rangle$ can be written as

$$w = \lambda_1 w_1 + \lambda_2 w_2 + \cdots + \lambda_r w_r,$$

for some $\lambda_1, \ldots, \lambda_r \in K$ and some $w_1, \ldots, w_r \in S$. Hence we have

$$g(v, w) = g\Big(v, \sum_{i=1}^{r} \lambda_i w_i\Big) = \sum_{i=1}^{r} \lambda_i \, g(v, w_i) = 0.$$

This means that v is orthogonal to all vectors of $\langle S \rangle$, hence $v \in \langle S \rangle^{\perp}$. □

Definition 5.3.16. Let V be a vector space over K and $g\colon V \times V \to K$ be a symmetric bilinear form. A basis $\{v_1, v_2, \dots, v_n\}$ of V formed by pairwise orthogonal vectors is called an *orthogonal basis* of V.

In a vector space V with a symmetric bilinear form g, we could define the square of the norm of a vector v by setting $\|v\|^2 = g(v, v)$. This definition is motivated by the fact that, in the vector space $\mathbb{R}^n$ with the usual scalar product, we have $\|v\|^2 = v \cdot v$, for every $v \in \mathbb{R}^n$. In general, however, it is not possible to define the *norm* of v; in fact it is not certain that in the field K there is a square root of $g(v, v)$.

As an example, we consider the vector space $V = \mathbb{Q}^2$, on the field $\mathbb{Q}$ of rational numbers, with the usual scalar product. The vector $v = (1, 1)$ is such that $\|v\|^2 = v \cdot v = 2$, therefore in the field $\mathbb{Q}$ it is not possible to define the norm of v, because it should be $\|v\| = \sqrt{2}$, but $\sqrt{2} \notin \mathbb{Q}$. An analogous example is obtained by considering the real vector space $V = \mathbb{R}^2$, with the symmetric bilinear form defined in Example 5.3.10. In this case, for the vector $v = (1, -1)$ we have $\|v\|^2 = g(v, v) = -2$, so it should be $\|v\| = \sqrt{-2}$, but $\sqrt{-2} \notin \mathbb{R}$.

Of course, in the special case where $g(v, v) = 1$, we also have $\|v\| = 1$. We can therefore give the following definition:

Definition 5.3.17. Let V be a vector space with a symmetric bilinear form g. A vector $v \in V$ is *normalized* if $g(v, v) = 1$.

Remark 5.3.18. If v is not isotropic, i.e., if $g(v, v) \neq 0$, and if in the field K there exists a square root of $g(v, v)$, it is possible to *normalize* the vector v by dividing it by $\sqrt{g(v, v)}$. In fact, by setting $v' = v/\sqrt{g(v, v)}$, we have:

$$\|v'\|^2 = g(v', v') = \frac{1}{\left(\sqrt{g(v, v)}\right)^2}\, g(v, v) = 1.$$

Definition 5.3.19. Let V be a vector space with a symmetric bilinear form g. A basis $\{v_1, v_2, \dots, v_n\}$ of V is *orthonormal* if it is an orthogonal basis and if all the vectors $v_1, \dots, v_n$ are normalized, i.e., if $g(v_i, v_j) = 0$ for any $i \neq j$, while $g(v_i, v_i) = 1$, for any $i = 1, \dots, n$.

Let us now consider in more detail the problem of defining the norm of a vector in a vector space V with a symmetric bilinear form g. As already mentioned above, this is possible if, for every $v \in V$, a square root of $g(v, v)$ exists in the field K. It is therefore natural to impose conditions on g that guarantee the existence of $\sqrt{g(v, v)}$, for every $v \in V$.

We note, for example, that if K is the field $\mathbb{C}$ of complex numbers, this request is satisfied for every bilinear symmetric form g. However, this is not true if K is the field of real numbers $\mathbb{R}$: in this case it is necessary to require that $g(v, v) \geq 0$, for each $v \in V$.

We therefore give the following definition:

Definition 5.3.20. Let V be a real vector space. A symmetric bilinear form g on V is called:

(*i*) *positive definite* if $g(v, v) > 0$, for any $v \in V$, $v \neq \mathbf{0}$;

(*ii*) *negative definite* if $g(v, v) < 0$, for any $v \in V$, $v \neq \mathbf{0}$;

(*iii*) *positive semidefinite* if $g(v, v) \geq 0$, for any $v \in V$;

(*iv*) *negative semidefinite* if $g(v, v) \leq 0$, for any $v \in V$;

(*v*) *indefinite* if there exist $v, w \in V$ such that $g(v, v) > 0$ and $g(w, w) < 0$.

If the symmetric bilinear form g is positive semidefinite it is possible to set $\|v\| = \sqrt{g(v, v)}$. However, we note that if we want the only vector with zero norm to be the zero vector, we need to require that g be positive definite. We therefore give the following definition:

Definition 5.3.21. Let V be a real vector space with a positive definite symmetric bilinear form g. The *norm* of a vector $v \in V$ is defined by setting

$$\|v\| = \sqrt{g(v, v)}.$$

This norm satisfies the following properties, whose proof is immediate:

(*i*) $\|v\| \geq 0$, for any $v \in V$, and $\|v\| = 0$ if and only if $v = \mathbf{0}$;

(*ii*) $\|\lambda v\| = |\lambda|\|v\|$, for any $\lambda \in \mathbb{R}$ and any $v \in V$.

As for the usual norm in $\mathbb{R}^n$, the following result holds:

Proposition 5.3.22 (Cauchy–Schwarz inequality). *Let g be a positive definite symmetric bilinear form on a real vector space V. For every pair of vectors $v, w \in V$, we have*

$$|g(v, w)| \leq \|v\|\|w\|. \tag{5.3.1}$$

Furthermore, equality holds if and only if the two vectors are linearly dependent.

Proof. The proof is exactly the same as that of Proposition 5.1.5. To adapt it to the present context it is sufficient to define the "scalar product" of two vectors $v, w \in V$ by setting $v \cdot w = g(v, w)$. □

An immediate consequence of the Cauchy–Schwarz inequality is the triangular inequality:

Proposition 5.3.23 (Triangular inequality). *Let V be a real vector space with a positive definite symmetric bilinear form g. For every $v, w \in V$, we have*

$$\|v + w\| \leq \|v\| + \|w\|.$$

Proof. The proof is quite similar to that of Proposition 5.1.7. □

The validity of the Cauchy–Schwarz inequality allows us, in turn, to define the angle between two vectors in a completely analogous way to what we have done for the vectors of $\mathbb{R}^n$:

Definition 5.3.24. Let V be a real vector space and g a positive definite symmetric bilinear form on V. Given two non-zero vectors $v, w \in V$, the (non-oriented) *angle* between them is $\phi \in [0, \pi]$ such that

$$\cos \phi = \frac{g(v, w)}{\|v\| \|w\|}. \tag{5.3.2}$$

The results described so far show how a positive definite symmetric bilinear form on a real vector space V allows us to define the lengths of the vectors and the angles between vectors in a completely analogous way to what happens for the vector space $\mathbb{R}^n$ equipped with the usual scalar product. For this reason, a real vector space with a positive definite symmetric bilinear form is called a *Euclidean vector space.* In a Euclidean vector space (V, g) we define the *scalar product* of two vectors $v, w \in V$ by setting $v \cdot w = g(v, w)$. With this definition, in the vector space V are valid, without any modification, (almost)[3] all the results on the areas and volumes described in Section 5.2.

[3] More precisely, all but those who use the fact that the canonical basis is an orthonormal basis.

5.4 Bilinear forms and matrices

Let V be a finite dimensional vector space over a field K and g a bilinear form on V. Let $\mathbf{v} = \{v_1, \dots, v_n\}$ be a basis of V and set $g_{ij} = g(v_i, v_j)$, for any $i, j = 1, \dots, n$. The scalars g_{ij} form a $n \times n$ matrix with entries in K.

Definition 5.4.1. With the previous notations, the matrix $G = \big(g_{ij}\big)$ is called the *matrix of the bilinear form* g, with respect to the basis $\mathbf{v}$ of V. When it is appropriate to explicitly indicate the dependence on the basis of V, we will write $G = M_{\mathbf{vv}}(g)$.

Knowing the matrix of g allows us to compute $g(v, w)$ for each pair of vectors $v, w \in V$. Indeed, expressing v and w as linear combinations of the vectors of the basis of V,

$$v = \sum_{i=1}^{n} \lambda_i v_i, \qquad w = \sum_{j=1}^{n} \mu_j v_j,$$

thanks to the bilinearity of g, we have:

$$\begin{aligned} g(v, w) &= g\Big(\sum_{i=1}^{n} \lambda_i v_i, \sum_{j=1}^{n} \mu_j v_j\Big) \\ &= \sum_{i,j=1}^{n} \lambda_i \mu_j \, g(v_i, v_j) = \sum_{i,j=1}^{n} \lambda_i \mu_j g_{ij}. \end{aligned}$$

If we consider the vectors $(\lambda_1, \lambda_2, \dots, \lambda_n)^{\mathbf{v}}$ and $(\mu_1, \mu_2, \dots, \mu_n)^{\mathbf{v}}$, formed by the components of v and w with respect to the basis $\mathbf{v}$ of V, it is possible to express $g(v, w)$ in terms of the product between matrices and vectors as follows:

$$g(v, w) = (\lambda_1, \dots, \lambda_n)^{\mathbf{v}} \, G \begin{pmatrix} \mu_1 \\ \vdots \\ \mu_n \end{pmatrix}^{\mathbf{v}} = (\lambda_1, \dots, \lambda_n)^{\mathbf{v}} \, M_{\mathbf{vv}}(g) \begin{pmatrix} \mu_1 \\ \vdots \\ \mu_n \end{pmatrix}^{\mathbf{v}}.$$

We can therefore associate a square matrix $G = M_{\mathbf{vv}}(g)$, with entries in K, to each bilinear form g on V; this matrix depends on the choice of a basis of V. It is immediate to verify that the bilinear form g is symmetric if and only if $g_{ij} = g_{ji}$, for every $i, j = 1, \dots, n$, i.e., if and only if G is a symmetric matrix. Similarly, g is antisymmetric if and only if $g_{ij} = -g_{ji}$, for any $i, j = 1, \dots, n$, i.e., if and only if G is an antisymmetric matrix.

Remark 5.4.2. Let U and W be two vector subspaces of a vector space V and $g_V = g_U \oplus g_W$, where g_U and g_W are bilinear forms on U and W, respectively. Let $\mathbf{u} = \{u_1, \ldots, u_r\}$ be a basis of U and let us denote by G_U the matrix of g_U with respect to the basis $\mathbf{u}$. Similarly, let $\mathbf{w} = \{w_1, \ldots, w_s\}$ be a basis of W and let us denote by G_W the matrix of g_W with respect to the basis $\mathbf{w}$. If we denote by G the matrix of g_V with respect to the basis $\mathbf{v}$ of V formed by the vectors $u_1, \ldots, u_r, w_1, \ldots, w_s$, it is immediate to verify that G has the following block form

$$G = \left(\begin{array}{c|c} G_U & \mathbf{0}_{r,s} \\ \hline \mathbf{0}_{s,r} & G_W \end{array}\right)$$

where $\mathbf{0}_{r,s}$ denotes the zero matrix with r rows and s columns.

We now describe the kernel of a symmetric bilinear form g in terms of its associated matrix:

Proposition 5.4.3. *Let V be a vector space over K, g a symmetric bilinear form on V and $\mathbf{v} = \{v_1, \ldots, v_n\}$ be a basis of V. Let us denote by $G \in M_n(K)$ the matrix of g with respect to the basis $\mathbf{v}$ of V. The kernel of g is the set of vectors $v = \lambda_1 v_1 + \lambda_2 v_2 + \cdots + \lambda_n v_n$, where the n-tuple $(\lambda_1, \ldots, \lambda_n)$ is a solution of the homogeneous linear system $GX = \mathbf{0}$.*

Proof. A vector $v = \lambda_1 v_1 + \lambda_2 v_2 + \cdots + \lambda_n v_n$ belongs to the kernel of g if and only if $g(w, v) = 0$ for every $w \in V$. Since the vectors $v_1, v_2, \ldots, v_n$ form a basis of V, this is equivalent to require that $g(v_i, v) = 0$, for $i = 1, \ldots, n$. Hence, for each vector v_i of the basis of V, we have

$$\begin{aligned} g(v_i, v) &= g(v_i, \lambda_1 v_1 + \lambda_2 v_2 + \cdots + \lambda_n v_n) \\ &= \lambda_1 g(v_i, v_1) + \lambda_2 g(v_i, v_2) + \cdots + \lambda_n g(v_i, v_n) \\ &= g_{i1}\lambda_1 + g_{i2}\lambda_2 + \cdots + g_{in}\lambda_n = 0. \end{aligned}$$

These are precisely the n equations of the linear system

$$G \begin{pmatrix} x_1 \\ \vdots \\ x_n \end{pmatrix} = \mathbf{0}.$$

□

Corollary 5.4.4. *Let V be a finite dimensional vector space over K. A symmetric bilinear form g on V is non-degenerate if and only if its matrix G, with respect to any basis of V, is non-singular, i.e., if and only if* $\det G \neq 0$.

Proof. Let us recall that g is non-degenerate if and only if $\operatorname{Ker}(g) = \{\mathbf{0}\}$. According to the previous proposition, this is equivalent to requiring that the linear system $GX = \mathbf{0}$ has the zero vector as the only solution. According to Cramer's theorem this happens if and only if the matrix G is non-singular. □

Proposition 5.4.5. *Let V be a n-dimensional vector space over K and g a non-degenerate symmetric bilinear form on V. For every vector subspace U of V, we have*

$$\dim U^{\perp} = n - \dim U.$$

Proof. Let U be a r-dimensional vector subspace of V and $v_1, v_2, \dots, v_r$ be a basis of U. Let us complete this basis so that we obtain a basis $v_1, \dots, v_r, v_{r+1}, \dots, v_n$ of V. A vector $v = \lambda_1 v_1 + \lambda_2 v_2 + \cdots + \lambda_n v_n$ belongs to $U^{\perp}$ if and only if $g(u, v) = 0$, for every $u \in U$. Since the vectors $v_1, v_2, \dots, v_r$ are a basis of U, this is equivalent to require that $g(v_i, v) = 0$ for $i = 1, \dots, r$. We obtain the following equations:

$$\begin{aligned} g(v_i, v) &= g(v_i, \lambda_1 v_1 + \lambda_2 v_2 + \cdots + \lambda_n v_n) \\ &= \lambda_1 g(v_i, v_1) + \lambda_2 g(v_i, v_2) + \cdots + \lambda_n g(v_i, v_n) \\ &= g_{i1}\lambda_1 + g_{i2}\lambda_2 + \cdots + g_{in}\lambda_n = 0, \end{aligned} \tag{5.4.1}$$

for $i = 1, \dots, r$. In this way we obtain a homogeneous system whose matrix G' is precisely the submatrix formed by the first r rows of the matrix G. Since by hypothesis g is non-degenerate, the matrix G is non-singular, hence all its rows are linearly independent. It follows that the submatrix G' has full rank, equal to r. From Rouché–Capelli's theorem it follows that the space of solutions of the linear system (5.4.1) has dimension $n-r$. Since the solutions of this system give the components, with respect to the basis $v_1, \dots, v_n$, of the vectors of the subspace $U^{\perp}$ of V, it follows that $\dim U^{\perp} = n - r$. □

Corollary 5.4.6. *Let V be a finite dimensional vector space over K and g a non-degenerate symmetric bilinear form on V. For every vector subspace U of V, we have*

$$(U^{\perp})^{\perp} = U.$$

Proof. If $n = \dim V$, by the previous proposition we have:

$$\dim(U^{\perp})^{\perp} = n - \dim U^{\perp} = n - (n - \dim U) = \dim U.$$

Hence U and $(U^{\perp})^{\perp}$ are two vector subspaces of the same dimension. Since we know that $U \subseteq (U^{\perp})^{\perp}$, we must have $U = (U^{\perp})^{\perp}$. □

Remark 5.4.7. Note that the proof of the preceding result makes essential use of the fact that V has finite dimension. In fact, it could be shown that, in the case of an infinite dimensional vector space V, for a subspace U of V there is, in general, only a proper inclusion $U \subset (U^{\perp})^{\perp}$.

Corollary 5.4.8. *Let V be a finite dimensional vector space over $\mathbb{R}$ and g a non-degenerate symmetric bilinear form on V. Let us also assume that g be positive (or negative) definite. Then for every vector subspace U of V, we have*

$$V = U \oplus U^{\perp}.$$

Proof. If $v \in U \cap U^{\perp}$, we must have $g(v, v) = 0$. Since g is positive (or negative) definite it follows that $v = \mathbf{0}$, hence $U \cap U^{\perp} = \{\mathbf{0}\}$.

If $n = \dim V$ we have $\dim U^{\perp} = n - \dim U$, hence from Grassmann's formula, we have

$$\dim(U + U^{\perp}) = \dim U + \dim U^{\perp} = \dim V,$$

hence $U + U^{\perp} = V$. □

Remark 5.4.9. With the same hypotheses of the previous corollary, every vector $v \in V$ can be written as the sum of a vector $u \in U$ and a vector $w \in U^{\perp}$:

$$v = u + w, \qquad u \in U, \quad w \in U^{\perp}.$$

The vectors u and w are called, respectively, the *orthogonal projections* of v on the subspaces U and $U^{\perp}$.

We end this section by proving the following result:

Proposition 5.4.10. *Let V be a vector space over the field K and g a non-degenerate symmetric bilinear form on V. For every vector subspaces W_1, W_2 of V, we have:*

(*i*) $(W_1 + W_2)^{\perp} = W_1^{\perp} \cap W_2^{\perp}$;

(*ii*) $(W_1 \cap W_2)^{\perp} \supseteq W_1^{\perp} + W_2^{\perp}$.

Moreover, if V is finite dimensional, in (ii) equality holds.

Proof. (*i*) Let $v \in (W_1 + W_2)^{\perp}$. We have $g(v, w) = 0$ for any $w \in W_1 + W_2$ hence $g(v, w_1) = 0$ and $g(v, w_2) = 0$, for any $w_1 \in W_1$ and any $w_2 \in W_2$. This shows that $v \in W_1^{\perp} \cap W_2^{\perp}$.

Vice versa, let us assume that $v \in W_1^\perp \cap W_2^\perp$, i.e., that $g(v, w_1) = 0$ and $g(v, w_2) = 0$, for any $w_1 \in W_1$ and any $w_2 \in W_2$. Let us recall that every vector $w \in W_1 + W_2$ can be written as $w = w_1 + w_2$, with $w_1 \in W_1$ and $w_2 \in W_2$. Hence we have $g(v, w) = g(v, w_1) + g(v, w_2) = 0$, which shows that $v \in (W_1 + W_2)^\perp$.

(*ii*) Let $v \in W_1^\perp + W_2^\perp$. Then we have $v = v_1 + v_2$, with $v_1 \in W_1^\perp$ and $v_2 \in W_2^\perp$. For every $w \in W_1 \cap W_2$ we have $g(v, w) = g(v_1, w) + g(v_2, w) = 0$, hence $v \in (W_1 \cap W_2)^\perp$, as claimed.

Finally, if V is finite dimensional, the equality follows by a dimensional computation. Let $n = \dim V$, $m_1 = \dim W_1$ and $m_2 = \dim W_2$; from Proposition 5.4.5 it follows that

$$\begin{aligned} \dim W_1^\perp &= n - m_1 \\ \dim W_2^\perp &= n - m_2 \\ \dim(W_1 \cap W_2)^\perp &= n - \dim(W_1 \cap W_2). \end{aligned}$$

By Grassmann's formula, we have

$$\begin{aligned} \dim(W_1^\perp + W_2^\perp) &= \dim W_1^\perp + \dim W_2^\perp - \dim(W_1^\perp \cap W_2^\perp) \\ &= 2n - m_1 - m_2 - \dim(W_1^\perp \cap W_2^\perp). \end{aligned} \tag{5.4.2}$$

By (*i*) we have

$$\dim(W_1^\perp \cap W_2^\perp) = \dim(W_1 + W_2)^\perp = n - \dim(W_1 + W_2)$$

hence, using once more Grassmann's formula, we have

$$\dim(W_1^\perp \cap W_2^\perp) = n - m_1 - m_2 + \dim(W_1 \cap W_2).$$

By substituting in (5.4.2) we finally obtain

$$\dim(W_1^\perp + W_2^\perp) = n - \dim(W_1 \cap W_2) = \dim(W_1 \cap W_2)^\perp.$$

From the inclusion (*ii*) and the equality of the dimensions of the two subspaces we deduce the equality $(W_1 \cap W_2)^\perp = W_1^\perp + W_2^\perp$. □

Remark 5.4.11. An alternative proof of the point (*ii*) of the proposition above, in the case in which the vector space V is finite dimensional, can be obtained by simply replacing, in the equality of the point (*i*), the vector subspaces W_1 and W_2 with their orthogonals and noting that, by Corollary 5.4.6, we have $(W_1^\perp)^\perp = W_1$ and $(W_2^\perp)^\perp = W_2$. We obtain

$$(W_1^\perp + W_2^\perp)^\perp = W_1 \cap W_2,$$

from which, by taking orthogonals and recalling once more Corollary 5.4.6, we get

$$W_1^{\perp} + W_2^{\perp} = (W_1 \cap W_2)^{\perp}.$$

5.4.1 Orthogonal projections and least squares

In the vector space $\mathbb{R}^n$ with the usual scalar product let $U \subset \mathbb{R}^n$ be a vector subspace of dimension $r \geq 1$, and let $\{u_1, u_2, \dots, u_r\}$ be a basis of U. By Corollary 5.4.8 we know that $\mathbb{R}^n = U \oplus U^{\perp}$, hence every vector $v \in \mathbb{R}^n$ can be written as $v = u + w$, with $u \in U$ and $w \in U^{\perp}$. Let us denote by $\mathrm{pr}_U \colon \mathbb{R}^n \to \mathbb{R}^n$ the linear function[4] that associates to any $v \in \mathbb{R}^n$ its orthogonal projection $u = \mathrm{pr}_U(v)$ on the subspace U.

In this section we want to determine the matrix of the function pr_U with respect to the canonical basis of $\mathbb{R}^n$. Then we will apply this result to the so-called *least squares method.*

Given $v \in \mathbb{R}^n$, the vector $u = \mathrm{pr}_U(v)$ can be written as a linear combination of the basis vectors of U

$$u = x_1 u_1 + x_2 u_2 + \cdots + x_r u_r.$$

If we denote by A the matrix whose *columns* are $u_1, u_2, \dots, u_r$, the previous equality can be written as follows

$$u = AX, \quad \text{where } X = \begin{pmatrix} x_1 \\ \vdots \\ x_r \end{pmatrix}$$

We have to determine the unknown coefficients $x_1, \dots, x_r$ by requiring that the vector $w = v - u = v - AX$ belong to the subspace $U^{\perp}$; hence we must have $u_i \cdot w = 0$, for any $i = 1, \dots, r$.

By recalling that the scalar product of the vectors u_i and w is equal to the product of the row vector u_i by the column vector w, the r equations $u_1 \cdot w = 0$, $u_2 \cdot w = 0$, $\dots$, $u_r \cdot w = 0$ can be written as ${}^tAw = \mathbf{0}$, i.e., ${}^tA(v - AX) = {}^tAv - {}^tAAX = \mathbf{0}$.

It follows that the coefficients $x_1, \dots, x_r$ are the solutions of the following system of linear equations:

$$({}^tAA)X = {}^tAv$$

[4] As an exercise the reader can check that the function pr_U is linear.

Lemma 5.4.12. *For every vector subspace $U \subseteq \mathbb{R}^n$ of dimension $r \geq 1$, with basis $\{u_1, u_2, \ldots, u_r\}$, the matrix ${}^tAA$ is invertible.*

Proof. Let $G = {}^tAA$. For every $i, j = 1, \ldots, r$, the entry g_{ij} of the matrix G is the scalar product of the vectors u_i and u_j, $g_{ij} = u_i \cdot u_j$.

This means that G is the matrix of the usual scalar product on U, with respect to the basis $\{u_1, u_2, \ldots, u_r\}$. Since the usual scalar product is a non-degenerate symmetric bilinear form, we have $\det G = \det({}^tAA) \neq 0$, hence the matrix ${}^tAA$ is invertible. □

The fact that the matrix ${}^tAA$ is invertible implies that the linear system $({}^tAA)X = {}^tAv$ has a unique solution, given by $X = ({}^tAA)^{-1}\,{}^tAv$. Therefore we have

$$u = \mathrm{pr}_U(v) = AX = A\,({}^tAA)^{-1}\,{}^tAv.$$

It follows that

$$P = A\,({}^tAA)^{-1}\,{}^tA$$

is the matrix of the orthogonal projection $\mathrm{pr}_U \colon \mathbb{R}^n \to \mathbb{R}^n$, with respect to the canonical basis of $\mathbb{R}^n$.

Remark 5.4.13. Note that the matrix P has the property that $P^2 = P$. In fact

$$\begin{aligned} P^2 = P \cdot P &= \big(A\,({}^tAA)^{-1}\,{}^tA\big)\big(A\,({}^tAA)^{-1}\,{}^tA\big) \\ &= A\,({}^tAA)^{-1}\,({}^tAA)\,({}^tAA)^{-1}\,{}^tA \\ &= A\,({}^tAA)^{-1}\,{}^tA = P. \end{aligned}$$

Remark 5.4.14. From the definition of $\mathrm{pr}_U \colon \mathbb{R}^n \to \mathbb{R}^n$ it follows that the only eigenvalues of $P = A\,({}^tAA)^{-1}\,{}^tA$ are 0 and 1. The eigenspace corresponding to the eigenvalue 1 is the subspace U, while the eigenspace corresponding to the eigenvalue 0 (i.e., the kernel of pr_U) is the susbpace $U^\perp$.

Example 5.4.15. Let $U = \langle u_1, u_2 \rangle$ be the vector subspace of $\mathbb{R}^4$ spanned by the vectors $u_1 = (1, 0, -1, 1)$ and $u_2 = (2, 1, 0, 1)$. The matrix A, whose columns are the vectors u_1 and u_2, is

$$A = \begin{pmatrix} 1 & 2 \\ 0 & 1 \\ -1 & 0 \\ 1 & 1 \end{pmatrix}$$

Hence we have

$$
{}^tAA = \begin{pmatrix} 1 & 0 & -1 & 1 \\ 2 & 1 & 0 & 1 \end{pmatrix} \begin{pmatrix} 1 & 2 \\ 0 & 1 \\ -1 & 0 \\ 1 & 1 \end{pmatrix} = \begin{pmatrix} 3 & 3 \\ 3 & 6 \end{pmatrix}
$$

The inverse of the matrix ${}^tAA$ is

$$
({}^tAA)^{-1} = \begin{pmatrix} \frac{2}{3} & -\frac{1}{3} \\ -\frac{1}{3} & \frac{1}{3} \end{pmatrix}
$$

The matrix P of the orthogonal projection on U is

$$
P = A\,({}^tAA)^{-1}\,{}^tA = \begin{pmatrix} 1 & 2 \\ 0 & 1 \\ -1 & 0 \\ 1 & 1 \end{pmatrix} \begin{pmatrix} \frac{2}{3} & -\frac{1}{3} \\ -\frac{1}{3} & \frac{1}{3} \end{pmatrix} \begin{pmatrix} 1 & 0 & -1 & 1 \\ 2 & 1 & 0 & 1 \end{pmatrix}
$$

$$
= \begin{pmatrix} \frac{2}{3} & \frac{1}{3} & 0 & \frac{1}{3} \\ \frac{1}{3} & \frac{1}{3} & \frac{1}{3} & 0 \\ 0 & \frac{1}{3} & \frac{2}{3} & -\frac{1}{3} \\ \frac{1}{3} & 0 & -\frac{1}{3} & \frac{1}{3} \end{pmatrix}
$$

The least squares method

Let's imagine we want to determine the equation of a straight line, in the Cartesian plane, passing through n assigned points $P_1 = (a_1, b_1)$, $P_2 = (a_2, b_2)$, $\dots$, $P_n = (a_n, b_n)$. Considering a line of equation $y = mx + q$ and imposing the conditions for the assigned points to lie on this line, we obtain the following system of linear equations, in the unknowns m and q:

$$
\begin{cases} ma_1 + q = b_1 \\ ma_2 + q = b_2 \\ \cdot \quad \cdot \quad \cdot \quad \cdot \\ ma_n + q = b_n \end{cases}
$$

In matrix notation this system is rewritten as follows:

$$
\begin{pmatrix} a_1 & 1 \\ a_2 & 1 \\ \vdots & \vdots \\ a_n & 1 \end{pmatrix} \begin{pmatrix} m \\ q \end{pmatrix} = \begin{pmatrix} b_1 \\ b_2 \\ \vdots \\ b_n \end{pmatrix} \tag{5.4.3}
$$

Of course, if the points assigned are more than two, there is no line that passes through them (unless these points are perfectly aligned).

From the point of view of the linear system (5.4.3), the reason why this system does not admit a solution is that the vector B formed by the column of constant terms does not belong to the vector subspace of $\mathbb{R}^n$ spanned by the columns of the matrix A.

In such a situation the best we can do is to look for a line that *approximates*, in the best possible way, the given points.

Motivated by this simple example, we consider a system of n linear equations in r unknowns (with $n > r$)

$$AX = B. \tag{5.4.4}$$

Let's assume that the matrix A has maximal rank, equal to r. This means that the r columns $A_{(1)}, A_{(2)}, \dots, A_{(r)}$ of A are linearly independent. Let us denote by $U \subset \mathbb{R}^n$ the vector subspace spanned by the columns of A. We know that the system (5.4.4) has a solution if and only if the column vector B belongs to the subspace U. If, on the other hand, $B \notin U$, then there does not exist a vector $X \in \mathbb{R}^r$ such that $AX = B$. In this case we can try to find a vector X that minimizes the difference $B - AX$.

Note that, as X varies in $\mathbb{R}^r$, the set of vectors AX span the whole subset U. Our problem is then to find a vector $AX \in U$ such that the vector $B - AX$ has minimum norm.[5] This situation is described in the following figure:

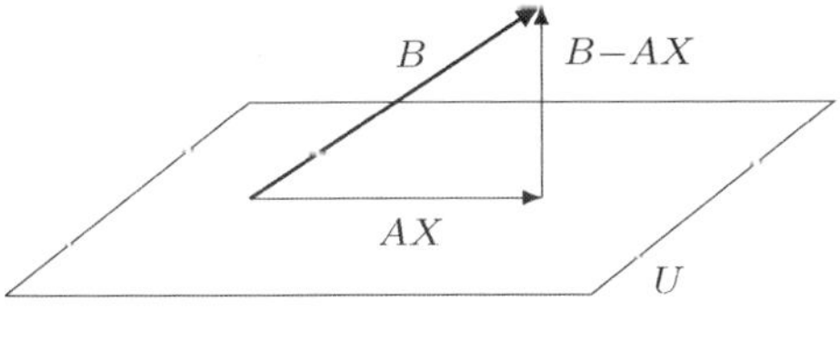

It is immediate to verify that the vector $B - AX$ has minimum norm precisely when it is orthogonal to the subspace U; in this case the vector AX is precisely the orthogonal projection of B on U.

Since the subspace U is spanned by the columns $A_{(1)}, A_{(2)}, \dots, A_{(r)}$ of the matrix A, the orthogonality condition between $B - AX$ and U, i.e., between

[5]Since the norm of a vector of $\mathbb{R}^n$ is the square root of the sum of squares of its components, minimizing the norm of the vector $B - AX$ is equivalent to minimizing the sum of squares of its components. Hence the name "least squares method."

$B-AX$ and the vectors $A_{(i)}$, is given by the equation ${}^tA(B-AX)=\mathbf{0}$. This equation is equivalent to the following system:

$$({}^tAA)X = {}^tAB. \tag{5.4.5}$$

The hypothesis that the matrix A has rank r, i.e., that its r columns are linearly independent, implies, by Lemma 5.4.12, that the matrix ${}^tAA$ is invertible, so the system (5.4.5) admits a single solution given by $X = ({}^tAA)^{-1}\,{}^tAB$. As we said, this is the vector $X \in \mathbb{R}^r$ that minimizes the difference between the vector B of constant terms and the vector AX.

Example 5.4.16. In the Cartesian plane, we want to determine the equation of the parabola, of the form $y = ax^2+bx+c$, which passes through the following points (or which better approximates them): $A_1 = (-2, 6.3)$, $A_2 = (-1, 1.7)$, $A_3 = (0, 1.3)$, $A_4 = (1, 3.1)$, $A_5 = (2, 8.6)$, $A_6 = (3, 16.2)$.

By requiring the points to lie on the parabola we obtain the following equations:

$$\begin{cases} 4a - 2b + c = 6.3 \\ a - b + c = 1.7 \\ c = 1.3 \\ a + b + c = 3.1 \\ 4a + 2b + c = 8.6 \\ 9a + 3b + c = 16.2 \end{cases}$$

It can be verified that this system has no solutions, therefore there is no parabola that passes exactly through the assigned points.

Rewriting the previous system in matrix notation, we obtain

$$\begin{pmatrix} 4 & -2 & 1 \\ 1 & -1 & 1 \\ 0 & 0 & 1 \\ 1 & 1 & 1 \\ 4 & 2 & 1 \\ 9 & 3 & 1 \end{pmatrix} \begin{pmatrix} a \\ b \\ c \end{pmatrix} = \begin{pmatrix} 6.3 \\ 1.7 \\ 1.3 \\ 3.1 \\ 8.6 \\ 16.2 \end{pmatrix}$$

To find the "least squares" solution it is sufficient to multiply both sides of the previous equality by the transpose of the matrix A. With a simple computation, we find

$${}^tAA = \begin{pmatrix} 115 & 27 & 19 \\ 27 & 19 & 3 \\ 19 & 3 & 6 \end{pmatrix}, \qquad {}^tAB = \begin{pmatrix} 210.2 \\ 54.6 \\ 37.2 \end{pmatrix}$$

Thus we obtain the system

$$\begin{cases} 115\,a + 27\,b + 19\,c = 210.2 \\ 27\,a + 19\,b + 3\,c = 54.6 \\ 19\,a + 3\,b + 6\,c = 37.2 \end{cases}$$

whose solution is $a \simeq 1.511$, $b \simeq 0.546$, $c \simeq 1.143$. The parabola thus found, with equation $y = 1.511\,x^2 + 0.546\,x + 1.143$, is the one that best approximates the given points. The situation is illustrated in the graph below:

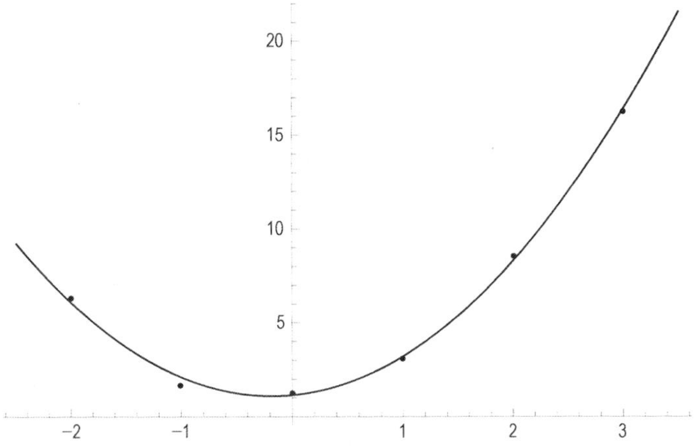

5.4.2 Change of basis

The matrix associated to a bilinear form g, defined on a finite dimensional vector space V, depends on the choice of a basis of V. In this section we will describe how the matrix of g changes in correspondence with a change in the basis of V.

Let V be a vector space of dimension n over K and g a bilinear form on V. Let $\mathbf{v} = \{v_1, \dots, v_n\}$ and $\mathbf{v}' = \{v'_1, \dots, v'_n\}$ be two bases of V and let us denote by $G = (g_{ij}) = M_{\mathbf{vv}}(g)$ the matrix of g with respect to the basis $\mathbf{v}$ and by $G' = (g'_{ij}) = M_{\mathbf{v}'\mathbf{v}'}(g)$ the matrix of g with respect to the basis $\mathbf{v}'$. We recall that this means that

$$g_{ij} = g(v_i, v_j), \quad \text{and} \quad g'_{ij} = g(v'_i, v'_j),$$

for any $i, j = 1, \dots, n$.

We shall denote by $\alpha_{\mathbf{v}} : V \xrightarrow{\sim} K^n$ the isomorphism that associates to each vector $v \in V$ the n-tuple $(x_1, \dots, x_n)^{\mathbf{v}}$ of its components with respect to the basis $\mathbf{v}$ and by $\alpha_{\mathbf{v}'}: V \xrightarrow{\sim} K^n$ the isomorphism that associates to each $v \in V$ the n-tuple $(x'_1, \ldots, x'_n)^{\mathbf{v}'}$ of its components with respect to the basis $\mathbf{v}'$.

By composing $\alpha_{\mathbf{v}}$ with the inverse of the isomorphism $\alpha_{\mathbf{v}'}$ we obtain an isomorphism of K^n with itself that corresponds to the multiplication by some invertible matrix $P \in \mathrm{GL}_n(K)$. We shall denote this isomorphism by $F_P: K^n \xrightarrow{\sim} K^n$. Hence we get the following commutative diagram:

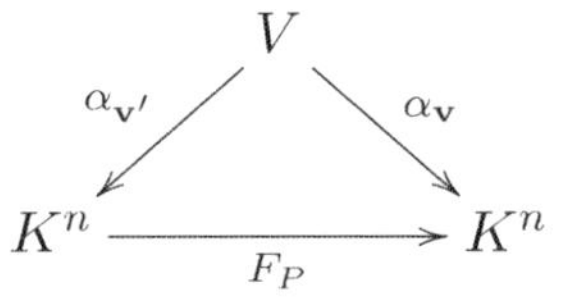

Let us now try to obtain a more explicit description of the matrix P. We recall that the columns of P are given by the images, by the isomorphism F_P, of the vectors of the canonical basis of K^n. Let $e_j = {}^t(0, \ldots, 0, 1, 0, \ldots, 0)$ be the j-th vector of the canonical basis of K^n (the coordinates are all zero except the j-th coordinate that is 1). By the isomorphism $\alpha_{\mathbf{v}'}$ the vector $e_j \in K^n$ corresponds to the j-th vector v'_j of the basis $\mathbf{v}'$ of V. Hence we have

$$F_P(e_j) = \alpha_{\mathbf{v}}(\alpha_{\mathbf{v}'}^{-1}(e_j)) = \alpha_{\mathbf{v}}(v'_j),$$

where we recall that $\alpha_{\mathbf{v}}(v'_j) \in K^n$ is the vector given by the components of v'_j computed with respect to the basis $\mathbf{v}$; this vector is the j-th column of P.

In conclusion, we can say that the *columns* of the matrix P are the components of the vectors $v'_1, \ldots, v'_n$ of the basis $\mathbf{v}'$ of V computed with respect to the first basis $\mathbf{v}$. In other words, this means that P is the matrix associated to the identity $\mathrm{id}_V: V \to V$, with respect to the basis $\mathbf{v}'$ of the domain and the basis $\mathbf{v}$ of the codomain, $P = M^{\mathbf{v}}_{\mathbf{v}'}(\mathrm{id}_V)$. Hence we have

$$v'_j = \sum_{i=1}^{n} p_{ij}\, v_i.$$

We are now able to determine the relationship between the two matrices $G = M_{\mathbf{vv}}(g)$ and $G' = M_{\mathbf{v}'\mathbf{v}'}(g)$. We have:

$$\begin{aligned} g'_{ij} = g(v'_i, v'_j) &= g\Big(\sum_{h=1}^{n} p_{hi}\, v_h, \sum_{k=1}^{n} p_{kj}\, v_k\Big) \\ &= \sum_{h,k=1}^{n} p_{hi}\, p_{kj}\, g(v_h, v_k) = \sum_{h,k=1}^{n} p_{hi}\, g_{hk}\, p_{kj}. \end{aligned}$$

This equality can be rewritten, in terms of matrices, as follows:

$$G' = {}^tPGP, \quad \text{that is} \quad M_{\mathbf{v}'\mathbf{v}'}(g) = {}^tM^{\mathbf{v}}_{\mathbf{v}'}(\mathrm{id}_V)M_{\mathbf{vv}}(g)M^{\mathbf{v}}_{\mathbf{v}'}(\mathrm{id}_V). \qquad (5.4.6)$$

Let us state the following definition:

Definition 5.4.17. Two square matrices G and G' of order n with entries in K are called *congruent* if there exists an invertible matrix $P \in \mathrm{GL}_n(K)$ such that

$$G' = {}^tPGP.$$

From the above, the following result is deduced:

Corollary 5.4.18. *Two matrices $G, G' \in M_n(K)$ represent the same bilinear form g defined on a vector space V of dimension n over K, with respect to different bases, if and only if they are congruent.*

Remark 5.4.19. It is easy to verify that the congruence relation is an equivalence relation on the set $M_n(K)$ of square matrices of order n with entries in K.

We remark that if $\mathbf{v} = \{v_1, \dots, v_n\}$ is an orthonormal basis of V, the corresponding matrix G of the bilinear form g is the identity matrix. In particular, if $\mathbf{v}$ and $\mathbf{v}'$ are both orthonormal bases of V, we have $G = G' = \mathbf{1}$ hence the formula (5.4.6) reduces to ${}^tPP = \mathbf{1}$, i.e., ${}^tP = P^{-1}$. This motivates the following definition:

Definition 5.4.20. A matrix $P \in M_n(K)$ is called an *orthogonal* matrix if ${}^tPP = \mathbf{1}$.

It is easy to verify that the set of orthogonal matrices

$$\mathrm{O}(n) = \{P \in M_n(K) \mid {}^tPP = \mathbf{1}\}$$

is a group with respect to the multiplication; this group is called the *orthogonal group* of order n.

We can now summarize the above as follows:

Proposition 5.4.21. *The change of basis matrix between two orthonormal bases of V is an orthogonal matrix.*

We end this section by extending to the matrices the Definition 5.3.20 given for symmetric bilinear forms:

Definition 5.4.22. Let $G \in M_n(\mathbb{R})$ be a symmetric matrix and $g\colon \mathbb{R}^n \times \mathbb{R}^n \to \mathbb{R}$ be the corresponding symmetric bilinear form (i.e., the bilinear form whose matrix, with respect to the canonical basis of $\mathbb{R}^n$, is G). The matrix G is:

(*i*) *positive definite* if g is positive definite;

(*ii*) *negative definite* if g negative definite;

(*iii*) *positive semidefinite* if g is positive semidefinite;

(*iv*) *negative semidefinite* if g is negative semidefinite;

(*v*) *indefinite* if g is indefinite.

5.4.3 Orthogonal and orthonormal bases

In this section we will prove that every real vector space V of finite dimension, endowed with a positive definite symmetric bilinear form, admits an orthonormal basis. We will also describe a procedure that allows to construct an orthonormal basis starting from any basis of V.

We start by proving the following result:

Proposition 5.4.23. *Let V be a real vector space and g be a positive definite symmetric bilinear form on V. If the vectors $v_1, v_2, \dots, v_r$ are pairwise orthogonal they are also linearly independent.*

Proof. Let's consider a linear combination

$$\lambda_1 v_1 + \lambda_2 v_2 + \cdots + \lambda_r v_r = \mathbf{0}.$$

For every $i = 1, \dots, r$, we have

$$g\Big(v_i, \sum_{j=1}^{r} \lambda_j v_j\Big) = \sum_{j=1}^{r} \lambda_j\, g(v_i, v_j) = \lambda_i\, g(v_i, v_i) = 0,$$

because $g(v_i, v_j) = 0$ for every $i \neq j$. Since g is positive definite we have $g(v_i, v_i) > 0$, hence $\lambda_i = 0$. This proves that the vectors $v_1, \dots, v_r$ are linearly independent. □

Now we can prove the following result:

Theorem 5.4.24. *Every finite dimensional real vector space V, endowed with a positive definite symmetric bilinear form g, has an orthonormal basis.*

Proof. We proceed by induction on $n = \dim V$. If $n = 1$ let us take a non-zero vector $v \in V$. Since g is positive definite we have $g(v, v) > 0$, hence we can set $w = v/\sqrt{g(v, v)}$. The vector w is normalized, i.e., $g(w, w) = 1$, hence it is an orthonormal basis of V. We therefore suppose that the theorem holds for any real vector space of dimension $n-1$ and prove that it also holds for vector spaces of dimension n.

Let V be a vector space of dimension n, and let $v \in V$ be a non-zero vector. Let's consider the subspace $W = \langle v\rangle^{\perp}$ endowed with the positive definite symmetric bilinear form induced by g. By Proposition 5.4.5 we have $\dim W = n - 1$ hence, by the induction hypothesis, W has an orthonormal basis $w_1, \ldots, w_{n-1}$. Since, by hypothesis, g is positive definite, we have $g(v, v) > 0$; hence we can set $w_n = v/\sqrt{g(v, v)}$. It is now easy to verify that $\{w_1, \ldots, w_{n-1}, w_n\}$ is an orthonormal basis of V. □

We now describe a method, known as the *Gram–Schmidt orthonormalization procedure*, which allows to construct an orthonormal basis starting from any basis of V.

Let V be a real vector space of dimension n endowed with a positive definite symmetric bilinear form g. Let $\mathbf{v} = \{v_1, \ldots, v_n\}$ be a basis of V. We set $w_1 = v_1$ and look for a vector w_2, orthogonal to w_1, of the form $w_2 = \alpha_1 w_1 + v_2$ (note that this implies that the vector subspace generated by w_1 and w_2 is equal to the subspace generated by v_1 and v_2). The orthogonality condition between w_1 and w_2 is expressed by setting $g(w_1, w_2) = 0$. Hence we have

$$g(w_1, w_2) = g(w_1, \alpha_1 w_1 + v_2) = \alpha_1 g(w_1, w_1) + g(w_1, v_2) = 0,$$

from which we get

$$\alpha_1 = -\frac{g(w_1, v_2)}{g(w_1, w_1)}.$$

The vector we are looking for is therefore

$$w_2 = v_2 - \frac{g(w_1, v_2)}{g(w_1, w_1)} w_1. \tag{5.4.7}$$

Now we look for a vector w_3, orthogonal to the subspace spanned by w_1 and w_2, of the form $w_3 = \alpha_1 w_1 + \alpha_2 w_2 + v_3$. Requiring w_3 to be orthogonal to w_1 and w_2 we get the equations $g(w_1, w_3) = 0$ and $g(w_2, w_3) = 0$. From these equations we obtain:

$$\begin{aligned} g(w_1, w_3) &= g(w_1, \alpha_1 w_1 + \alpha_2 w_2 + v_3) \\ &= \alpha_1 g(w_1, w_1) + \alpha_2 g(w_1, w_2) + g(w_1, v_3) \\ &= \alpha_1 g(w_1, w_1) + g(w_1, v_3) = 0, \end{aligned}$$

$$\begin{aligned} g(w_2, w_3) &= g(w_2, \alpha_1 w_1 + \alpha_2 w_2 + v_3) \\ &= \alpha_1 g(w_2, w_1) + \alpha_2 g(w_2, w_2) + g(w_2, v_3) \\ &= \alpha_2 g(w_2, w_2) + g(w_2, v_3) = 0, \end{aligned}$$

from which it follows

$$\alpha_1 = -\frac{g(w_1, v_3)}{g(w_1, w_1)}, \qquad \alpha_2 = -\frac{g(w_2, v_3)}{g(w_2, w_2)}.$$

The vector we are looking for is therefore

$$w_3 = v_3 - \frac{g(w_1, v_3)}{g(w_1, w_1)} w_1 - \frac{g(w_2, v_3)}{g(w_2, w_2)} w_2. \tag{5.4.8}$$

Continuing in this way, we obtain an orthogonal basis $\{w_1, w_2, \ldots, w_n\}$ of V. More precisely, let's assume (by induction) that we have already constructed the pairwise orthogonal vectors $w_1, w_2, \ldots, w_{r-1}$. Now we look for a vector w_r, of the form

$$w_r = \alpha_1 w_1 + \alpha_2 w_2 + \cdots + \alpha_{r-1} w_{r-1} + v_r,$$

that is orthogonal to the subspace spanned by $w_1, \ldots, w_{r-1}$ (note that this implies that the vector subspace generated by $w_1, \ldots, w_r$ is equal to the subspace spanned by the vectors $v_1, \ldots, v_r$). For every $i = 1, \ldots, r-1$, the orthogonality condition between w_r and w_i gives the following equation

$$\begin{aligned} g(w_i, w_r) &= g\Big(w_i, \sum_{j=1}^{r-1} \alpha_j w_j + v_r\Big) \\ &= \sum_{j=1}^{r-1} \alpha_j g(w_i, w_j) + g(w_i, v_r) \\ &= \alpha_i g(w_i, w_i) + g(w_i, v_r) = 0, \end{aligned}$$

from which we obtain

$$\alpha_i = -\frac{g(w_i, v_r)}{g(w_i, w_i)}.$$

The vector w_r we are looking for is therefore

$$w_r = v_r - \frac{g(w_1, v_r)}{g(w_1, w_1)} w_1 - \frac{g(w_2, v_r)}{g(w_2, w_2)} w_2 - \cdots - \frac{g(w_{r-1}, v_r)}{g(w_{r-1}, w_{r-1})} w_{r-1}.$$

The n vectors $w_1, \ldots, w_n$ thus constructed are pairwise orthogonal, hence they are linearly independent (see Proposition 5.4.23); this means that they are an orthogonal basis of V. To get an orthonormal basis all we need to do now is normalize these vectors by setting

$$w_i' = \frac{w_i}{\sqrt{g(w_i, w_i)}},$$

for any $i = 1, \ldots, n$ (this is possible because g is positive definite). The vectors $w_1', \ldots, w_n'$ are an orthonormal basis of V. If we denote by G the matrix of g with respect to the basis $\mathbf{v} = \{v_1, \ldots, v_n\}$ of V and by G' the matrix of g with respect to the basis $\mathbf{w}' = \{w_1', \ldots, w_n'\}$, we have

$$G' = {}^tPGP,$$

where P is the change of basis matrix, i.e., the matrix whose columns are formed by the components of the vectors $w_1', \ldots, w_n'$ of the new basis, with respect to the vectors $v_1, \ldots, v_n$ of the original basis of V. Since $\mathbf{w}' = \{w_1', \ldots, w_n'\}$ is an orthonormal basis, we have

$$g_{ij}' = g(w_i', w_j') = \begin{cases} 1 & \text{if } i = j, \\ 0 & \text{if } i \neq j, \end{cases}$$

hence G' is the identity matrix.

Remark 5.4.25. Note that each vector w_j' of the orthonormal basis $\mathbf{w}'$ is a linear combination of the vectors $v_1, v_2, \ldots, v_j$ of the basis $\mathbf{v}$. This means that, in the change of basis matrix P, all elements below the principal diagonal are zero, hence P is an upper triangular matrix.

We can summarize what we have seen so far in the following proposition:

Proposition 5.4.26. *Let $G \in M_n(\mathbb{R})$ be a positive definite symmetric matrix. Then there exists an invertible matrix $P \in \mathrm{GL}_n(\mathbb{R})$ such that ${}^tPGP = \mathbf{1}$. Moreover, the matrix P can be chosen to be upper triangular.*

Example 5.4.27. Let us now apply the Gram–Schmidt orthonormalization method described above to a concrete example. Let V be a real vector space of dimension 4 and g be the symmetric bilinear form on V with matrix

$$G = \begin{pmatrix} 4 & 2 & -2 & 2 \\ 2 & 10 & -7 & -2 \\ -2 & -7 & 6 & 3 \\ 2 & -2 & 3 & 10 \end{pmatrix}$$

with respect to the basis $\mathbf{v} = \{v_1, v_2, v_3, v_4\}$ of V. We want to construct an orthonormal basis of V.

We start by setting $w_1 = v_1$. We have

$$g(w_1, w_1) = g(v_1, v_1) = 4, \qquad g(w_1, v_2) = g(v_1, v_2) = 2,$$

hence, from the formula (5.4.7), we get

$$w_2 = v_2 - \frac{g(w_1, v_2)}{g(w_1, w_1)} w_1 = v_2 - \frac{1}{2} v_1.$$

Now we have:

$$\begin{aligned} g(w_1, v_3) &= g(v_1, v_3) = -2, \\ g(w_2, v_3) &= g\Big(v_2 - \frac{1}{2} v_1, v_3\Big) = g(v_2, v_3) - \frac{1}{2} g(v_1, v_3) = -6, \\ g(w_2, w_2) &= g\Big(v_2 - \frac{1}{2} v_1, v_2 - \frac{1}{2} v_1\Big) = g(v_2, v_2) - g(v_2, v_1) + \frac{1}{4} g(v_1, v_1) = 9. \end{aligned}$$

From the formula (5.4.8) we get

$$w_3 = v_3 - \frac{g(w_1, v_3)}{g(w_1, w_1)} w_1 - \frac{g(w_2, v_3)}{g(w_2, w_2)} w_2 = v_3 + \frac{1}{6} v_1 + \frac{2}{3} v_2.$$

Finally, in a completely analogous way, we have

$$w_4 = v_4 - \frac{g(w_1, v_4)}{g(w_1, w_1)} w_1 - \frac{g(w_2, v_4)}{g(w_2, w_2)} w_2 - \frac{g(w_3, v_4)}{g(w_3, w_3)} w_3.$$

Developing the computations, one finds

$$\begin{aligned} g(w_1, v_4) &= 2 \qquad & g(w_2, v_4) &= -3 \\ g(w_3, v_4) &= 2 \qquad & g(w_3, w_3) &= 1, \end{aligned}$$

from which we get

$$w_4 = -v_1 - v_2 - 2v_3 + v_4.$$

Now let us compute $g(w_4, w_4)$:

$$g(w_4, w_4) = g(-v_1 - v_2 - 2v_3 + v_4, -v_1 - v_2 - 2v_3 + v_4) = 4.$$

Now all that remains to do is to normalize these vectors:

$$\begin{aligned}
w_1' &= \frac{w_1}{\sqrt{g(w_1, w_1)}} = \frac{w_1}{\sqrt{4}} = \frac{1}{2} v_1, \\
w_2' &= \frac{w_2}{\sqrt{g(w_2, w_2)}} = \frac{w_2}{\sqrt{9}} = -\frac{1}{6} v_1 + \frac{1}{3} v_2, \\
w_3' &= \frac{w_3}{\sqrt{g(w_3, w_3)}} = \frac{w_3}{\sqrt{1}} = \frac{1}{6} v_1 + \frac{2}{3} v_2 + v_3, \\
w_4' &= \frac{w_4}{\sqrt{g(w_4, w_4)}} = \frac{w_4}{\sqrt{4}} = -\frac{1}{2} v_1 - \frac{1}{2} v_2 - v_3 + \frac{1}{2} v_4.
\end{aligned}$$

The change of basis matrix is

$$P = \begin{pmatrix} \frac{1}{2} & -\frac{1}{6} & \frac{1}{6} & -\frac{1}{2} \\ 0 & \frac{1}{3} & \frac{2}{3} & -\frac{1}{2} \\ 0 & 0 & 1 & -1 \\ 0 & 0 & 0 & \frac{1}{2} \end{pmatrix}$$

that is indeed upper triangular. Now it is easy to check that ${}^tPGP = \mathbf{1}$.

5.5 Symmetric Linear Functions

In this section we will study some properties of symmetric linear functions and of the corresponding symmetric matrices. In particular, we will prove that every symmetric matrix A is *orthogonally* diagonalizable, i.e., it can be transformed into a diagonal matrix by means of an orthogonal change of basis matrix.

Let (V, g) be a Euclidean vector space. This means that V is a real vector space endowed with a positive definite symmetric bilinear form g. In this situation we define the scalar product of two vectors $v, w \in V$ by setting $v \cdot w = g(v, w)$.

Definition 5.5.1. A linear function $f\colon V \to V$ is said to be *symmetric* if

$$f(v) \cdot w = v \cdot f(w), \tag{5.5.1}$$

for every $v, w \in V$.

The following result states that, to verify if a linear function is symmetric, it is sufficient to verify that the equality (5.5.1) holds when v and w are chosen among the vectors of a basis of V.

Lemma 5.5.2. *A linear function $f: V \to V$ is symmetric if and only if $f(v_i) \cdot v_j = v_i \cdot f(v_j)$, for every $i, j = 1, \dots, n$, where $\{v_1, v_2, \dots, v_n\}$ is a basis of V.*

Proof. The necessity of the stated condition is evident. Let us suppose that, for every $i, j = 1, \dots, n$, the equality $f(v_i) \cdot v_j = v_i \cdot f(v_j)$ holds, and let us take any two vectors $v, w \in V$. Since $\{v_1, v_2, \dots, v_n\}$ is a basis of V, we have

$$v = \sum_{i=1}^{n} \lambda_i v_i, \qquad w = \sum_{j=1}^{n} \mu_j v_j.$$

From the linearity of f and the bilinearity of the scalar product it follows that

$$\begin{aligned} f(v) \cdot w &= \Big(\sum_{i=1}^{n} \lambda_i f(v_i)\Big) \cdot \Big(\sum_{j=1}^{n} \mu_j v_j\Big) \\ &= \sum_{i,j=1}^{n} \lambda_i \mu_j f(v_i) \cdot v_j \\ &= \sum_{i,j=1}^{n} \lambda_i \mu_j v_i \cdot f(v_j) \\ &= \Big(\sum_{i=1}^{n} \lambda_i v_i\Big) \cdot \Big(\sum_{j=1}^{n} \mu_j f(v_j)\Big) \\ &= v \cdot f(w), \end{aligned}$$

which proves that f is symmetric. □

The link between symmetric linear functions and symmetric matrices is clarified by the following result:

Proposition 5.5.3. *Let $\mathbf{v} = \{v_1, v_2, \dots, v_n\}$ be an orthonormal basis of V. Then a linear function $f: V \to V$ is symmetric if and only if its matrix with respect to the basis $\mathbf{v}$ is symmetric.*

Proof. Let $f: V \to V$ be a linear function and $A = (a_{ij})$ be its matrix with respect to the orthonormal basis $\{v_1, v_2, \dots, v_n\}$. We have

$$f(v_j) = \sum_{l=1}^{n} a_{lj} v_l,$$

for every $j = 1, \dots, n$.

Let us recall that $v_i \cdot v_j = 0$, for every $i \neq j$, while $v_i \cdot v_i = 1$, hence we have

$$v_i \cdot f(v_j) = v_i \cdot \Big(\sum_{l=1}^{n} a_{lj} v_l\Big) = \sum_{l=1}^{n} a_{lj} v_i \cdot v_l = a_{ij}.$$

By the previous lemma, f is symmetric if and only if, for every $i, j = 1, \dots, n$, we have $v_i \cdot f(v_j) = f(v_i) \cdot v_j$, i.e., if and only if $a_{ij} = a_{ji}$, which means that the matrix A is symmetric. □

We are now able to prove the following important result:

Theorem 5.5.4 (Spectral Theorem). *Let V be a Euclidean vector space and $f\colon V \to V$ be a linear function. Then f is* orthogonally diagonalizable, *i.e., there exists an orthonormal basis of V formed by eigenvectors of f, if and only if f is symmetric.*

Proof. Let us suppose that f is orthogonally diagonalizable, i.e., that there exists an orthonormal basis $\mathbf{v} = \{v_1, v_2, \dots, v_n\}$ of V where the v_i's are eigenvectors of f. The matrix of f with respect to this basis is a diagonal matrix, hence it is symmetric. By the previous proposition this implies that f is a symmetric linear function.

Conversely, we will prove that every symmetric linear function f is orthogonally diagonalizable, reasoning by induction on the dimension of the vector space V.

If $\dim V = 1$, let v_1 be an orthonormal basis of V. We have $f(v_1) \in V$, hence $f(v_1) = \lambda_1 v_1$, which shows that v_1 is an eigenvector of f.

Now we assume that V has dimension $n > 1$ and that the result holds for vector spaces of dimension $n-1$. If we arbitrarily choose an orthonormal basis $\{u_1, u_2, \dots, u_n\}$ of V, from the previous proposition we know that the matrix A of f with respect to this basis is a symmetric matrix, hence it has n real eigenvalues (cf. Theorem 4.3.4). Since the eigenvalues of A are the eigenvalues of f, it follows that all the eigenvalues of f are real.

Let λ_1 be an eigenvalue of f and v_1 be a corresponding eigenvector. Let us denote by W the vector subspace of V orthogonal to the subspace spanned by v_1, $W = \langle v_1 \rangle^{\perp}$. We want to prove that, for every $w \in W$, we have $f(w) \in W$, i.e., that f induces, by restriction, a linear function $f|_W\colon W \to W$.

From the symmetry of f it follows that, for every $w \in W$, $v_1 \cdot f(w) = f(v_1) \cdot w = \lambda_1 v_1 \cdot w = 0$, hence $f(w) \in \langle v_1 \rangle^{\perp} = W$. The function $f|_W : W \to W$ is a symmetric linear function, and since $\dim W = n - 1$, we can apply the induction hypothesis to deduce that there exists an orthonormal basis $\{w_2, \dots, w_n\}$ of W formed by eigenvectors of $f|_W$, i.e., by eigenvectors of f. Now it is enough to set $w_1 = v_1/\|v_1\|$ and note that the vectors $w_1, w_2, \dots, w_n$ are an orthonormal basis of V formed by eigenvectors of f. □

An immediate consequence of this theorem is the following:

Corollary 5.5.5. *A real symmetric matrix A of order n is orthogonally diagonalizable, i.e., there exists an orthonormal basis of $\mathbb{R}^n$ formed by eigenvectors of A.*

Proof. It is sufficient to apply the previous theorem to the Euclidean vector space $V = \mathbb{R}^n$ endowed with the usual scalar product and to the linear function $f\colon \mathbb{R}^n \to \mathbb{R}^n$ defined by $f(v) = Av$. □

Remark 5.5.6. The previous corollary states that, given a real symmetric matrix A of order n, it is always possible to find an orthogonal matrix P (i.e., P is an invertible matrix such that $P^{-1} = {}^tP$), such that the matrix $P^{-1}AP = {}^tPAP$ is diagonal. The columns of such a matrix P are an orthonormal basis of $\mathbb{R}^n$ formed by eigenvectors of A.

5.6 Classification of symmetric bilinear forms

In this section we will deal with the study and classification of symmetric bilinear forms defined on a real vector space of finite dimension.

Let V be a vector space of dimension n over the field $\mathbb{R}$ of real numbers and g be a symmetric bilinear form on V. By Proposizione 5.3.7, there exists a vector subspace U of V such that $V = \operatorname{Ker}(g) \oplus U$ and the restriction of g to $U \times U$ is non-degenerate. This means that it is always possible to find a basis $\{v_1, \dots, v_k, v_{k+1}, \dots, v_n\}$ of V (where $\{v_1, \dots, v_k\}$ is a basis of $\operatorname{Ker}(g)$) with respect to which the matrix G of g is a block matrix of the form

$$G = \left(\begin{array}{c|c} \mathbf{0}_{k,k} & \mathbf{0}_{k,n-k} \\ \hline \mathbf{0}_{n-k,k} & G_U \end{array}\right)$$

where $\mathbf{0}_{r,s}$ denotes the zero matrix with r rows and s columns and G_U is the matrix of the restriction of g to the subspace U spanned by the vectors $\{v_{k+1}, \dots, v_n\}$. To classify the symmetric bilinear forms it is therefore sufficient to classify the non-degenerate ones. We will now describe the classification of these forms over the real numbers.

Let g be a non-degenerate symmetric bilinear form on a real vector space V. If g is positive definite the Gram–Schmidt procedure allows to construct an orthonormal basis of V. With respect to such basis the matrix of g is the identity matrix. If g is negative definite then $-g$ is positive definite. It follows that there exists a basis of V with respect to which the matrix of g is the opposite of the identity matrix. Since g is supposed to be non-degenerate, it remains only to study the case in which g is indefinite, i.e., the case in which there exist a vector v such that $g(v, v) > 0$ and a vector w such that $g(w, w) < 0$.

The idea to deal with this case is the following: we start from any basis of V and use a method similar to the Gram–Schmidt procedure to construct an orthogonal basis of V. The matrix of g with respect to such a basis will be diagonal. Each positive diagonal element $\alpha > 0$ can be made equal to 1 by dividing the corresponding basis vector by $\sqrt{\alpha}$, while every negative diagonal element $\beta < 0$ can be made equal to -1 by dividing the corresponding basis vector by $\sqrt{-\beta}$; it is not possibile to have diagonal elements equal to zero because g is non-degenerate. In this way we obtain a basis of V with respect to which the matrix of g is a diagonal matrix in which the elements on the diagonal are 1 or -1.

Theorem 5.6.1 (Sylvester's theorem). *Let V be a real vector space of dimension n and $g\colon V \times V \to \mathbb{R}$ be a non-degenerate symmetric bilinear form. There exists a basis of V with respect to which the matrix of g is a block matrix of the form*

$$\left(\begin{array}{c|c} \mathbf{1}_{r,r} & \mathbf{0}_{r,s} \\ \hline \mathbf{0}_{s,r} & -\mathbf{1}_{s,s} \end{array}\right)$$

where $r + s = n$, $\mathbf{1}_{r,r}$ denotes the identity matrix of order r and $-\mathbf{1}_{s,s}$ is the opposite of the identity matrix of order s. Furthermore, the numbers r and s (the numbers of 1's and -1's on the diagonal) are uniquely determined (they depend only on g and not on the chosen basis). The pair (r, s) is called the signature *and the difference $r - s$ is called the* inertia index *of g.*

Proof. As previously said, the statement of the theorem is trivially verified if g is positive or negative definite. So we suppose that g is indefinite. In this case there exists a vector $v \in V$ such that $g(v, v) = \alpha > 0$. By setting $v_1 = v/\sqrt{\alpha}$, we have $g(v_1, v_1) = 1$. Let $W = \langle v_1 \rangle^{\perp}$; we have $\dim W = n - 1$. If $\{w_2, \ldots, w_n\}$ is a basis of W, the vectors $\{v_1, w_2, \ldots, w_n\}$ form a basis of

V with respect to which the matrix G of g is a block matrix of the form

$$G = \left(\begin{array}{c|c} 1 & \mathbf{0}_{1,n-1} \\ \hline \mathbf{0}_{n-1,1} & G_W \end{array}\right)$$

where G_W is the matrix of the restriction of g to W (with respect to the basis of W stated above). Now it is enough to repeat the reasoning just described applying it to the restriction of g to W (or we can use induction on the dimension of V).

This proves the existence of a basis of V with respect to which the matrix of g has the desired form. It now remains to prove the uniqueness of the numbers r and s, i.e., their independence from the particular basis chosen.

Let us suppose that there exist two bases $\{v_1, \dots, v_n\}$ and $\{v'_1, \dots, v'_n\}$ of V with respect to which the matrices of g are

$$G = \left(\begin{array}{c|c} \mathbf{1}_{r,r} & \mathbf{0}_{r,s} \\ \hline \mathbf{0}_{s,r} & -\mathbf{1}_{s,s} \end{array}\right), \qquad G' = \left(\begin{array}{c|c} \mathbf{1}_{r',r'} & \mathbf{0}_{r',s'} \\ \hline \mathbf{0}_{s',r'} & -\mathbf{1}_{s',s'} \end{array}\right)$$

with $r + s = r' + s' = n$.

We set $W^+ = \langle v_1, \dots, v_r \rangle$, $W^- = \langle v_{r+1}, \dots, v_n \rangle$, $Z^+ = \langle v'_1, \dots, v'_{r'} \rangle$, $Z^- = \langle v'_{r'+1}, \dots, v'_n \rangle$. We have $\dim W^+ = r$, $\dim W^- = s$, $\dim Z^+ = r'$, $\dim Z^- = s'$,

$$V = W^+ \oplus W^- = Z^+ \oplus Z^-,$$

moreover the restriction of g to W^+ and Z^+ is positive definite, while g restricted to W^- and Z^- is negative definite.

Let us suppose that it is $r < r'$. Every vector v'_i $(i = 1, \dots, r')$ of the basis of Z^+ can be (uniquely) written in the form

$$v'_i = u_i + w_i,$$

with $u_i \in W^+$ and $w_i \in W^-$.

Since $r' > r = \dim W^+$, the vectors $u_1, \dots, u_{r'}$ must be linearly dependent. Hence we have

$$\lambda_1 u_1 + \cdots + \lambda_{r'} u_{r'} = \mathbf{0},$$

with $\lambda_1, \dots, \lambda_{r'}$ not all zero.

Let $z = \lambda_1 v'_1 + \cdots + \lambda_{r'} v'_{r'}$. Then $z \neq \mathbf{0}$, because the λ_j's are not all zero and the vectors $\{v'_1, \dots, v'_{r'}\}$ are a basis of Z^+, hence $g(z, z) > 0$ (because the restriction of g to Z^+ is positive definite). However, we have

$$\begin{aligned} z &= \lambda_1 v_1' + \cdots + \lambda_{r'} v_{r'}' \\ &= \lambda_1(u_1 + w_1) + \cdots + \lambda_{r'}(u_{r'} + w_{r'}) \\ &= \lambda_1 w_1 + \cdots + \lambda_{r'} w_{r'} \in W^- \end{aligned}$$

hence $g(z, z) < 0$, because the restriction of g to W^- is negative definite. We have found a contradiction that is due to the fact that we have assumed that $r' > r$, hence we must have $r' \leq r$.

Repeating the reasoning after switching the roles of r and r', we deduce that we must also have $r \leq r'$, hence the equalities $r = r'$ and $s = s'$ hold. □

We can summarize what we have seen so far in the following result:

Theorem 5.6.2. *Let V be a vector space of dimension n over the field $\mathbb{R}$ of real numbers and $g\colon V \times V \to \mathbb{R}$ be a symmetric bilinear form.*

There exists a basis of V with respect to which the matrix of g is a block matrix of the form

$$\left(\begin{array}{c|c|c} \mathbf{0}_{k,k} & \mathbf{0}_{k,r} & \mathbf{0}_{k,s} \\ \hline \mathbf{0}_{r,k} & \mathbf{1}_{r,r} & \mathbf{0}_{r,s} \\ \hline \mathbf{0}_{s,k} & \mathbf{0}_{s,r} & -\mathbf{1}_{s,s} \end{array}\right)$$

where $k = \dim \operatorname{Ker}(g)$ and $k+r+s = n$. The numbers k, r and s are uniquely determined and $r + s$ is equal to the rank of the matrix of g with respect to any basis of V.

Proof. The only statement that has not yet been proved is that which identifies $r + s$ with the rank of the matrix of g with respect to any basis of V. This follows from the fact that the rank of the matrix in the statement of the theorem is precisely $r + s$ and two matrices G and G' of g are related by the equality $G' = {}^tPGP$, for some invertible matrix P. Now it is sufficient to observe that multiplying a matrix G to the right or to the left by an invertible matrix does not change its rank. □

Let's now see some concrete examples of what we have previously described.

Exercise 1. Let V be a real vector space and g be the symmetric bilinear form given by the matrix

$$G = \begin{pmatrix} 0 & 2 & 1 \\ 2 & 3 & -1 \\ 1 & -1 & 2 \end{pmatrix}$$

with respect to the basis $\{v_1, v_2, v_3\}$ of V. Find a basis of V with respect to which the matrix of g is diagonal, with only 1's and -1's on the diagonal.

Solution. We will use an analogue of the Gram–Schmidt procedure. We remark that the vector v_1 is isotropic, hence we cannot choose as the first vector v_1' of the new basis some multiple of it. To solve this problem it suffices to exchange the vectors v_1 and v_2 of the given basis and apply the orthogonalization procedure starting with v_2. Since $g(v_2, v_2) = 3$, we set $v_1' = v_2/\sqrt{3}$. Hence we have $g(v_1', v_1') = 1$. Now we set $v_2'' = \alpha v_1' + v_1$ and require that $g(v_1', v_2'') = 0$. We find that $\alpha = -g(v_1', v_1) = -2/\sqrt{3}$, hence $v_2'' = -2/\sqrt{3}\, v_1' + v_1 = -\frac{2}{3}\, v_2 + v_1$. We have $g(v_2'', v_2'') = -\frac{4}{3}$, and $\sqrt{-\frac{4}{3}}$ does not exist in $\mathbb{R}$. Hence we set $v_2' = v_2''/\sqrt{\frac{4}{3}} = \frac{\sqrt{3}}{2}\, v_2'' = -\frac{\sqrt{3}}{3}\, v_2 + \frac{\sqrt{3}}{2}\, v_1$, but now we get $g(v_2', v_2') = -1$.

To find the third, and last, vector of the basis, we set $v_3'' = \alpha v_1' + \beta v_2' + v_3$ and require that $g(v_1', v_3'') = 0$ and $g(v_2', v_3'') = 0$. From this we get $\alpha = -g(v_1', v_3) = 1/\sqrt{3}$ and $\beta = g(v_2', v_3) = 5\sqrt{3}/6$ (beware that in this case we have $g(v_2', v_2') = -1$). It follows that $v_3'' = \frac{1}{\sqrt{3}}\, v_1' + \frac{5\sqrt{3}}{6}\, v_2' + v_3 = -\frac{1}{2}\, v_2 + \frac{5}{4}\, v_1 + v_3$. Now we get $g(v_3'', v_3'') = 15/4$, hence we set $v_3' = \frac{2}{\sqrt{15}}\, v_3'' = \frac{5}{2\sqrt{15}}\, v_1 - \frac{1}{\sqrt{15}}\, v_2 + \frac{2}{\sqrt{15}}\, v_3$, and it follows that $g(v_3', v_3') = 1$.

The matrix of g with respect to the basis $\{v_1', v_2', v_3'\}$ is

$$G' = \begin{pmatrix} 1 & 0 & 0 \\ 0 & -1 & 0 \\ 0 & 0 & 1 \end{pmatrix}.$$

Finally we can note that we have

$$G' = {}^tPGP,$$

where the change of basis matrix P is the matrix whose columns are the components of the vectors v_1', v_2' and v_3' with respect to the basis $\{v_1, v_2, v_3\}$.

Hence we have

$$P = \begin{pmatrix} 0 & \frac{\sqrt{3}}{2} & \frac{5}{2\sqrt{15}} \\ \frac{1}{\sqrt{3}} & -\frac{\sqrt{3}}{3} & -\frac{1}{\sqrt{15}} \\ 0 & 0 & \frac{2}{\sqrt{15}} \end{pmatrix} = \begin{pmatrix} 0 & \frac{\sqrt{3}}{2} & \frac{\sqrt{15}}{6} \\ \frac{\sqrt{3}}{3} & -\frac{\sqrt{3}}{3} & -\frac{\sqrt{15}}{15} \\ 0 & 0 & \frac{2\sqrt{15}}{15} \end{pmatrix}.$$

Exercise 2. Let V be a real vector space of dimension 3, and $\{v_1, v_2, v_3\}$ be a basis. Let g be the symmetric bilinear form given by the matrix

$$G = \begin{pmatrix} -3 & 1 & 0 \\ 1 & 2 & -1 \\ 0 & -1 & -1 \end{pmatrix},$$

with respect to the given basis.

(i) Verify that g is non-degenerate and find an orthogonal basis of V, relative to g.

(ii) Compute the inertia index $i(g)$.

(iii) Check whether there exists non-zero isotropic vectors for g and, in case, determine a totally isotropic subspace of V of maximum dimension.

Solution. (i) We have

$$\det G = -3 \begin{vmatrix} 2 & -1 \\ -1 & -1 \end{vmatrix} - \begin{vmatrix} 1 & -1 \\ 0 & -1 \end{vmatrix} = 10,$$

hence g is non-degenerate.

Let's find an orthogonal basis. We set $w_1 = v_1$; we have $g(w_1, w_1) = -3$. Note that it is not necessary to normalize the vectors of the new basis because we are looking for an *orthogonal* basis.

By looking at the matrix G we observe that the vector v_3 is orthogonal to w_1, hence as the second vector of the orthogonal basis we can take $w_2 = v_3$; we have $g(w_2, w_2) = -1$.

It remains only to find a third vector w_3 orthogonal to w_1 and w_2. We set $w_3 = \lambda_1 v_1 + \lambda_2 v_2 + \lambda_3 v_3$ and require w_3 to be orthogonal to w_1 and w_2. We get the following system:

$$\begin{cases} g(w_1, w_3) = -3\lambda_1 + \lambda_2 = 0 \\ g(w_2, w_3) = -\lambda_2 - \lambda_3 = 0, \end{cases}$$

from which it follows

$$\begin{cases} \lambda_1 = -\frac{1}{3}\lambda_3 \\ \lambda_2 = -\lambda_3. \end{cases}$$

Setting, for instance, $\lambda_3 = 3$, we get the vector $w_3 = -v_1 - 3v_2 + 3v_3$. Now we find that $g(w_3, w_3) = 30$, and it follows that the matrix of g with respect to the orthogonal basis $\{w_1, w_2, w_3\}$ is

$$\begin{pmatrix} -3 & 0 & 0 \\ 0 & -1 & 0 \\ 0 & 0 & 30 \end{pmatrix}.$$

(ii) By looking at the matrix just found we can see that the restriction of g to the subspace $\langle w_1, w_2 \rangle$, spanned by w_1 and w_2, is negative definite, while the restriction of g to the subspace spanned by w_3 is positive definite. It follows that the inertia index of g is $i(g) = 1 - 2 = -1$.

(iii) Since g is indefinite, there certainly exist non-zero isotropic vectors. For instance, if we take a vector $w = \lambda_1 w_1 + \lambda_2 w_2 + \lambda_3 w_3$, the requirement that w be isotropic is

$$g(w, w) = -3\lambda_1^2 - \lambda_2^2 + 30\lambda_3^2 = 0,$$

that obviously has non-zero real solutions.

Let U be a totally isotropic subspace; we have $U \subset U^\perp$ and $\dim U^\perp = 3 - \dim U$. It follows that U can have, at most, dimension 1. Hence to find a totally isotropic subspace of maximum dimension it suffices to find a non-zero isotropic vector; for instance the vector $w = \sqrt{30}\, w_2 + w_3$. The subspace $U = \langle w \rangle$ is a totally isotropic subspace of maximum dimension (obviously such a totally isotropic subspace is not unique).

Exercise 3. On the vector space $\mathbb{R}^4$ we consider the symmetric bilinear form g given by the matrix

$$G = \begin{pmatrix} 0 & 2 & -2 & 0 \\ 2 & 0 & 1 & -3 \\ -2 & 1 & 0 & -1 \\ 0 & -3 & -1 & 0 \end{pmatrix}$$

with respect to the canonical basis.

(i) Find the inertia index of g.

(ii) Find, if possible, a basis of $\mathbb{R}^4$ with respect to which g has matrix

$$\begin{pmatrix} 1 & 0 & 0 & 0 \\ 0 & 1 & 0 & 0 \\ 0 & 0 & 0 & 1 \\ 0 & 0 & 1 & 0 \end{pmatrix}$$

(iii) Find, if possible, a basis of $\mathbb{R}^4$ with respect to which g has matrix

$$\begin{pmatrix} 0 & 1 & 0 & 0 \\ 1 & 0 & 0 & 0 \\ 0 & 0 & 0 & 1 \\ 0 & 0 & 1 & 0 \end{pmatrix}$$

(iv) Find, if possible, a basis of $\mathbb{R}^4$ with respect to which g has matrix

$$\begin{pmatrix} 0 & 1 & 0 & 0 \\ 1 & 0 & 0 & 0 \\ 0 & 0 & -1 & 0 \\ 0 & 0 & 0 & -1 \end{pmatrix}$$

Solution. (i) To determine the inertia index we construct an orthogonal basis. We denote by $\{e_1, \dots, e_4\}$ the canonical basis of $\mathbb{R}^4$. Note that in this case all the vectors of the canonical basis are isotropic, hence we cannot take one of these as the first vector of an orthogonal basis. For this reason we set $v_1 = e_1 + e_2$; this vector is not isotropic, in fact we have $g(v_1, v_1) = 2g(e_1, e_2) = 4$.

Is it easy to see that as the second vector we can take $v_2 = e_1 - e_2$, since it is orthogonal to v_1: $g(v_1, v_2) = 0$. We have $g(v_2, v_2) = -4$.

Now we can set

$$v_3 = \lambda_1 v_1 + \lambda_2 v_2 + e_3,$$

and require that $g(v_1, v_3) = g(v_2, v_3) = 0$. We find $\lambda_1 = \frac{1}{4}$ and $\lambda_2 = -\frac{3}{4}$. Hence we have

$$v_3 = \tfrac{1}{4} v_1 - \tfrac{3}{4} v_2 + e_3 = -\tfrac{1}{2} e_1 + e_2 + e_3,$$

from which it follows that $g(v_3, v_3) = 2$. Similarly, by setting

$$v_4 = \lambda_1 v_1 + \lambda_2 v_2 + \lambda_3 v_3 + e_4,$$

and requiring that $g(v_1, v_4) = g(v_2, v_4) = g(v_3, v_4) = 0$, we find $\lambda_1 = \frac{3}{4}$, $\lambda_2 = \frac{3}{4}$ and $\lambda_3 = 2$, from which it follows that

$$v_4 = \tfrac{3}{4}\, v_1 + \tfrac{3}{4}\, v_2 + 2v_3 + e_4 = \tfrac{1}{2}\, e_1 + 2e_2 + 2e_3 + e_4,$$

and $g(v_4, v_4) = -8$. The matrix of g with respect to the basis $\{v_1, \dots, v_4\}$ is therefore

$$\begin{pmatrix} 4 & 0 & 0 & 0 \\ 0 & -4 & 0 & 0 \\ 0 & 0 & 2 & 0 \\ 0 & 0 & 0 & -8 \end{pmatrix}$$

from which it follows that the inertia index of g is $i(g) = 2 - 2 = 0$.

(ii) It is easy to verify that the matrix $\begin{pmatrix} 0 & 1 \\ 1 & 0 \end{pmatrix}$ has inertia index 0. It follows that the matrix

$$\begin{pmatrix} 1 & 0 & 0 & 0 \\ 0 & 1 & 0 & 0 \\ 0 & 0 & 0 & 1 \\ 0 & 0 & 1 & 0 \end{pmatrix}$$

has inertia index $2 + 0 = 2$, which is different from the inertia index of g. By Sylvester's theorem it follows that there does not exist a basis of $\mathbb{R}^4$ with respect to which g has this matrix.

(iii) By what we have already seen it follows that the matrix

$$\begin{pmatrix} 0 & 1 & 0 & 0 \\ 1 & 0 & 0 & 0 \\ 0 & 0 & 0 & 1 \\ 0 & 0 & 1 & 0 \end{pmatrix}$$

has inertia index $0+0 = 0$, that is equal to the inertia index of g. Sylvester's theorem implies that there exists a basis of $\mathbb{R}^4$ with respect to which g has this matrix. By recalling the matrix of g with respect to the basis $\{v_1, \dots, v_4\}$ it is easy to see that by setting, for instance, $w_1 = v_1 + v_2$, $w_2 = \frac{1}{4}\,(v_1 - v_2)$, $w_3 = v_3 + \frac{1}{2}\, v_4$ and $w_4 = \frac{1}{4}\,(v_3 - \frac{1}{2}\, v_4)$, we get a basis with respect to which the matrix of g is the desired one.

(iv) By what we have previously seen it follows that the matrix

$$\begin{pmatrix} 0 & 1 & 0 & 0 \\ 1 & 0 & 0 & 0 \\ 0 & 0 & -1 & 0 \\ 0 & 0 & 0 & -1 \end{pmatrix}$$

has inertia index $0 - 2 = -2$, that is different from the inertia index of g. By Sylvester's theorem it follows that there does not exist a basis of $\mathbb{R}^4$ with respect to which g has this matrix.

We end this section by stating a useful criterion for establishing whether a real symmetric matrix is positive definite, negative definite or indefinite.

Let $G = (g_{ij}) \in M_n(\mathbb{R})$ be a symmetric matrix. For every $r = 1, \dots, n$ we denote by G_r the square submatrix of G formed by the elements g_{ij}, with $1 \leq i, j \leq r$:

$$G_1 = (g_{11}),\ G_2 = \begin{pmatrix} g_{11} & g_{12} \\ g_{21} & g_{22} \end{pmatrix},\ G_3 = \begin{pmatrix} g_{11} & g_{12} & g_{13} \\ g_{21} & g_{22} & g_{23} \\ g_{31} & g_{32} & g_{33} \end{pmatrix}, \dots,\ G_n = G.$$

The determinants of the matrices G_r are called the *principal minors* of G.

Theorem 5.6.3. *Let $G \in M_n(\mathbb{R})$ be a symmetrix matrix, with* $\det G \neq 0$.

(i) G is positive definite if and only if $\det G_r > 0$ *for every $r = 1, \dots, n$;*

(ii) G is negative definite if and only if $\det G_r < 0$ *for r odd and* $\det G_r > 0$ *for r even;*

(iii) in all other cases G is indefinite.

In the proof of this theorem we will use the following result:

Lemma 5.6.4. *Let $G \in M_n(\mathbb{R})$ be a symmetric matrix and $P \in \mathrm{GL}_n(\mathbb{R})$ an upper triangular invertible matrix. Let $G' = {}^tP\,G\,P$. Then, for every $r = 1, \dots, n$, we also have $G'_r = {}^tP_r\,G_r\,P_r$. In particular, each principal minor of the matrix G has the same sign of the corresponding principal minor of G'.*

Proof. Let $G = (g_{ij})$, $G' = (g'_{ij})$ and $P = (p_{ij})$. The assumption that P is upper triangular means that $p_{ij} = 0$ if $i > j$. With these notations, the equality $G' = {}^tP\,G\,P$ can be written as

$$g'_{ij} = \sum_{h,k=1}^{n} p_{hi}\, g_{hk}\, p_{kj}.$$

For each element g'_{ij} of the submatrix G'_r, we have $i, j \leq r$, hence $p_{hi} = 0$ and $p_{kj} = 0$, for every $h, k > r$. If $i, j \leq r$, we have therefore

$$g'_{ij} = \sum_{h,k=1}^{n} p_{hi}\, g_{hk}\, p_{kj} = \sum_{h,k=1}^{r} p_{hi}\, g_{hk}\, p_{kj}$$

that is equivalent to $G'_r = {}^tP_r\, G_r\, P_r$. It follows that $\det G'_r = \det G_r (\det P_r)^2$, hence $\det G'_r$ and $\det G_r$ have the same sign. □

Now we can prove Theorem 5.6.3.

(i) Let V be a real vector space with basis $\{v_1, \ldots, v_n\}$ and g be the symmetric bilinear form whose matrix, with respect to this basis, is G. If G is positive definite, Proposition 5.4.26 implies that there exists an upper triangular invertible matrix P such that ${}^tP\, G\, P = \mathbf{1}$. By the previous lemma, the principal minors of G have the same sign of the principal minors of the identity matrix, hence they are all positive.

Conversely, let us suppose that all the principal minors of the matrix G are positive. We must prove that G is positive definite. We make induction on the order n of the matrix G.

If $n = 1$ this is trivially true. Let us assume that this result holds for matrices of order $n - 1$. By inductive hypothesis, the submatrix G_{n-1} is positive definite. We note that G_{n-1} is the matrix of the restriction of the bilinear form g to the subspace $W = \langle v_1, \ldots, v_{n-1} \rangle$ spanned by the first $n-1$ vectors of the basis of V. Since the restriction of g to W is positive definite, we can apply the Gram–Schmidt procedure to the basis $\{v_1, \ldots, v_{n-1}\}$ of W in order to construct an orthonormal basis $\{w_1, \ldots, w_{n-1}\}$. If we denote by $\tilde{P}$ the corresponding change of basis matrix (which is upper triangular), we have ${}^t\tilde{P}\, G_{n-1}\, \tilde{P} = \mathbf{1}_{n-1}$.

Now we set

$$u = v_n - g(w_1, v_n)\, w_1 - g(w_2, v_n)\, w_2 - \cdots - g(w_{n-1}, v_n)\, w_{n-1}.$$

It is easy to verify that the vectors $w_1, \ldots, w_{n-1}, u$ are linearly independent and, by recalling that $g(w_i, w_j) = \delta_{ij}$, we have $g(w_i, u) = g(w_i, v_n) - g(w_i, v_n) = 0$, for $i = 1, \ldots, n-1$. Hence the vectors $w_1, \ldots, w_{n-1}, u$ form an orthogonal basis of V. The matrix G' of the symmetric bilinear form g with respect to the basis $\{w_1, \ldots, w_{n-1}, u\}$ is a block matrix of the form

$$G' = \begin{pmatrix} \mathbf{1}_{n-1} & \mathbf{0} \\ \mathbf{0} & \alpha \end{pmatrix}$$

where $\alpha = g(u, u)$. Hence we have $G' = {}^tP\, G\, P$ (by recalling how the basis $\{w_1, \ldots, w_{n-1}, u\}$ was constructed, we deduce that the change of basis matrix P is also upper triangular, moreover we have $P_{n-1} = \tilde{P}$). It follows that $\alpha = \det G' = \det G (\det P)^2 > 0$. Hence we can normalize the vector u dividing it by $\sqrt{\alpha}$. By setting $w_n = \frac{1}{\sqrt{\alpha}}\, u$, we get an orthonormal basis

$\{w_1, \ldots, w_{n-1}, w_n\}$ of V. The existence of such a basis allows us to conclude that the matrix G is congruent to the identity matrix, hence it is positive definite.

(*ii*) G is negative definite if and only if $-G$ is positive definite, hence, by (*i*), if and only if $\det(-G_r) > 0$ for $r = 1, \ldots, n$. To conclude it is enough to observe that $\det(-G_r) = (-1)^r \det G_r$.

(*iii*) This last case is an immediate consequence of Sylvester's theorem (Theorem 5.6.1). From this theorem it follows that, if $\det G \neq 0$, the matrix G can only be positive definite, negative definite or indefinite. □

5.7 Isometries

Let (V, g) and (W, h) be two vector spaces over the field K endowed with symmetric bilinear forms

$$g\colon V \times V \to K, \qquad h\colon W \times W \to K.$$

Definition 5.7.1. A linear function $f\colon V \to W$ is *compatible* with the symmetric bilinear forms g and h if, for every $v_1, v_2 \in V$, we have

$$h\big(f(v_1), f(v_2)\big) = g(v_1, v_2). \tag{5.7.1}$$

In this case we say that $f\colon (V, g) \to (W, h)$ is a *homomorphism of vector spaces endowed with symmetric bilinear forms.*

The compatibility requirement (5.7.1) implies the existence of relationships between the kernels of the functions f, g and h. More precisely, we have:

Lemma 5.7.2. *Let $f\colon (V, g) \to (W, h)$ be a homomorphism of vector spaces endowed with symmetric bilinear forms. The following inclusions hold:*

$$\operatorname{Ker} f \subseteq f^{-1}(\operatorname{Ker} h) \subseteq \operatorname{Ker} g. \tag{5.7.2}$$

Proof. The first inclusion is trivial: if $v \in \operatorname{Ker} f$ we have $f(v) = \mathbf{0}$, hence $h\big(f(v), w\big) = 0$ for every $w \in W$, hence $f(v) \in \operatorname{Ker} h$.

Let's prove the second inclusion. For every $v \in f^{-1}(\operatorname{Ker} h)$ and every $v' \in V$, we have $g(v, v') = h\big(f(v), f(v')\big) = 0$, because $f(v) \in \operatorname{Ker} h$. It follows that $v \in \operatorname{Ker} g$. □

From this result some interesting consequences descend:

Corollary 5.7.3. *Let* $f\colon (V,g) \to (W,h)$ *be a homomorphism of vector spaces endowed with symmetric bilinear forms. Then:*

(*i*) *if* h *is non-degenerate,* $\operatorname{Ker} f = f^{-1}(\operatorname{Ker} h)$*;*

(*ii*) *if* g *is non-degenerate, we have* $\operatorname{Ker} f = f^{-1}(\operatorname{Ker} h) = \{\mathbf{0}\}$. *In particular,* f *is injective and the restriction of* h *to* $\operatorname{Im}(f) \times \operatorname{Im}(f)$ *is non-degenerate;*

(*iii*) *if* g *is non-degenerate and* f *is surjective, then also* h *is non-degenerate. Moreover, in this case,* f *is an isomorphism of vector spaces and the inverse fuction* $f^{-1}\colon (W,h) \to (V,g)$ *is compatible with the bilinear forms* h *and* g.

Proof. To prove (*i*) it suffices to recall that h is non-degenerate if and only if $\operatorname{Ker} h = \{\mathbf{0}\}$.

(*ii*) If g is non-degenerate, i.e., if $\operatorname{Ker} g = \{\mathbf{0}\}$, (5.7.2) reduces to $\operatorname{Ker} f = f^{-1}(\operatorname{Ker} h) = \{\mathbf{0}\}$, hence f is injective. Denoting by $\tilde{h}$ the restriction of h to $\operatorname{Im}(f) \times \operatorname{Im}(f)$, for every $w_1, w_2 \in \operatorname{Im}(f)$ we have $\tilde{h}(w_1, w_2) = g(v_1, v_2)$, where $v_1, v_2 \in V$ are such that $w_1 = f(v_1)$ and $w_2 = f(v_2)$. Since g is non-degenerate it follows that $\tilde{h}$ is also non-degenerate.

Finally we prove (*iii*). If f is surjective, we have $\operatorname{Im} f = W$ hence, since g is non-degenerate, from (*ii*) it follows that h is non-degenerate and f is injective. Hence the function $f\colon V \to W$ is bijective and, therefore, it is an isomorphism of vector spaces. It only remains to prove that its inverse $f^{-1}\colon W \to V$ is compatible with the bilinear forms h and g. Let $v_1, v_2 \in V$ and set $w_1 = f(v_1)$ and $w_2 = f(v_2)$. We have $v_1 = f^{-1}(w_1)$, $v_2 = f^{-1}(w_2)$, hence

$$g\big(f^{-1}(w_1), f^{-1}(w_2)\big) = g(v_1, v_2) = h\big(f(v_1), f(v_2)\big) = h(w_1, w_2).$$

This proves that also f^{-1} is a homomorphism of vector spaces endowed with symmetric bilinear forms. □

We also note that, under appropriate hypotheses, the compatibility condition (5.7.1) implies the linearity of f.

Proposition 5.7.4. *Let* (V,g) *and* (W,h) *be two vector spaces endowed with symmetric bilinear forms and* $f\colon V \to W$ *be any function (note that we are not assuming* f *to be linear) that verifies the equality* (5.7.1)*, for every* $v_1, v_2 \in V$. *If* h *is non-degenerate and* f *is surjective, then* f *is linear.*

Proof. Let $v_1, v_2 \in V$ and $\lambda_1, \lambda_2 \in K$. We want to prove that

$$f(\lambda_1 v_1 + \lambda_2 v_2) = \lambda_1 f(v_1) + \lambda_2 f(v_2).$$

For every $v \in V$, we have:

$$\begin{aligned} &h\big(f(\lambda_1 v_1 + \lambda_2 v_2) - \lambda_1 f(v_1) - \lambda_2 f(v_2), f(v)\big) = \\ &\quad = h\big(f(\lambda_1 v_1 + \lambda_2 v_2), f(v)\big) - \lambda_1 h\big(f(v_1), f(v)\big) - \lambda_2 h\big(f(v_2), f(v)\big) \\ &\quad = g(\lambda_1 v_1 + \lambda_2 v_2, v) - \lambda_1 g(v_1, v) - \lambda_2 g(v_2, v) = 0, \end{aligned}$$

by the bilinearity of h and g. Since f is surjective and h is non-degenerate, it follows that

$$f(\lambda_1 v_1 + \lambda_2 v_2) - \lambda_1 f(v_1) - \lambda_2 f(v_2) = \mathbf{0},$$

which proves the linearity of f. □

Let's now see how the compatibility condition (5.7.1) is expressed in terms of matrices.

Proposition 5.7.5. *Let (V, g) and (W, h) be two vector spaces endowed with symmetric bilinear forms and $f\colon V \to W$ a linear function. Let $\mathbf{v} = \{v_1, \dots, v_n\}$ and $\mathbf{w} = \{w_1, \dots, w_m\}$ be bases of V and W, respectively. Let us denote by A the matrix of f with respect to the bases $\mathbf{v}$ and $\mathbf{w}$, by G the matrix of g with respect to the basis $\mathbf{v}$ and by H the matrix of h with respect to the basis $\mathbf{w}$. Then f is compatible with the bilinear forms g and h if and only if*

$$G = {}^tAHA. \tag{5.7.3}$$

Proof. From the definition of the matrix associated to a linear function it follows that, for each vector v_i of the basis $\mathbf{v}$, we have

$$f(v_i) = \sum_{r=1}^{m} a_{ri}\, w_r.$$

For every $i, j = 1, \dots, n$, the compatibility condition (5.7.1) is equivalent to:

$$\begin{aligned} g_{ij} = g(v_i, v_j) &= h\big(f(v_i), f(v_j)\big) = h\Big(\sum_{r=1}^{m} a_{ri}\, w_r, \sum_{s=1}^{m} a_{sj}\, w_s\Big) \\ &= \sum_{r,s=1}^{m} a_{ri}\, a_{sj}\, h(w_r, w_s) = \sum_{r,s=1}^{m} a_{ri}\, a_{sj}\, h_{rs}. \end{aligned}$$

In terms of matrices this equality is precisely (5.7.3). □

By specializing this result to the case $(W,h)=(V,g)$, we get:

Corollary 5.7.6. *Let V be a vector space over the field K and g be a symmetric bilinear form on V. Let us denote by G the matrix of g with respect to a basis $\mathbf{v}=\{v_1,\dots,v_n\}$ of V. An endomorphism f of V, with matrix A with respect to the basis $\mathbf{v}$, is compatible with the bilinear form g if and only if A satisfies the following equality:*

$$G={}^tAGA.$$

Let us now restrict our attention to the case of vector spaces with non-degenerate symmetric bilinear forms.

Definition 5.7.7. Let (V,g) and (W,h) be two vector spaces endowed with non-degenerate symmetric bilinear forms. A linear function $f\colon V\to W$ compatible with the bilinear forms g and h is called an *isometry.*

We remark that, by Corollary 5.7.3, f is necessarily injective, hence it determines an isomorphism of V with the subspace $\operatorname{Im}(f)$ of W.

Definition 5.7.8. Two vector spaces endowed with non-degenerate symmetric bilinear forms (V,g) and (W,h) are said to be *isometric* if there exists a surjective isometry of V onto W, i.e., if there exists an isomorphism $f\colon V\xrightarrow{\sim}W$ compatible with the bilinear forms g and h.

Remark 5.7.9. If $f\colon(V,g)\to(W,h)$ is an isometry of V in W, the vector space V is isometric to the vector subspace $\operatorname{Im}(f)$ of W endowed with the non-degenerate symmetric bilinear form given by the restriction of h.

Remark 5.7.10. In the case of Euclidean vector spaces (i.e., real vector spaces endowed with positive definite symmetric bilinear forms) the isometries preserve the length of vectors and the angles between vectors. In fact, let (V,g) and (W,h) be two Euclidean vector spaces and $f\colon V\to W$ be an isometry. For every $v\in V$, we have:

$$\|f(v)\|=\sqrt{h\big(f(v),f(v)\big)}=\sqrt{g(v,v)}=\|v\|.$$

Moreover, if we denote by ϕ the angle between the vectors $v_1,v_2\in V$ and by ϕ' the angle between their images $f(v_1),f(v_2)\in W$, we have

$$\cos\phi'=\frac{h\big(f(v_1),f(v_2)\big)}{\|f(v_1)\|\|f(v_2)\|}=\frac{g(v_1,v_2)}{\|v_1\|\|v_2\|}=\cos\phi.$$

It is easy to prove that a linear function $f\colon V \to W$ that preserves the norm of vectors is an isometry. In fact, given $v_1, v_2 \in V$, we have

$$\begin{aligned}\|v_1+v_2\|^2 &= g(v_1+v_2, v_1+v_2)\\ &= g(v_1,v_1)+2g(v_1,v_2)+g(v_2,v_2)\\ &= \|v_1\|^2+2g(v_1,v_2)+\|v_2\|^2.\end{aligned}$$

Similarly,

$$\begin{aligned}\|f(v_1+v_2)\|^2 &= \|f(v_1)+f(v_2)\|^2\\ &= h\big(f(v_1)+f(v_2), f(v_1)+f(v_2)\big)\\ &= h\big(f(v_1),f(v_1)\big)+2h\big(f(v_1),f(v_2)\big)+h\big(f(v_2),f(v_2)\big)\\ &= \|f(v_1)\|^2+2h\big(f(v_1),f(v_2)\big)+\|f(v_2)\|^2.\end{aligned}$$

Since, by hypothesis, f preserves the norm of vectors, we have

$$\|f(v_1)\| = \|v_1\|, \qquad \|f(v_2)\| = \|v_2\|, \qquad \|f(v_1+v_2)\| = \|v_1+v_2\|.$$

It follows that we also have $h\big(f(v_1), f(v_2)\big) = g(v_1, v_2)$, hence f is an isometry.

An immediate consequence of the existence of orthonormal bases in Euclidean vector spaces is the following:

Proposition 5.7.11. *Every Euclidean vector space (V, g) of dimension n is isometric to the Euclidean vector space $\mathbb{R}^n$, endowed with the usual scalar product.*

Proof. Let $\mathbf{v} = \{v_1, v_2, \dots, v_n\}$ be an orthonormal basis of V, whose existence is garanteed by Theorem 5.4.24. We know that the function $\alpha_{\mathbf{v}}\colon V \to \mathbb{R}^n$, that associates to every vector $v \in V$ the n-tuple $(\lambda_1, \dots, \lambda_n)$ of its coordinates with respect to the basis $\mathbf{v}$, is an isomorphism of vector spaces; it remains only to prove that it is also an isometry.

Let's take two vectors $v, w \in V$ and write

$$v = \sum_{i=1}^{n} \lambda_i v_i, \qquad w = \sum_{j=1}^{n} \mu_j v_j.$$

We have $\alpha_{\mathbf{v}}(v) = (\lambda_1, \dots, \lambda_n)$, $\alpha_{\mathbf{v}}(w) = (\mu_1, \dots, \mu_n)$ hence

$$\alpha_{\mathbf{v}}(v) \cdot \alpha_{\mathbf{v}}(w) = \sum_{i=1}^{n} \lambda_i \mu_i.$$

On the other hand, remembering that $\mathbf{v}$ is an orthonormal basis, we have

$$\begin{aligned} g(v,w) &= g\Big(\sum_{i=1}^{n} \lambda_i v_i, \sum_{j=1}^{n} \mu_j v_j\Big) \\ &= \sum_{i,j=1}^{n} \lambda_i \mu_j g(v_i, v_j) \\ &= \sum_{i=1}^{n} \lambda_i \mu_i. \end{aligned}$$

This shows that, for every $v, w \in V$, we have $\alpha_{\mathbf{v}}(v) \cdot \alpha_{\mathbf{v}}(w) = g(v,w)$, i.e., $\alpha_{\mathbf{v}}$ is an isometry. □

We have already remarked in Corollary 5.7.3 that if an isometry $f\colon V \to W$ is bijective then its inverse $f^{-1}\colon W \to V$ is also an isometry. Moreover, it is easy to verify that the composition of two isometries is an isometry. It follows that the set $\mathrm{O}(V,g)$ of the isometries $f\colon V \to V$ of a vector space V endowed with a non-degenerate symmetric bilinear form g, with respect to the composition of functions, is a group. This group, called the *isometry group* or the *orthogonal group* of (V,g), is a subgroup of the group $\mathrm{GL}(V)$ of all automorphisms of V.

Let V be a vector space endowed with a non-degenerate symmetric bilinear form g, let's choose a basis $\mathbf{v} = \{v_1, \dots, v_n\}$ of V and denote by G the matrix of g with respect to the basis $\mathbf{v}$. As we have already seen, an endomorphism $f\colon V \to V$ is an isometry if and only if its matrix A, with respect to the basis $\mathbf{v}$, satisfies the equality ${}^t\!AGA = G$. It follows that, in terms of matrices, the isometry group of (V,g) has the following expression:

$$\mathrm{O}(G) = \{A \in \mathrm{GL}(n) \mid {}^t\!AGA = G\}.$$

From Binet's theorem it follows that, for every $A \in \mathrm{O}(G)$, we have

$$\det(G) = \det({}^t\!AGA) = \det(G)\det(A)^2.$$

Since g is non-degenerate, we have $\det(G) \neq 0$, hence $\det(A)^2 = 1$, i.e., $\det(A) = \pm 1$. It is easy to verify that the isometries with determinant 1 form a subgroup (of index 2) of the group of all isometries of (V,g). This group, denoted by $\mathrm{SO}(V,g)$, is called the *group of direct isometries*, or the *special orthogonal group*, of the vector space V endowed with the symmetric bilinear form g. In terms of matrices this group has the following description:

$$\mathrm{SO}(G) = \{A \in \mathrm{GL}(n) \mid {}^t\!AGA = G,\ \det(A) = 1\}.$$

If (V, g) is a Euclidean vector space, Theorem 5.4.24 guarantees the existence of an orthonormal basis $\mathbf{v} = \{v_1, \dots, v_n\}$ of V. With respect to this basis the matrix of g is the identity, hence the matrices of the isometries of (V, g) are characterized by the following equation:

$$ {}^tAA = \mathbf{1}. $$

The group of such matrices is the *orthogonal group of order n*

$$ \mathrm{O}(n) = \{A \in \mathrm{GL}(n) \,|\, {}^tAA = \mathbf{1}\}. $$

A matrix $A \in \mathrm{O}(n)$, i.e., a square matrix A of order n such that ${}^tAA = \mathbf{1}$ or, equivalently, such that ${}^tA = A^{-1}$, is called an *orthogonal matrix*. If we denote by $A_{(1)}, \dots, A_{(n)}$ the columns of A, the equality ${}^tAA = \mathbf{1}$ is equivalent to

$$ A_{(i)} \cdot A_{(j)} = \begin{cases} 1 & \text{if } i = j \\ 0 & \text{if } i \neq j, \end{cases} $$

which means that the columns of A are an orthonormal basis of $\mathbb{R}^n$, with respect to the usual scalar product.

5.7.1 Isometries of $\mathbb{R}^2$

Let us now study in detail the structure of the isometries of the Euclidean vector space $\mathbb{R}^2$, endowed with the usual scalar product. The group of such isometries is the orthogonal group of order 2

$$ \mathrm{O}(2) = \{A \in \mathrm{GL}(2) \,|\, {}^tAA = \mathbf{1}\}. $$

Given a matrix

$$ A = \begin{pmatrix} a & b \\ c & d \end{pmatrix}, $$

the equality ${}^tAA = \mathbf{1}$ is equivalent to the following system:

$$ \begin{cases} a^2 + c^2 = 1 \\ b^2 + d^2 = 1 \\ ab + cd = 0. \end{cases} $$

From the equation $a^2 + c^2 = 1$ it follows that there exists an angle α such that

$$ a = \cos\alpha, \qquad c = \sin\alpha. $$

Similarly, from $b^2 + d^2 = 1$ it follows that there exists an angle β such that

$$b = \sin\beta, \qquad d = \cos\beta.$$

The third equation, $ab + cd = 0$, becomes

$$\cos\alpha\sin\beta + \sin\alpha\cos\beta = \sin(\alpha+\beta) = 0,$$

from which it follows that $\alpha+\beta = k\pi$, i.e., $\beta = k\pi - \alpha$, for some $k \in \mathbb{Z}$. Thus we get two different types of isometries, depending on the parity of k. More precisely, if k is even the matrix A has the following form:

$$\begin{pmatrix} \cos\alpha & -\sin\alpha \\ \sin\alpha & \cos\alpha \end{pmatrix}. \tag{5.7.4}$$

This matrix has determinant 1, hence it is a direct isometry. In geometrical terms, this is the matrix of a counterclockwise rotation of the plane $\mathbb{R}^2$, around the origin, by an angle α. The two columns

$$u_1 = \begin{pmatrix} \cos\alpha \\ \sin\alpha \end{pmatrix} \qquad u_2 = \begin{pmatrix} -\sin\alpha \\ \cos\alpha \end{pmatrix}$$

of the matrix (5.7.4) are the images of the vectors e_1 and e_2 of the canonical basis of $\mathbb{R}^2$ by this rotation, as represented in the following figure:

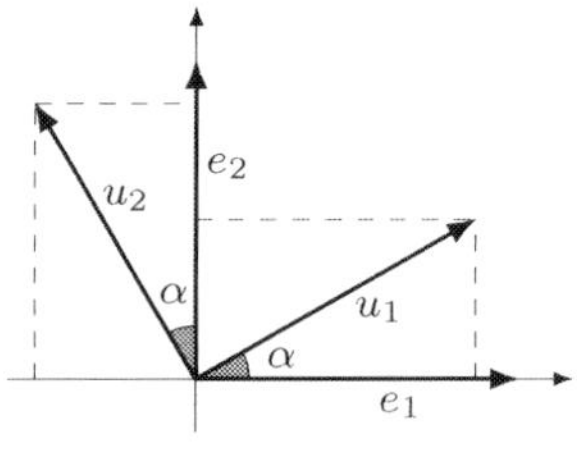

Instead, if k is odd, we get the following matrices:

$$\begin{pmatrix} \cos\alpha & \sin\alpha \\ \sin\alpha & -\cos\alpha \end{pmatrix} \tag{5.7.5}$$

These matrices have determinant -1, hence they represent inverse isometries. Geometrically they correspond to a reflection of the plane with respect to the X-axis, followed by a counterclockwise rotation, around the origin, by an angle α. This is described in the following figure:

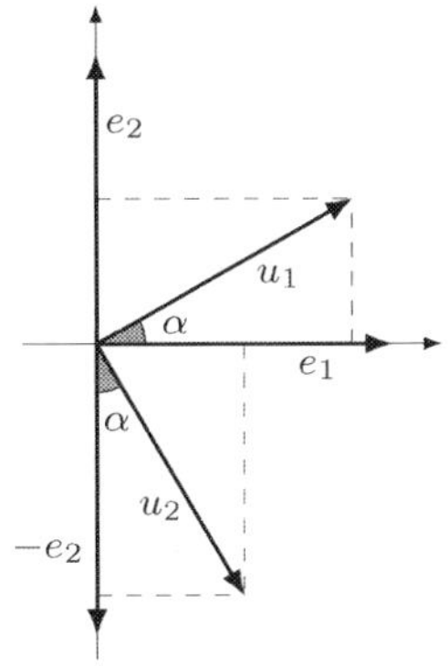

As we can see in this figure, the vectors u_1 and u_2 (that are the images of the vectors e_1 and e_2 of the canonical basis), have the following coordinates:

$$u_1 = \begin{pmatrix} \cos\alpha \\ \sin\alpha \end{pmatrix} \qquad u_2 = \begin{pmatrix} \sin\alpha \\ -\cos\alpha \end{pmatrix}$$

Hence they are equal to the columns of the matrix (5.7.5).

5.8 The cross product

To end this chapter on Euclidean vector spaces, we will describe a construction that has many useful applications, especially in physics: the *cross product* (or *vector product*) of two vectors. It is an operation that associates to two vectors of a 3-dimensional Euclidean vector space V another vector of V. As we will see below, the definition of this product will depend not only on the Euclidean structure but also on the choice of an *orientation* of the vector space V.

Recall that any Euclidean vector space of dimension n is isometric to the vector space $\mathbb{R}^n$, endowed with the usual scalar product; hence we can assume that $V = \mathbb{R}^3$.

Definition 5.8.1. Given two vectors $u = (a_1, a_2, a_3)$, $v = (b_1, b_2, b_3) \in \mathbb{R}^3$, their *cross product* is the vector $w = u \times v$ defined as follows:

$$w = u \times v = \det \begin{pmatrix} e_1 & e_2 & e_3 \\ a_1 & a_2 & a_3 \\ b_1 & b_2 & b_3 \end{pmatrix} \tag{5.8.1}$$

where $e_1 = (1,0,0)$, $e_2 = (0,1,0)$, $e_3 = (0,0,1)$ are the vectors of the canonical basis of $\mathbb{R}^3$ (this determinant must be understood in a purely formal sense).

The Laplace expansion of the determinant (5.8.1), along the first row, gives the following explicit formula:

$$\begin{aligned} u \times v &= (a_2b_3 - a_3b_2)e_1 + (a_3b_1 - a_1b_3)e_2 + (a_1b_2 - a_2b_1)e_3 \\ &= (a_2b_3 - a_3b_2, a_3b_1 - a_1b_3, a_1b_2 - a_2b_1). \end{aligned} \tag{5.8.2}$$

Let's describe now the main properties of the cross product:

Proposition 5.8.2. *For every $u, u_1, u_2, v, v_1, v_2 \in \mathbb{R}^3$ and $\lambda_1, \lambda_2, \mu_1, \mu_2 \in \mathbb{R}$, we have:*

(i) $(\lambda_1 u_1 + \lambda_2 u_2) \times v = \lambda_1(u_1 \times v) + \lambda_2(u_2 \times v)$,

(ii) $u \times (\mu_1 v_1 + \mu_2 v_2) = \mu_1(u \times v_1) + \mu_2(u \times v_2)$,

(iii) $u \times v = -v \times u$,

(iv) $u \times v = \mathbf{0}$ *if and only if u and v are linearly dependent.*

Proof. All the properties listed above are an immediate consequence of the properties of the determinants, studied in Chapter 3. □

Remark 5.8.3. From the properties listed in the previous proposition it follows that the cross product

$$\mathbb{R}^3 \times \mathbb{R}^3 \to \mathbb{R}^3, \qquad (u, v) \mapsto u \times v$$

is an alternating bilinear function.

Proposition 5.8.4. *Given three vectors $u = (a_1, a_2, a_3)$, $v = (b_1, b_2, b_3)$, $w = (c_1, c_2, c_3) \in \mathbb{R}^3$, the* mixed product *$u \cdot (v \times w)$ is given by*

$$u \cdot (v \times w) = \det \begin{pmatrix} a_1 & a_2 & a_3 \\ b_1 & b_2 & b_3 \\ c_1 & c_2 & c_3 \end{pmatrix}$$

Proof. The proof consists in a simple computation. From the formula (5.8.2) we have

$$v \times w = (b_2c_3 - b_3c_2, b_3c_1 - b_1c_3, b_1c_2 - b_2c_1)$$

hence

$$u \cdot (v \times w) = a_1(b_2c_3 - b_3c_2) + a_2(b_3c_1 - b_1c_3) + a_3(b_1c_2 - b_2c_1).$$

This is precisely the expression that we get by using the Laplace expansion along the first row to compute the determinant

$$\det \begin{pmatrix} a_1 & a_2 & a_3 \\ b_1 & b_2 & b_3 \\ c_1 & c_2 & c_3 \end{pmatrix}$$

□

Proposition 5.8.5. *If $u, v \in \mathbb{R}^3$ are linearly independent, then $u \times v$ is a non-zero vector orthogonal to the plane spanned by u and v.*

Proof. If u and v are linearly independent we have $u \times v \neq \mathbf{0}$, by (iv) of Proposition 5.8.2. Moreover, we have $u \cdot (u \times v) = 0$ and $v \cdot (u \times v) = 0$ because, according to the previous proposition, the mixed products are given by the determinant of a matrix with two equal rows. This means that the vector $u \times v$ is orthogonal to both u and v, hence to the plane spanned by these vectors. □

As for the geometric meaning of the cross product of two vectors, or of the mixed product of three vectors, it is expressed by the following result:

Proposition 5.8.6. *Given $u, v, w \in \mathbb{R}^3$, we have:*

(i) $\|u \times v\| = \text{Area}\,\mathscr{P}(u, v)$,

(ii) $|u \cdot (v \times w)| = \text{Vol}\,\mathscr{P}(u, v, w)$,

where $\mathscr{P}(u, v)$ denotes the parallelogram determined by the vectors u and v, while $\mathscr{P}(u, v, w)$ denotes the parallelepiped whose sides are u, v and w.

Proof. (i) If $u = (a_1, a_2, a_3)$ and $v = (b_1, b_2, b_3)$, we have

$$u \times v = (a_2b_3 - a_3b_2, a_3b_1 - a_1b_3, a_1b_2 - a_2b_1)$$

hence

$$\|u \times v\| = \sqrt{(a_2b_3 - a_3b_2)^2 + (a_3b_1 - a_1b_3)^2 + (a_1b_2 - a_2b_1)^2}.$$

To compute the area of the parallelogram $\mathscr{P}(u, v)$ we use the formula (5.2.1). We have:

$$u \cdot u = a_1^2 + a_2^2 + a_3^2,\ v \cdot v = b_1^2 + b_2^2 + b_3^2,\ u \cdot v = v \cdot u = a_1b_1 + a_2b_2 + a_3b_3,$$

hence

$$\text{Area}\,\mathscr{P}(u,v)=\sqrt{(a_1^2+a_2^2+a_3^2)(b_1^2+b_2^2+b_3^2)-(a_1b_1+a_2b_2+a_3b_3)^2}.$$

Now it is enough to develop the computations to verify the equality $\|u\times v\| =$ Area $\mathscr{P}(u,v)$.

(*ii*) If $u=(a_1,a_2,a_3)$, $v=(b_1,b_2,b_3)$, $w=(c_1,c_2,c_3)$, by using the formula (5.2.4) we have

$$\text{Vol}\,\mathscr{P}(u,v,w)=\left|\det\begin{pmatrix}a_1 & a_2 & a_3\\ b_1 & b_2 & b_3\\ c_1 & c_2 & c_3\end{pmatrix}\right|,$$

that, according to Proposition 5.8.4, can be written as

$$\text{Vol}\,\mathscr{P}(u,v,w)=|u\cdot(v\times w)|.$$

□

Proposition 5.8.7. *If $u=(a_1,a_2,a_3)$ and $v=(b_1,b_2,b_3)$ are linearly independent, the vectors u, v and $u\times v$, in this order, form a basis of $\mathbb{R}^3$ that has the same orientation of the canonical basis e_1, e_2, e_3.*

Proof. Since u and v are linearly independent, the vector $u\times v$ is orthogonal to the plane spanned by u and v (cf. Proposition 5.8.5); this implies that the vectors u, v and $u\times v$ are linearly independent, hence they are a basis of $\mathbb{R}^3$. The change of basis matrix P (from the canonical basis e_1, e_2, e_3 to the basis u, v, $u\times v$) is the matrix whose columns are formed by the components of the vectors u, v, $u\times v$ with respect to the vectors of the canonical basis:

$$P=\begin{pmatrix}a_1 & b_1 & a_2b_3-a_3b_2\\ a_2 & b_2 & a_3b_1-a_1b_3\\ a_3 & b_3 & a_1b_2-a_2b_1\end{pmatrix}$$

Computing the determinant of P using Laplace expansion along the third column, we get

$$\det P=(a_2b_3-a_3b_2)^2+(a_3b_1-a_1b_3)^2+(a_1b_2-a_2b_1)^2,$$

from which it follows that P has positive determinant, hence the bases $\{u,v,u\times v\}$ and $\{e_1,e_2,e_3\}$ have the same orientation. □

Remark 5.8.8. From what we have previously seen it follows that the cross product of two linearly independent vectors $u,\ v \in \mathbb{R}^3$ can be characterized as follows: $u \times v$ is the vector orthogonal to the plane spanned by u and v, whose norm is equal to the area of the parallelogram determined by u and v and whose orientation is such that the basis $\{u, v, u \times v\}$ has the same orientation of the canonical basis $\{e_1, e_2, e_3\}$. This is actually the usual definition of the cross product used by physicists.

Note that this definition depends on the choice of an orientation of the Euclidean vector space $\mathbb{R}^3$. Because of this, in physics the cross product of two vectors is usually called a *pseudovector* or an *axial vector.*

Finally, we list a series of identities involving the cross product and the scalar product of vectors of $\mathbb{R}^3$:

Proposition 5.8.9. *Let $u, v, w, z \in \mathbb{R}^3$. The following identities hold:*

(*i*) $u \times (v \times w) = (u \cdot w)v - (u \cdot v)w$ (*Lagrange's identity*);

(*ii*) $u \times (v \times w) + v \times (w \times u) + w \times (u \times v) = \mathbf{0}$ (*Jacobi's identity*);

(*iii*) $\|u \times v\|^2 + |u \cdot v|^2 = \|u\|^2 \, \|v\|^2$ (*Lagrange's identity*);

(*iv*) $(u \times v) \times (u \times w) = (u \cdot (v \times w))u$;

(*v*) $(u \times v) \cdot (w \times z) = (u \cdot w)(v \cdot z) - (u \cdot z)(v \cdot w)$.

Proof. Proofs of all previous identities can be obtained by direct calculation, using the definition of cross product given by the formula (5.8.2). As an example, we prove (*i*).

Let $u = (a_1, a_2, a_3)$, $v = (b_1, b_2, b_3)$ and $w = (c_1, c_2, c_3)$. We have

$$\begin{aligned} u \times (v \times w) = \big(& a_2(b_1c_2 - b_2c_1) - a_3(b_3c_1 - b_1c_3), \\ & a_3(b_2c_3 - b_3c_2) - a_1(b_1c_2 - b_2c_1), \\ & a_1(b_3c_1 - b_1c_3) - a_2(b_2c_3 - b_3c_2) \big). \end{aligned}$$

Then we have

$$\begin{aligned} (u \cdot w)v &= \big((a_1c_1 + a_2c_2 + a_3c_3)b_1, \\ &\qquad (a_1c_1 + a_2c_2 + a_3c_3)b_2, \\ &\qquad (a_1c_1 + a_2c_2 + a_3c_3)b_3 \big) \\ (u \cdot v)w &= \big((a_1b_1 + a_2b_2 + a_3b_3)c_1, \\ &\qquad (a_1b_1 + a_2b_2 + a_3b_3)c_2, \\ &\qquad (a_1b_1 + a_2b_2 + a_3b_3)c_3 \big) \end{aligned}$$

from which, developing the calculations, we obtain the equality $u \times (v \times w) = (u \cdot w)v - (u \cdot v)w$.

The proof of other identities is left as an exercise. □

Exercises

Exercise 5.1. Let V be a vector space over the field of complex numbers $\mathbb{C}$, with basis $\{v_1, v_2, v_3\}$, and g be the symmetric bilinear form whose matrix, with respect to the given basis, is

$$G = \begin{pmatrix} 2 & 3 & -1 \\ 3 & 4 & 0 \\ -1 & 0 & 1 \end{pmatrix}.$$

Determine an invertible matrix P such that ${}^tPGP = \mathbf{1}$.

Exercise 5.2. Let V be a vector space over the field of real numbers, with basis $\{v_1, v_2, v_3\}$, and g be the symmetric bilinear form whose matrix, with respect to the given basis, is

$$G = \begin{pmatrix} 0 & 2 & 1 \\ 2 & 3 & -1 \\ 1 & -1 & 2 \end{pmatrix}.$$

Determine a basis of V with respect to which the matrix of g is diagonal, with only 1's and -1's on the diagonal.

Exercise 5.3. Let V be a vector space over the field of real numbers, with basis $\{v_1, v_2, v_3\}$, and g be the symmetric bilinear form whose matrix, with respect to the given basis, is

$$G = \begin{pmatrix} 0 & 2 & -1 \\ 2 & 0 & 3 \\ -1 & 3 & 0 \end{pmatrix}.$$

Determine a basis of V with respect to which the matrix of g is diagonal, with only 1's and -1's on the diagonal.

Exercise 5.4. Let V be a real vector space of dimension 3, and $\{v_1, v_2, v_3\}$ be a basis. Let g be the symmetric bilinear form whose matrix, with respect to the given basis, is

$$G = \begin{pmatrix} -3 & 1 & 0 \\ 1 & 2 & -1 \\ 0 & -1 & -1 \end{pmatrix}$$

(1) Verify that g is non-degenerate and find an orthogonal basis of V for the bilinear form g.

(2) Compute the inertia index $i(g)$.

(3) Find if there exist non-zero isotropic vectors for the form g and, if they exist, find a totally isotropic subspace of maximum dimension.

Exercise 5.5. Let $V = \mathbb{R}^4$ and g be the symmetric bilinear form of matrix

$$G = \begin{pmatrix} 3 & -2 & 1 & 1 \\ -2 & 1 & -1 & -1 \\ 1 & -1 & 1 & 4 \\ 1 & -1 & 4 & 2 \end{pmatrix},$$

with respect to the canonical basis.

(1) Verify that g is non-degenerate, compute the inertia index and find an orthogonal basis of V for the form g.

(2) Is $(\mathbb{R}^4, g)$ isometric to the vector space $\mathbb{R}^4$ endowed with the usual scalar product?

Exercise 5.6. Let V be a finite dimensional real vector space and $g\colon V \times V \to \mathbb{R}$ be a positive definite symmetric bilinear form. If $\phi\colon V \to V$ is an isometry, prove that we have

$$\operatorname{Im}(\phi - \mathrm{id}) = \operatorname{Ker}(\phi - \mathrm{id})^{\perp}.$$

Exercise 5.7. Let's consider $\mathbb{C}$ as a vector space over $\mathbb{R}$ and let us define a function $g\colon \mathbb{C} \times \mathbb{C} \to \mathbb{R}$ by setting

$$g(z_1, z_2) = 2\Re(z_1 z_2),$$

where $\Re(z)$ denotes the real part of the complex number z.

(1) Prove that g is bilinear and non-degenerate.

(2) Write the matrix of g with respect to the basis $\{1, i\}$ of $\mathbb{C}$.

(3) Let $z_1, z_2 \in \mathbb{C}$ be linearly independent over $\mathbb{R}$ (i.e., z_1 and z_2 are a basis of $\mathbb{C}$, considered as an $\mathbb{R}$-vector space). Find the matrix G of g with respect to the basis $\{z_1, z_2\}$ and prove that

$$\det G = \left[\det \begin{pmatrix} z_1 & z_2 \\ \bar{z}_1 & \bar{z}_2 \end{pmatrix}\right]^2,$$

where $\bar{z}$ is the complex conjugate of $z \in \mathbb{C}$.

Exercise 5.8. Let A be a symmetric matrix with entries in $\mathbb{R}$. It is true that the matrix $\mathbf{1} + A^2$ is invertible?

6

Affine Geometry

In this chapter we will see how linear algebra can be used for the study of geometric problems. Starting from the notions of vector and vector space, we will introduce the notion of affine space as a natural environment for the study of Euclidean geometry. We will then study the main geometric properties of these spaces and their subspaces.

6.1 Affine spaces

Vector spaces, which underlie the development of linear algebra, prove inadequate to provide a model of "space" for the study of Euclidean geometry.[1] As we have seen in Chapter 1, vectors were originally introduced to represent displacements; it is therefore necessary to have a "space" (i.e., a set of points) of which the vectors can represent the translations.

These simple considerations lead us to conclude that a model of space suitable for the study of geometry must consist of a set of points and a set of vectors, that is, of a vector space, which will represent the set of translations of these points.

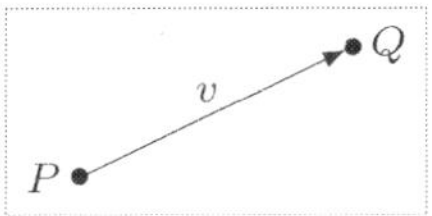

Two equivalent ways of expressing this fact are as follows: (1) establish that an ordered pair of points P and Q determines a vector v (graphically represented as an oriented segment that originates at the point P with the tip of the arrow in Q), or (2) establish that a point P and a vector v determine another point Q (which represents the effect of the translation represented by the vector v, applied to the point P).

[1] An example of this inadequacy is given by the fact that in every vector space there is a privileged element, the null vector $\mathbf{0}$. On the contrary, in the space of Euclidean geometry, there is no privileged point O that can be made to correspond to the null vector.

From this second interpretation it follows that our space must be endowed with a composition law, which we will denote with the symbol +, which associates the point Q to the point-vector pair (P, v); we will therefore write $Q = P + v$.

Now we can formalize the notion of affine space by following the idea just described.

Definition 6.1.1. An *affine space* $\mathbb{A}$ over the field K is a non-empty set $\mathscr{A}$, called the *set of points* of $\mathbb{A}$, together with a vector space V over K and an operation

$$+_{\mathbb{A}} : \mathscr{A} \times V \to \mathscr{A}, \qquad (P, v) \mapsto Q = P +_{\mathbb{A}} v,$$

that satisfies the following properties:

(*i*) $(P +_{\mathbb{A}} v_1) +_{\mathbb{A}} v_2 = P +_{\mathbb{A}} (v_1 + v_2)$, for every $P \in \mathscr{A}$ and every $v_1, v_2 \in V$;

(*ii*) $P +_{\mathbb{A}} \mathbf{0} = P$, for every $P \in \mathscr{A}$;

(*iii*) for every $P, Q \in \mathscr{A}$ there exists a (unique) vector $v \in V$ such that $Q = P +_{\mathbb{A}} v$.

The vector space V is called the *underlying vector space* of the affine space $\mathbb{A}$ and the dimension of V is called the *dimension* of the affine space $\mathbb{A}$. An affine space is *finite dimensional* if the underlying vector space is finitely generated.

For simplicity, throughout this chapter we will consider only finite dimensional affine spaces.

In the sequel, an affine space $\mathbb{A} = (\mathscr{A}, V, +_{\mathbb{A}})$ will often be identified with its set of points $\mathscr{A}$. To indicate that P is a point of $\mathbb{A}$ we will therefore write $P \in \mathbb{A}$ instead of $P \in \mathscr{A}$. Similarly, to indicate that v is a vector belonging to the vector space V underlying the affine space $\mathbb{A}$, we can write $v \in \mathbb{A}$. The meaning of such expressions will always be clear from the context.

Remark 6.1.2. The operation of "sum between points and vectors" in an affine space $\mathbb{A} = (\mathscr{A}, V, +_{\mathbb{A}})$ has been denoted by the symbol $+_{\mathbb{A}}$ in order to distinguish it from the operation of addition of vectors in the vector space V, denoted simply with +. However, since it will always be clear from the context which of the two addition operations we are using, in the sequel we will denote with the symbol + both the addition operation of vectors and the addition operation between a point and a vector.

Remark 6.1.3. For each pair of points P and Q of an affine space $\mathbb{A}$, the property (*iii*) of Definition 6.1.1 guarantees the existence and uniqueness of a vector v such that $Q = P + v$; this vector will be denoted by $\overrightarrow{PQ}$. Formally deriving v from the expression $Q = P + v$, we will also write

$$v = \overrightarrow{PQ} = Q - P,$$

so that we have the identity $P + (Q - P) = Q$. As already mentioned, the vector $\overrightarrow{PQ} = Q - P$ will be represented graphically as an oriented segment having the origin in P and the tip of the arrow at the point Q.

Remark 6.1.4. In an affine space $\mathbb{A} = (\mathscr{A}, V, +_{\mathbb{A}})$ the additive group of the vector space V acts on the set $\mathscr{A}$ of the points of $\mathbb{A}$; this action is free and transitive. It is easy to verify that an affine space can be defined, in a completely equivalent way, as a non-empty set endowed with a free and transitive action of the additive group of a vector space.

For every point P of an affine space $\mathbb{A} = (\mathscr{A}, V, +)$ we can define the function

$$f_P : V \to \mathscr{A}, \qquad v \mapsto f_P(v) = P + v.$$

An immediate consequence of Definition 6.1.1 is the following result:

Proposition 6.1.5. *For every point P of an affine space $\mathbb{A}$, the function $f_P : V \to \mathscr{A}$ is bijective.*

Proof. The injectivity and the surjectivity of f_P follow from the property (iii) of Definition 6.1.1. □

Remark 6.1.6. The existence of a bijection between the set of points $\mathscr{A}$ and the vector space V of an affine space $\mathbb{A} = (\mathscr{A}, V, +)$ allows us to conclude that $\mathscr{A}$ can be identified with V. However, note that there is no canonical bijection between V and $\mathscr{A}$; in fact, such a bijection depends on the choice of a point P of $\mathscr{A}$. We also remark that in the bijection $f_P : V \to \mathscr{A}$ the point P corresponds to the null vector of V. We can therefore conclude that, in any affine space $\mathbb{A}$, the set of points can be identified with the underlying vector space only after having (arbitrarily) fixed a point of $\mathbb{A}$.

Remark 6.1.7. Given an affine space $\mathbb{A} = (\mathscr{A}, V, +)$, for every vector $v \in V$ the function

$$\tau_v : \mathscr{A} \to \mathscr{A}, \qquad P \mapsto P + v,$$

is called the *translation* parallel to the vector v. If v is the zero vector, $\tau_{\mathbf{0}}$ is the identity, while for any $v \neq \mathbf{0}$, the translation τ_v is a bijection without fixed points, i.e., $\tau_v(P) \neq P$, for any $P \in \mathscr{A}$ and any $v \neq \mathbf{0}$.

Example 6.1.8. We will now show how every vector space has a canonical structure of affine space.

Given a vector space V over the field K, we define the affine space $\mathbb{V} = (\mathscr{V}, V, +_{\mathbb{V}})$ associated with it by setting $\mathscr{V} = V$. In this way the operation $+_{\mathbb{V}}$ coincides with the operation of adding elements of V:

$$+_{\mathbb{V}}\colon \mathscr{V} \times V \to \mathscr{V}, \qquad (P, v) \mapsto P +_{\mathbb{V}} v = P +_V v.$$

It is easy to verify that, with these definitions, $\mathbb{V}$ turns out to be an affine space over K, having V as the underlying vector space. We note that this affine space structure depends exclusively on the vector space structure of V: it is therefore called the *canonical affine structure* of the vector space V.

Example 6.1.9. The basic example of affine space is provided by the standard n-dimensional affine space over the field K, which will be denoted by $\mathbb{A}^n_K$. It is the affine space associated, in a canonical way, to the vector space K^n, as described in the previous example. $\mathbb{A}^n_K$ is therefore the affine space whose set of points is the *set* K^n and whose underlying vector space is the *vector space* K^n. The operation of sum of points and vectors is defined as the usual componentwise sum of two n-tuples of elements of K; more precisely, if $P = (p_1, p_2, \ldots, p_n)$ is a point and $v = (a_1, a_2, \ldots, a_n)$ is a vector of $\mathbb{A}^n_K$, we have

$$Q = P + v = (p_1 + a_1, p_2 + a_2, \ldots, p_n + a_n).$$

It is immediate to verify that the properties (i), (ii) and (iii) of Definition 6.1.1 are satisfied, hence $\mathbb{A}^n_K$ is an affine space. Note also that, from the definition, it immediately follows that if $P = (p_1, p_2, \ldots, p_n)$ and $Q = (q_1, q_2, \ldots, q_n)$ are two points of $\mathbb{A}^n_K$, the vector $v = Q - P$ is given by

$$v = (q_1 - p_1, q_2 - p_2, \ldots, q_n - p_n).$$

As we will see later, this example is of particular importance: in fact every affine space of dimension n over the field K turns out to be *isomorphic* (although not canonically) to the standard affine space $\mathbb{A}^n_K$.

Remark 6.1.10. In the affine space $\mathbb{A}^n_K$ both points and vectors are simply n-tuples of elements of K. However, we observe that while the sum of a point and a vector, or the difference of two points, are legal operations, the sum of two points, although algebraically possible, is not a legal operation.

In this regard, we note that to distinguish the n-tuples of elements of K that represent points from those that represent vectors, it is possible to adopt the following convention: to every n-tuple $(a_1, a_2, \ldots, a_n) \in K^n$ we add an element $a_0 \in \{0, 1\}$, with the convention that if $a_0 = 0$ the n-tuple $(a_1, a_2, \ldots, a_n)$ represents a vector, while if $a_0 = 1$ such n-tuple represents a point of the affine space. In other words, the vectors of $\mathbb{A}^n_K$ are written in the form

$$v = (0, a_1, a_2, \ldots, a_n),$$

while the points are written in the form

$$A = (1, a_1, a_2, \ldots, a_n).$$

Note that this convention is compatible with the definition of the operations between points and vectors of an affine space: a linear combination of vectors is still a vector (the first component is 0), the sum of a point and a vector gives a point (in fact, if we observe the first component, we have $1 + 0 = 1$) and the difference between two points is a vector (in the first component we have $1 - 1 = 0$). The sum of two points, on the other hand, is not a legitimate operation; this is evidenced by the fact that, by adding two points, one would obtain a $(n + 1)$-tuple having as first component $a_0 = 2$, which is not a permitted value.

6.2 Affine subspaces

Let $\mathbb{A} = (\mathscr{A}, V, +_{\mathbb{A}})$ be an affine space over the field K.

Definition 6.2.1. An *affine subspace* (or *linear subspace*) of $\mathbb{A}$ is an affine space $\mathbb{B} = (\mathscr{B}, W, +_{\mathbb{B}})$, where $\mathscr{B}$ is a subset of $\mathscr{A}$, W is a vector subspace of V and the operation $+_{\mathbb{B}} : \mathscr{B} \times W \to \mathscr{B}$ is obtained, by restriction, from the operation $+_{\mathbb{A}} : \mathscr{A} \times V \to \mathscr{A}$ defined on the affine space $\mathbb{A}$.

The structure of the affine subspaces of an affine space $\mathbb{A}$ is further specified by the following result:

Proposition 6.2.2. *Let $\mathbb{B} = (\mathscr{B}, W, +_{\mathbb{B}})$ be an affine subspace of the affine space $\mathbb{A} = (\mathscr{A}, V, +_{\mathbb{A}})$. For every point $P \in \mathscr{B}$, we have*

$$\mathscr{B} = P +_{\mathbb{B}} W = \{P +_{\mathbb{B}} w \mid w \in W\}.$$

Proof. Given $P \in \mathscr{B}$, from the definition of affine space it follows that, for each vector $w \in W$, the point $Q = P +_{\mathbb{B}} w$ belongs to the set $\mathscr{B}$. Conversely, from the property (*iii*) of Definition 6.1.1, it follows that, for each $Q \in \mathscr{B}$, there exists only one vector $w \in W$ such that $Q = P +_{\mathbb{B}} w$. □

Remark 6.2.3. If $\mathbb{A} = (\mathscr{A}, V, +_{\mathbb{A}})$ is an affine space and W is a vector subspace of V, for every point $P \in \mathscr{A}$ we can construct an affine subspace $\mathbb{B} = (\mathscr{B}, W, +_{\mathbb{B}})$ of $\mathbb{A}$ by setting

$$\mathscr{B} = P +_{\mathbb{A}} W = \{P +_{\mathbb{A}} w \,|\, w \in W\}$$

and defining the operation $+_{\mathbb{B}}$ as the restriction of the sum of points and vectors defined in $\mathbb{A}$. It is immediate to verify that $\mathbb{B}$ is an affine space. As an example we verify that the function

$$+_{\mathbb{B}} \colon \mathscr{B} \times W \to \mathscr{B}$$

is well defined. More precisely, we check that, for any $A \in \mathscr{B}$ and any $w \in W$, we have $A +_{\mathbb{B}} w \in \mathscr{B}$.

Let $A \in \mathscr{B}$ and $u \in W$ be such that $A = P +_{\mathbb{A}} u$. Then, for every $w \in W$, we have:

$$A +_{\mathbb{B}} w = A +_{\mathbb{A}} w = (P +_{\mathbb{A}} u) +_{\mathbb{A}} w = P +_{\mathbb{A}} (u + w) \in \mathscr{B},$$

given that $u + w \in W$, because W is a vector subspace of V.

The affine subspace $\mathbb{B}$ thus defined will often be denoted simply by $\mathbb{B} = P + W$: it is called the *affine subspace of* $\mathbb{A}$ *passing through the point* P *and parallel to the subspace* W.

Remark 6.2.4. An affine subspace of $\mathbb{A}$ of dimension 0 is simply a point of $\mathbb{A}$. An affine subspace of dimension 1 is called a *line*, an affine subspace of dimension 2 is called a *plane* and, if $\dim \mathbb{A} = n$, an affine subspace of $\mathbb{A}$ of dimension $n - 1$ is called a *hyperplane.*

Definition 6.2.5. Let $\mathscr{S}$ be a non-empty subset (of the set of points) of an affine space $\mathbb{A}$. The smallest affine subspace of $\mathbb{A}$ containing $\mathscr{S}$ is called the *affine subspace generated by* $\mathscr{S}$, and is denoted by $\langle \mathscr{S} \rangle$.

Proposition 6.2.6. *Let* $\mathbb{A} = (\mathscr{A}, V, +)$ *be an affine space and* $\mathscr{S}$ *a non-empty subset of* $\mathscr{A}$. *Let us fix a point* $P \in \mathscr{S}$. *We will denote by* W *the vector subspace of* V *generated by the set of vectors* $v_Q = Q - P$, *as the point* Q *varies in* $\mathscr{S}$. *Then we have* $\langle \mathscr{S} \rangle = P + W$.

Proof. Let us denote by $\mathbb{B}$ the affine subspace $\mathbb{B} = P + W$ of $\mathbb{A}$. For every $Q \in \mathscr{S}$, we have $Q = P + (Q - P) = P + v_Q \in \mathbb{B}$, which proves that $\mathscr{S} \subseteq \mathbb{B}$. To prove that $\langle \mathscr{S} \rangle = \mathbb{B}$ we must show that $\mathbb{B}$ is the smallest affine subspace of $\mathbb{A}$ that contains $\mathscr{S}$, i.e., that every affine subspace of $\mathbb{A}$ containing $\mathscr{S}$ also contains $\mathbb{B}$. Let $\mathbb{A}' = (\mathscr{A}', V', +)$ be an affine subspace of $\mathbb{A}$ that contains $\mathscr{S}$. For any $P, Q \in \mathscr{S}$ we also have $P, Q \in \mathbb{A}'$ hence $v_Q = Q - P \in V'$. It follows that the vector subspace V' contains the set of vectors $v_Q = Q - P$, as Q varies in $\mathscr{S}$, hence it also contains the vector subspace W spanned by these vectors. It follows that for each vector $w \in W$ we have $P + w \in \mathbb{A}'$, which proves that $\mathbb{B} \subseteq \mathbb{A}$. □

As already happened in the case of vector spaces, the union of two affine subspaces of an affine space $\mathbb{A}$ is not, in general, an affine subspace of $\mathbb{A}$. We define the *sum* of two affine subspaces $\mathbb{L}$ and $\mathbb{M}$, denoted[2] by $\mathbb{L} \vee \mathbb{M}$, as the affine subspace generated by the union of $\mathbb{L}$ and $\mathbb{M}$, i.e., as the smallest affine subspace of $\mathbb{A}$ that contains $\mathbb{L}$ and $\mathbb{M}$. In the particular case in which the affine subspaces are two points, P and Q, the smallest affine subspace containing the points P and Q, denoted with the notation $P \vee Q$, is the straight line through P and Q (if the points P and Q are distinct). Similarly, given three points P, Q and R of an affine space $\mathbb{A}$, the notation $P \vee Q \vee R$ denotes the smallest affine subspace of $\mathbb{A}$ containing these three points, i.e., the plane passing through P, Q and R, if the three points are not aligned.

For the sum of two affine subspaces the following result holds:

Proposition 6.2.7. *Let $\mathbb{L} = (\mathscr{L}, L, +)$ and $\mathbb{M} = (\mathscr{M}, M, +)$ be two affine subspaces of an affine space $\mathbb{A}$. If $\mathscr{L} \cap \mathscr{M} \neq \varnothing$, then*

$$\mathbb{L} \vee \mathbb{M} = P + (L + M) = \{P + v \,|\, v \in L + M\},$$

for every point $P \in \mathscr{L} \cup \mathscr{M}$. If $\mathscr{L} \cap \mathscr{M} = \varnothing$, then

$$\mathbb{L} \vee \mathbb{M} = P + (L + M + \langle u \rangle) = \{P + v \,|\, v \in L + M + \langle u \rangle\},$$

where $P \in \mathscr{L} \cup \mathscr{M}$ and $u = B - A$, with $A \in \mathscr{L}$ and $B \in \mathscr{M}$.

Proof. Let us first consider the case in which $\mathscr{L} \cap \mathscr{M} \neq \varnothing$. Let $A \in \mathscr{L} \cap \mathscr{M}$ and let us fix a point $P \in \mathscr{L} \cup \mathscr{M}$. Possibly exchanging the roles of $\mathscr{L}$ and $\mathscr{M}$, we can assume that $P \in \mathscr{L}$. Let us consider the affine subspace $\mathbb{B} = P + (L + M)$. To prove that $\mathbb{B} = \mathbb{L} \vee \mathbb{M}$ we must show that $\mathbb{B}$ is the smallest affine subspace of $\mathbb{A}$ that contains $\mathbb{L}$ and $\mathbb{M}$. For every point $Q \in \mathbb{L}$ the vector $v = Q - P$ belongs to the vector space L, hence $Q = P + v \in P + L \subseteq \mathbb{B}$; this proves that $\mathbb{L} \subseteq \mathbb{B}$. To prove that we also have $\mathbb{M} \subseteq \mathbb{B}$, let us consider a point $R \in \mathbb{M}$ and set $w = R - A$ (we remark that $w \in M$, because the point

[2]Another commonly used notation for the sum of two affine suspaces $\mathbb{L}$ and $\mathbb{M}$ is $\mathbb{L} + \mathbb{M}$.

A belongs to $\mathbb{M}$). Given that the points A and P belong to $\mathbb{L}$, the vector $u = A - P$ belongs to the vector subspace L. It follows that

$$R = P + (A - P) + (R - A) = P + u + w \in P + (L + M),$$

hence $R \in \mathbb{B}$. This proves that $\mathbb{B}$ contains $\mathbb{L}$ and $\mathbb{M}$.

Let's now consider an affine subspace $\mathbb{B}'$ of $\mathbb{A}$ containing $\mathbb{L}$ and $\mathbb{M}$. If we denote by W the underlying vector space to $\mathbb{B}'$, we have $W \supseteq L$ and $W \supseteq M$, hence $W \supseteq L + M$. Since $P \in \mathbb{B}'$, we have $\mathbb{B} = P + (L + M) \subseteq \mathbb{B}'$, which proves that $\mathbb{B}$ is the smallest affine subspace of $\mathbb{A}$ that contains $\mathbb{L}$ and $\mathbb{M}$, hence $\mathbb{B} = \mathbb{L} \vee \mathbb{M}$.

Let's now consider the case in which $\mathscr{L} \cap \mathscr{M} = \varnothing$. Let $A \in \mathscr{L}$, $B \in \mathscr{M}$ and let $u = B - A$. Let us fix a point $P \in \mathscr{L} \cup \mathscr{M}$. Possibly exchanging the roles of $\mathscr{L}$ and $\mathscr{M}$, we can assume that $P \in \mathscr{L}$. Let us consider the affine subspace $\mathbb{B} = P + (L + M + \langle u \rangle)$. To prove that $\mathbb{B} = \mathbb{L} \vee \mathbb{M}$ we must show that $\mathbb{B}$ is the smallest affine subspace of $\mathbb{A}$ that contains $\mathbb{L}$ and $\mathbb{M}$. As in the previous case, for every point $Q \in \mathbb{L}$ the vector $v = Q - P$ belongs to the vector space L, hence $Q = P + v \in P + L \subseteq \mathbb{B}$; this proves that $\mathbb{L} \subseteq \mathbb{B}$. To prove that $\mathbb{M} \subseteq \mathbb{B}$, let's consider a point $R \in \mathbb{M}$ and notice that we have

$$R = P + (A - P) + (B - A) + (R - B).$$

Now we remark that the vector $A - P$ belongs to L, because $A, P \in \mathbb{L}$, that $R - B \in M$ because the points B and R belong to $\mathbb{M}$, and that $B - A = u$. It follows that $R \in P + (L + M + \langle u \rangle) = \mathbb{B}$. We have thus proved that $\mathbb{B}$ contains $\mathbb{L}$ and $\mathbb{M}$.

Let us consider now an affine subspace $\mathbb{B}'$ containing $\mathbb{L}$ and $\mathbb{M}$. If we denote by W the underlying vector subspace to $\mathbb{B}'$, we have $W \supseteq L$ and $W \supseteq M$. Moreover, since $A, B \in \mathbb{B}'$, we also have $u = B - A \in W$, hence $W \supseteq L + M + \langle u \rangle$. Then we have $\mathbb{B} = P + (L + M + \langle u \rangle) \subseteq \mathbb{B}'$. It follows that $\mathbb{B}$ is the smallest affine subspace of $\mathbb{A}$ that contains $\mathbb{L}$ and $\mathbb{M}$, hence $\mathbb{B} = \mathbb{L} \vee \mathbb{M}$. □

Unlike union, the intersection of two or more affine subspaces is usually an affine subspace. In fact, we have:

Proposition 6.2.8. *Let $\mathbb{B}_i = (\mathscr{B}_i, W_i, +)$, $i \in I$, be a collection of affine subspaces of an affine space $\mathbb{A} = (\mathscr{A}, V, +)$. If $\bigcap_{i \in I} \mathscr{B}_i \neq \varnothing$, the intersection of the affine subspaces $\mathbb{B}_i$ is an affine subspace of $\mathbb{A}$, whose set of points is $\bigcap_{i \in I} \mathscr{B}_i$ and whose underlying vector space is $\bigcap_{i \in I} W_i$. Hence we have*

$$\bigcap_{i \in I} \mathbb{B}_i = \Big(\bigcap_{i \in I} \mathscr{B}_i, \bigcap_{i \in I} W_i, + \Big).$$

Proof. The proof is reduced to the verification that for the triple

$$\Big(\bigcap_{i\in I}\mathscr{B}_i, \bigcap_{i\in I} W_i, +\Big)$$

the properties (i), (ii) and (iii) of the definition of affine space hold, which is completely obvious. □

Remark 6.2.9. The only case in which the intersection of two or more affine subspaces is not an affine subspace occurs when this intersection is the empty set. It is therefore convenient to also consider the empty set as an affine space. Of course, the dimension of such an affine space is not defined. It is however possible to assign a dimension to the empty affine space by setting $\dim\varnothing = -1$ (in this regard, remember that an affine space of dimension zero is not empty, but consists of a single point).

Remark 6.2.10. Given a subset $\mathscr{S}$ of an affine space $\mathbb{A}$, we can now easily check that the affine subspace of $\mathbb{A}$ generated by $\mathscr{S}$ coincides with the intersection of all the affine subspaces of $\mathbb{A}$ containing $\mathscr{S}$.

As regards the reciprocal positions of two affine subspaces, we can give the following definition:

Definition 6.2.11. Let $\mathbb{L}$ and $\mathbb{M}$ be two affine subspaces of an affine space $\mathbb{A}$, whose underlying vector subspaces are L and M, respectively.

(i) $\mathbb{L}$ and $\mathbb{M}$ *intersect* if $\mathbb{L}\cap\mathbb{M}\neq\varnothing$;

(ii) $\mathbb{L}$ and $\mathbb{M}$ are *parallel* if $L\neq\{\mathbf{0}\}$, $M\neq\{\mathbf{0}\}$ and we have $L\subseteq M$ or $M\subseteq L$;

(iii) $\mathbb{L}$ and $\mathbb{M}$ are *skew* if $\mathbb{L}\cap\mathbb{M}=\varnothing$ and $L\cap M=\{\mathbf{0}\}$.

From the Propositions 6.2.7 and 6.2.8 we can deduce the following result:

Corollary 6.2.12. *Let $\mathbb{L}$ and $\mathbb{M}$ be two affine subspaces of $\mathbb{A}$. If $\mathbb{L}$ and $\mathbb{M}$ intersect or are skew, we have*

$$\dim(\mathbb{L}\vee\mathbb{M}) = \dim\mathbb{L}+\dim\mathbb{M}-\dim(\mathbb{L}\cap\mathbb{M}),$$

otherwise

$$\dim(\mathbb{L}\vee\mathbb{M}) < \dim\mathbb{L}+\dim\mathbb{M}-\dim(\mathbb{L}\cap\mathbb{M}).$$

Proof. Let us denote by L and M the underlying vector spaces of $\mathbb{L}$ and $\mathbb{M}$, respectively. If $\mathbb{L}$ and $\mathbb{M}$ intersect, $\mathbb{L}\cap\mathbb{M}$ is an affine subspace whose underlying vector space is $L\cap M$, while the underlying vector space of $\mathbb{L}\vee\mathbb{M}$

is the sum of the subspaces L and M. From Grassmann's formula (Chapter 1, Proposition 1.3.48) it follows that

$$\begin{aligned}\dim(\mathbb{L} \vee \mathbb{M}) &= \dim(L + M) \\ &= \dim L + \dim M - \dim(L \cap M) \\ &= \dim \mathbb{L} + \dim \mathbb{M} - \dim(\mathbb{L} \cap \mathbb{M}).\end{aligned}$$

If $\mathbb{L}$ and $\mathbb{M}$ are skew, we have $\mathbb{L} \cap \mathbb{M} = \varnothing$ and $L \cap M = \{\mathbf{0}\}$. Since by convention we have $\dim \varnothing = -1$ (see Remark 6.2.9), from Proposition 6.2.7 and Grassmann's formula we derive that

$$\begin{aligned}\dim(\mathbb{L} \vee \mathbb{M}) &= \dim(L + M + \langle u \rangle) \\ &= \dim L + \dim M + 1 \\ &= \dim \mathbb{L} + \dim \mathbb{M} - \dim(\mathbb{L} \cap \mathbb{M}).\end{aligned}$$

On the other hand, if $\mathbb{L}$ and $\mathbb{M}$ are not skew and do not intersect, we have $\mathbb{L} \cap \mathbb{M} = \varnothing$ and $L \cap M \neq \{\mathbf{0}\}$, hence $\dim(L \cap M) > 0$. In this case we easily deduce that

$$\begin{aligned}\dim(\mathbb{L} \vee \mathbb{M}) &= \dim(L + M + \langle u \rangle) \\ &= \dim L + \dim M - \dim(L \cap M) + 1 \\ &< \dim L + \dim M + 1 \\ &= \dim \mathbb{L} + \dim \mathbb{M} - \dim(\mathbb{L} \cap \mathbb{M}).\end{aligned}$$ □

6.3 Reference frames

A reference frame in an affine space is the analogue of a basis for a vector space. We recall that the choice of a basis of a n-dimensional vector space V allows us to associate to each vector $v \in V$ a n-tuple of elements of the field K. In a completely analogous way, the choice of a reference frame in a n-dimensional affine space $\mathbb{A}$ allows us to associate to each point (and to each vector) of $\mathbb{A}$ a n-tuple of elements of K.

Definition 6.3.1. Let $\mathbb{A} = (\mathscr{A}, V, +)$ be a n-dimensional affine space over the field K. A *reference frame* $\mathscr{R}$ in $\mathbb{A}$ is determined by a point $O \in \mathbb{A}$ (called *origin*) and a basis $v_1, v_2, \dots, v_n$ of V.

Equivalently, a reference frame in $\mathbb{A}$ is determined by $n + 1$ points, $P_0, P_1, \dots, P_n$, such that the vectors $v_1 = P_1 - P_0$, $v_2 = P_2 - P_0$, $\dots$, $v_n = P_n - P_0$, are a basis of V (in this case we can take as origin of the reference frame the point $O = P_0$).

Let $\mathscr{R} = \{O, v_1, v_2, \dots, v_n\}$ be a reference frame in an affine space $\mathbb{A}$ over the field K. For every point $P \in \mathbb{A}$ there exists a vector $v \in V$ such that $P = O + v$. Writing v as a linear combination of the basis vectors $v_1, v_2, \dots, v_n$,

$$v = \lambda_1 v_1 + \lambda_2 v_2 + \cdots + \lambda_n v_n,$$

we get

$$P = O + \lambda_1 v_1 + \lambda_2 v_2 + \cdots + \lambda_n v_n.$$

The scalars $\lambda_1, \dots, \lambda_n$ are called the *coordinates* of the point P in the given reference frame. Hence to every point $P \in \mathbb{A}$ we can associate a n-tuple $(\lambda_1, \dots, \lambda_n) \in K^n$; note that to the point O (the origin of the reference frame) we associate the n-tuple $(0, 0, \dots, 0)$. Similarly, to every vector $v \in V$ we associate the n-tuple of its components with respect to the basis $v_1, v_2, \dots, v_n$. It follows that the choice of a reference frame determines a bijection between the n-dimensional affine space $\mathbb{A}$ over the field K and the affine space $\mathbb{A}^n_K$; this bijection is obtained by associating to each point $P \in \mathbb{A}$ its coordinates $(\lambda_1, \dots, \lambda_n) \in K^n$ and to each vector $v \in V$ its components with respect to the given basis.

Definition 6.3.2. Let $\mathscr{R} = \{O, v_1, v_2, \dots, v_n\}$ be a reference frame in an affine space $\mathbb{A}$. The n lines passing through the origin O and parallel to the vectors $v_1, v_2, \dots, v_n$ are called the *axes* of the reference frame (or the *coordinate axes*). The n hyperplanes π_i $(i = 1, \dots, n)$ passing through O and parallel to the vector subspaces generated by the vectors $v_1, \dots, \hat{v}_i, \dots, v_n$ (the notation $\hat{v}_i$ means that such vector is omitted) are called the *coordinate hyperplanes*.

Definition 6.3.3. Given $n + 1$ points in an affine space $\mathbb{A}$, we say that they are *affinely independent* (or *in generic position*) if the affine subspace spanned by these points has dimension n.

Proposition 6.3.4. *A reference frame in an affine space $\mathbb{A} = (\mathscr{A}, V, +)$ of dimension n is given by $n + 1$ affinely independent points.*

Proof. Given $n + 1$ points $P_0, P_1, \dots, P_n$, we set $v_1 = P_1 - P_0$, $v_2 = P_2 - P_0$, $\dots$, $v_n = P_n - P_0$ and denote by W the vector subspace of V generated by the vectors v_i, for $i = 1, \dots, n$. By Proposition 6.2.6, the affine subspace spanned by the points $P_0, \dots, P_n$ is

$$\langle P_0, \dots, P_n \rangle = P_0 + W.$$

It follows that the points $P_0, \dots, P_n$ are affinely independent if and only if $\dim W = n$, i.e., if and only if the vectors $v_1, \dots, v_n$ are a basis of V. This is equivalent to the fact that the $n+1$ points $P_0, P_1, \dots, P_n$ form a reference frame in $\mathbb{A}$. □

Remark 6.3.5. Let $P_0, P_1, \dots, P_n$ be a reference frame in an affine space $\mathbb{A}$. It it easy to verify that, for every permutation $\sigma \in \mathfrak{S}_{n+1}$, the points $P_{\sigma(0)}, P_{\sigma(1)}, \dots, P_{\sigma(n)}$ form another reference frame in $\mathbb{A}$.

6.4 Equations of affine subspaces

As we saw in the previous section, the choice of a reference frame in an affine space allows us to introduce a coordinate system. In this way, any affine space of dimension n over the field K can be identified with the standard affine space $\mathbb{A}^n_K$.

In this section we will show how the affine subspaces of an affine space with a reference frame can be identified with the sets of solutions of systems of linear equations.

Let us denote by $\mathbb{A} = \mathbb{A}^n_K$ the n-dimensional affine space over the field K and let us consider an affine subspace $\mathbb{B}$, of dimension r, of $\mathbb{A}$. Let $W \subseteq K^n$ be the underlying vector subspace of $\mathbb{B}$ and let $w_1, w_2, \dots, w_r$ be a basis of W. Given a point $P \in \mathbb{B}$, we have:

$$\mathbb{B} = P + W = \{P + w \mid w \in W\}.$$

Since any vector $w \in W$ can be written as a linear combination of the basis vectors $w_1, w_2, \dots, w_r$, the generic point $X = (x_1, x_2, \dots, x_n)$ of $\mathbb{B}$ can be written in the form

$$X = P + \lambda_1 w_1 + \lambda_2 w_2 + \cdots + \lambda_r w_r, \tag{6.4.1}$$

for some $\lambda_1, \lambda_2, \dots, \lambda_r \in K$. Introducing coordinates, the equation (6.4.1) translates into the following system:

$$\begin{cases} x_1 = p_1 + \lambda_1 a_{11} + \lambda_2 a_{12} + \cdots + \lambda_r a_{1r} \\ x_2 = p_2 + \lambda_1 a_{21} + \lambda_2 a_{22} + \cdots + \lambda_r a_{2r} \\ \quad \cdot \quad \cdot \quad \cdot \\ x_n = p_n + \lambda_1 a_{n1} + \lambda_2 a_{n2} + \cdots + \lambda_r a_{nr} \end{cases} \tag{6.4.2}$$

where $P = (p_1, p_2, \ldots, p_n)$ and $w_i = (a_{1i}, a_{2i}, \ldots, a_{ni}) \in K^n$, for every $i = 1, \ldots, r$. The scalars $\lambda_1, \lambda_2, \ldots, \lambda_r$ are called the *parameters* and the previous equations are called the *parametric equations* of the affine subspace $\mathbb{B}$.

By eliminating $\lambda_1, \ldots, \lambda_r$ in the system of equations (6.4.2) we obtain a system of (non homogeneous) linear equations in the indeterminates $x_1, x_2, \ldots x_n$, of the form

$$S : \begin{cases} a'_{11}x_1 + a'_{12}x_2 + \cdots + a'_{1n}x_n = b_1 \\ a'_{21}x_1 + a'_{22}x_2 + \cdots + a'_{2n}x_n = b_2 \\ \quad \cdot \quad \cdot \quad \cdot \\ a'_{s1}x_1 + a'_{s2}x_2 + \cdots + a'_{sn}x_n = b_s \end{cases} \tag{6.4.3}$$

for some coefficients a'_{ij} and b_i in K. These are called the *cartesian equations* of $\mathbb{B}$.

Conversely, given a system S of linear equations as in (6.4.3), we denote by W the vector subspace of K^n formed by the solutions of the homogeneous system S_0 associated to S (see Chapter 2, Proposition 2.3.2):

$$S_0 : \begin{cases} a'_{11}x_1 + a'_{12}x_2 + \cdots + a'_{1n}x_n = 0 \\ a'_{21}x_1 + a'_{22}x_2 + \cdots + a'_{2n}x_n = 0 \\ \quad \cdot \quad \cdot \quad \cdot \\ a'_{s1}x_1 + a'_{s2}x_2 + \cdots + a'_{sn}x_n = 0. \end{cases}$$

If the system S is incompatible, the set Σ of the solutions of S is empty, hence it can be considered an affine subspace of $\mathbb{A}$ (see Remark 6.2.9). If S has solutions every such solution can be written as the sum of a special solution $P = (p_1, \ldots, p_n)$ of S and a solution $w \in W$ of the associated homogeneous system (see Chapter 2, Proposition 2.3.3). In this case the set Σ of the solutions of S is given by $\Sigma = P + W$, hence Σ is an affine subspace of $\mathbb{A}$.

We can therefore conclude that the affine subspaces of $\mathbb{A}^n_K$ can be described as the sets of solutions of systems of linear equations in n indeterminates. However, it should be noted that this correspondence between systems of linear equations and affine subspaces is not one-to-one; in fact there are different systems having the same set of solutions.

Let's now look at some particularly significant examples.

Example 6.4.1 (Equation of a line in the affine plane). Let r be a line in the affine plane $\mathbb{A}^2_K$. The underlying vector subspace of r has dimension 1 and is generated by a non-zero vector $v = (a_1, a_2) \in K^2$. If $P = (p_1, p_2)$ is a point of r and if we denote by $X = (x_1, x_2)$ a generic point of the plane, we have

$$X \in r \iff X = P + \lambda v, \ \lambda \in K.$$

The vector equation $X = P + \lambda v$ translates into the following system:

$$\begin{cases} x_1 = p_1 + \lambda a_1 \\ x_2 = p_2 + \lambda a_2. \end{cases}$$

These are the parametric equations of a line in the affine plane.

By computing λ from one of the two equations and replacing into the other equation, we get a linear equation in the indeterminates x_1 and x_2, of the form

$$ax_1 + bx_2 + c = 0,$$

where $a, b, c \in K$. We can conclude that a line in the affine plane is the zero locus of a polynomial of first degree in the indeterminates x_1 and x_2.

Example 6.4.2 (EQUATION OF A PLANE IN THE 3-DIMENSIONAL AFFINE SPACE). Let π be a plane in the affine space $\mathbb{A}^3_K$. The underlying vector subspace of π has dimension 2 and is generated by two linearly independent vectors $v_1 = (a_{11}, a_{21}, a_{31})$ and $v_2 = (a_{12}, a_{22}, a_{32})$. If $P = (p_1, p_2, p_3)$ is a point of the plane π and if we denote by $X = (x_1, x_2, x_3)$ a generic point of the affine space, we have

$$X \in \pi \iff X = P + \lambda_1 v_1 + \lambda_2 v_2, \ \lambda_1, \lambda_2 \in K.$$

The vector equation $X = P + \lambda_1 v_1 + \lambda_2 v_2$ translates into the following system:

$$\begin{cases} x_1 = p_1 + \lambda_1 a_{11} + \lambda_2 a_{12} \\ x_2 = p_2 + \lambda_1 a_{21} + \lambda_2 a_{22} \\ x_3 = p_3 + \lambda_1 a_{31} + \lambda_2 a_{32}. \end{cases}$$

These are the parametric equations of a plane in the 3-dimensional affine space.

By eliminating the parameters λ_1 and λ_2 from these equations we obtain a linear equation in the indeterminates x_1, x_2 and x_3, of the form

$$ax_1 + bx_2 + cx_3 + d = 0,$$

where $a, b, c, d \in K$. It follows that a plane in the 3-dimensional affine space is the zero locus of a polynomial of first degree in the indeterminates x_1, x_2, x_3.

Example 6.4.3 (EQUATION OF A HYPERPLANE IN n-DIMENSIONAL AFFINE SPACE). By generalizing the two previous examples it is possible to determine the equation of an affine subspace of dimension $n-1$ of the affine space $\mathbb{A}^n_K$.

Let π be a hyperplane in $\mathbb{A}^n_K$. The underlying vector subspace of π has dimension $n-1$ and is generated by $n-1$ linearly independent vectors $v_1 = (a_{11}, a_{21}, \dots, a_{n1})$, $\dots$, $v_{n-1} = (a_{1,n-1}, a_{2,n-1}, \dots, a_{n,n-1}) \in K^n$. If $P = (p_1, p_2, \dots, p_n)$ is a point of the hyperplane π and if we denote by $X = (x_1, x_2, \dots, x_n)$ a generic point of $\mathbb{A}^n_K$, we have

$$X \in \pi \iff X = P + \lambda_1 v_1 + \cdots + \lambda_{n-1} v_{n-1}, \ \ \lambda_1, \dots, \lambda_{n-1} \in K.$$

The vector equation $X = P + \lambda_1 v_1 + \cdots + \lambda_{n-1} v_{n-1}$ translates into the following system:

$$\begin{cases} x_1 = p_1 + \lambda_1 a_{11} + \lambda_2 a_{12} + \cdots + \lambda_{n-1} a_{1,n-1} \\ x_2 = p_2 + \lambda_1 a_{21} + \lambda_2 a_{22} + \cdots + \lambda_{n-1} a_{2,n-1} \\ \quad \cdot \quad \cdot \quad \cdot \\ x_n = p_n + \lambda_1 a_{n1} + \lambda_2 a_{n2} + \cdots + \lambda_{n-1} a_{n,n-1}. \end{cases}$$

These are the parametric equations of a hyperplane in the n-dimensional affine space.

By eliminating the parameters $\lambda_1, \lambda_2, \dots, \lambda_{n-1}$ from the previous equations, we obtain a linear equation in the indeterminates $x_1, x_2, \dots, x_n$, of the form

$$a_1 x_1 + a_2 x_2 + \cdots + a_n x_n + a_{n+1} = 0,$$

where $a_1, \dots, a_{n+1} \in K$. It follows that a hyperplane in the n-dimensional affine space is the zero locus of a polynomial of first degree in the indeterminates $x_1, x_2, \dots, x_n$.

Example 6.4.4 (EQUATIONS OF A LINE IN 3-DIMENSIONAL AFFINE SPACE). Let r be a line in the affine space $\mathbb{A}^3_K$. The underlying vector subspace of r has dimension 1 and is generated by a non-zero vector $v = (a_1, a_2, a_3) \in K^3$. If $P = (p_1, p_2, p_3)$ is a point of r and if we denote by $X = (x_1, x_2, x_3)$ a generic point of the affine space, we have

$$X \in r \iff X = P + \lambda v, \ \ \lambda \in K.$$

The vector equation $X = P + \lambda v$ translates into the following system:

$$\begin{cases} x_1 = p_1 + \lambda a_1 \\ x_2 = p_2 + \lambda a_2 \\ x_3 = p_3 + \lambda a_3. \end{cases}$$

These are the parametric equations of a line in the 3-dimensional affine space.

By eliminating λ from the previous equations we obtain a system of two linear equations in the indeterminates x_1, x_2 and x_3, of the form

$$\begin{cases} ax_1 + bx_2 + cx_3 + d = 0 \\ ex_1 + fx_2 + gx_3 + h = 0. \end{cases}$$

It follows that a line in the 3-dimensional affine space can be identified with the set of solutions of a system of two linear equations in the indeterminates x_1, x_2, x_3. Of course, it is not true that every linear system of two equations in three indeterminates determines a line: for instance, the set of solutions of such a system can be the empty set.

We observe that this result has an obvious geometric interpretation. In fact the set of solutions of a linear equation in the unknowns x_1, x_2, x_3 is a plane in the 3-dimensional affine space. Therefore the set of solutions of a system of two linear equations corresponds to the set of points common to two planes. If these planes are parallel and distinct, their intersection is the empty set; in this case the corresponding system of linear equations is incompatible. If, on the other hand, the two planes are not parallel, their intersection is a straight line.

Remark 6.4.5. From the previous considerations we can deduce that if the (non-empty) affine subspace $\mathbb{B} = (\mathscr{B}, W, +)$ of the affine space $\mathbb{A}$ is given by the solutions of the system of linear equations

$$S : AX = B,$$

then the underlying vector subspace W of $\mathbb{B}$ is given by the solutions of the associated homogeneous system

$$S_0 : AX = \mathbf{0}.$$

Moreover, if $\mathbb{B}$ is not empty, its dimension is given by

$$\dim \mathbb{B} = n - \operatorname{rk}(A),$$

where n is the dimension of the affine space $\mathbb{A}$. In particular, $\mathbb{B}$ is a hyperplane of $\mathbb{A}$ if and only if $\operatorname{rk}(A) = 1$, i.e., if and only if $\mathbb{B}$ can be described by one (non-trivial) equation.

Definition 6.4.6. Let $\mathbb{A}$ be a n-dimensional affine space and $\mathbb{L}$ a linear subspace of dimension r. The set of hyperplanes of $\mathbb{A}$ that contain $\mathbb{L}$ is called the *star of hyperplanes with center* $\mathbb{L}$ (if $r = n - 2$ it is called a *pencil of hyperplanes*). More generally, given m such that $r \leq m \leq n$, the *star of m-dimensional linear subspaces with center* $\mathbb{L}$ is the set of all m-dimensional affine subspaces of $\mathbb{A}$ that contain $\mathbb{L}$.

Definition 6.4.7. Let W be a vector subspace of dimension r of the underlying vector space of a n-dimensional affine space $\mathbb{A}$. The set of hyperplanes of $\mathbb{A}$ that contain W in their underlying vector space is called the *improper star of hyperplanes parallel to* W (if $r = n - 2$ it is called the *improper pencil of hyperplanes*). More generally, given m such that $r \leq m \leq n$, the *improper star of m-dimensional linear subspaces parallel to* W is the set of all m-dimensional affine subspaces of $\mathbb{A}$ that contain W in their underlying vector space.

Example 6.4.8. In the affine space $\mathbb{A}^n_K$ let $\mathbb{L}$ be the linear subspace given by the solutions of the system:

$$\begin{cases} a_{11}x_1 + a_{12}x_2 + \cdots + a_{1n}x_n + b_1 = 0 \\ a_{21}x_1 + a_{22}x_2 + \cdots + a_{2n}x_n + b_2 = 0 \\ \quad \cdot \quad \cdot \quad \cdot \quad \cdot \quad \cdot \quad \cdot \\ a_{m1}x_1 + a_{m2}x_2 + \cdots + a_{mn}x_n + b_m = 0 \end{cases}$$

Then the cartesian equations of the hyperplanes that form the star with center $\mathbb{L}$ are given by

$$\begin{aligned} &\lambda_1(a_{11}x_1 + a_{12}x_2 + \cdots + a_{1n}x_n + b_1) \\ &\qquad + \lambda_2(a_{21}x_1 + a_{22}x_2 + \cdots + a_{2n}x_n + b_2) \\ &\qquad + \cdots + \lambda_m(a_{m1}x_1 + a_{m2}x_2 + \cdots + a_{mn}x_n + b_m) = 0, \end{aligned}$$

for every $\lambda_1, \lambda_2, \ldots, \lambda_m \in K$ (not all zero).

6.5 Euclidean affine spaces

In this section we will study the affine spaces whose underlying vector space is a Euclidean space, i.e., a real vector space with a scalar product. As we will see, the presence of such a structure in the vector space V underlying an affine space $\mathbb{A} = (\mathscr{A}, V, +)$ will allow us to introduce in $\mathbb{A}$ notions of a metric nature such as distances, angles, areas, volumes.

Definition 6.5.1. A *Euclidean affine space* is an affine space $\mathbb{A}$ over the field of real numbers $\mathbb{R}$ such that the underlying vector space V is endowed with a Euclidean structure, i.e., a positive definite symmetric bilinear form $g\colon V \times V \to \mathbb{R}$. For simplicity, given two vectors $v, w \in V$, we will write $v \cdot w$ instead of $g(v, w)$.

Example 6.5.2. The fundamental example of a Euclidean affine space is provided by the affine space $\mathbb{A}^n_{\mathbb{R}}$, where the underlying vector space $\mathbb{R}^n$ is equipped with the usual scalar product. This space will be simply called the Euclidean affine space $\mathbb{A}^n_{\mathbb{R}}$.

Definition 6.5.3. In a Euclidean affine space $\mathbb{A}$ (of dimension n), a reference frame $\mathscr{R} = \{O, v_1, v_2, \dots, v_n\}$ is called *orthogonal* (resp., *orthonormal*) if the vectors $v_1, v_2, \dots, v_n$ are an orthogonal (resp., orthonormal) basis of the Euclidean vector space V. An orthonormal reference frame is also called *cartesian*.

Remark 6.5.4. Note that the existence of orthonormal reference frames in a Euclidean affine space is guaranteed by the existence of orthonormal bases in Euclidean vector spaces (see Chapter 5, Section 5.4.3). Moreover, the choice of an orthonormal reference frame $\mathscr{R}$ in a Euclidean affine space $\mathbb{A}$ of dimension n, determines an isomorphism of affine spaces

$$\Phi_{\mathscr{R}}\colon \mathbb{A} \xrightarrow{\sim} \mathbb{A}^n_{\mathbb{R}}$$

By means of this isomorphism the bilinear form g defined on the vector space V underlying the affine space $\mathbb{A}$ is identified with the usual scalar product of $\mathbb{R}^n$.

From Definition 6.5.1 it follows that all the results proved in Chapter 5 for Euclidean vector spaces can be restated for Euclidean affine spaces.

The presence of a scalar product in the vector space V underlying a Euclidean affine space $\mathbb{A}$, allows us to define the distance between two points of $\mathbb{A}$:

Definition 6.5.5. Let $\mathbb{A}$ be a Euclidean affine space and P, Q be two points of $\mathbb{A}$. The *distance* $d(P, Q)$ between P and Q is the norm of the vector $Q - P$:

$$d(P, Q) = \|Q - P\| = \sqrt{(Q - P) \cdot (Q - P)}.$$

From the properties of the norm of a vector studied in Chapter 5, it follows that, in a Euclidean affine space $\mathbb{A} = (\mathscr{A}, V, +)$, the distance function

$$d\colon \mathscr{A} \times \mathscr{A} \to \mathbb{R}$$

has the following properties:

(i) $d(P, Q) \geq 0$, for every $P, Q \in \mathscr{A}$, and $d(P, Q) = 0$ if and only if $P = Q$;

(ii) $d(P, Q) = d(Q, P)$, for every $P, Q \in \mathscr{A}$;

(iii) $d(P, R) \leq d(P, Q) + d(Q, R)$, for every $P, Q, R \in \mathscr{A}$.

The property (iii) is called the *triangular inequality*: it states that, in a triangle, the length of each side is smaller than or equal to the sum of the lengths of the other two sides (see Chapter 5, Propositions 5.1.7 and 5.3.23).

If $\mathscr{S}$ and $\mathscr{T}$ are two non-empty subsets (of the set of points) of a Euclidean affine space $\mathbb{A}$, it is natural to define the distance of $\mathscr{S}$ from $\mathscr{T}$ as the lower bound of the distances of the points of $\mathscr{S}$ from the points of $\mathscr{T}$:

$$d(\mathscr{S}, \mathscr{T}) = \inf\{d(P, Q) \mid P \in \mathscr{S}, Q \in \mathscr{T}\}.$$

From this definition it follows that

$$\mathscr{S} \cap \mathscr{T} \neq \varnothing \quad \Longrightarrow \quad d(\mathscr{S}, \mathscr{T}) = 0.$$

The following remark shows that the opposite implication is not true.

Remark 6.5.6. If $\mathscr{S}$ and $\mathscr{T}$ are two subsets of a Euclidean affine space $\mathbb{A}$, in general there do not exist points $P \in \mathscr{S}$ and $Q \in \mathscr{T}$ such that $d(P, Q) = d(\mathscr{S}, \mathscr{T})$. For instance, let us consider the real affine line $\mathbb{A}^1_{\mathbb{R}}$ and let us set

$$\mathscr{S} = \{x \in \mathbb{R} \mid x \leq 0\}, \qquad \mathscr{T} = \{x \in \mathbb{R} \mid x > 0\}.$$

Then we have $d(\mathscr{S}, \mathscr{T}) = 0$, but $d(P, Q) > 0$ for every $P \in \mathscr{S}$ and every $Q \in \mathscr{T}$.

In the special case where the subsets $\mathscr{S}$ and $\mathscr{T}$ are (the sets of points of) two affine subsets $\mathbb{L}$ and $\mathbb{M}$ of the Euclidean affine space $\mathbb{A}$, the distance function

$$d \colon \mathbb{L} \times \mathbb{M} \to \mathbb{R}, \qquad (P, Q) \mapsto d(P, Q)$$

has minimums, i.e., there exist points $P_0 \in \mathbb{L}$ and $Q_0 \in \mathbb{M}$ such that

$$d(\mathbb{L}, \mathbb{M}) = d(P_0, Q_0).$$

Before proving this result, we state and prove the following lemma:

Lemma 6.5.7. *Let $\mathbb{L} = (\mathscr{L}, L, +)$ and $\mathbb{M} = (\mathscr{M}, M, +)$ be two affine subspaces of a Euclidean affine space $\mathbb{A}$ and let $P_0 \in \mathbb{L}$ and $Q_0 \in \mathbb{M}$ be two points such that the vector $u = Q_0 - P_0$ is orthogonal to $\mathbb{L}$ and $\mathbb{M}$. Then, for every $P \in \mathbb{L}$ and every $Q \in \mathbb{M}$, we have $d(P_0, Q_0) \leq d(P, Q)$. Moreover, if $P_1 \in \mathbb{L}$ and $Q_1 \in \mathbb{M}$ are two points such that $d(P_1, Q_1) = d(P_0, Q_0)$, we necessarily have $Q_1 - P_1 = Q_0 - P_0$.*

Proof. For every $P \in \mathbb{L}$ and every $Q \in \mathbb{M}$, we have $Q - P = (Q - Q_0) + (Q_0 - P_0) + (P_0 - P)$, hence $Q - P = u + v + w$, where $u = Q_0 - P_0$, $v = P_0 - P \in L$ and $w = Q - Q_0 \in M$.

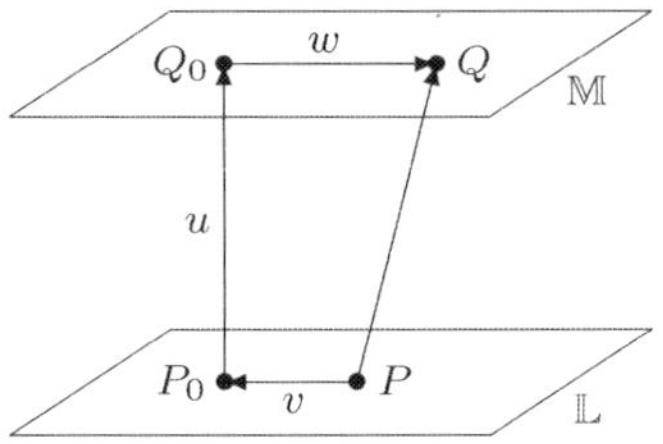

Since u is orthogonal to $\mathbb{L}$ and $\mathbb{M}$, we have $u \cdot v = u \cdot w = 0$. It follows that

$$\begin{aligned}\|Q - P\|^2 &= (u + v + w) \cdot (u + v + w) \\ &= \|u\|^2 + \|v\|^2 + \|w\|^2 + 2v \cdot w \\ &= \|u\|^2 + \|v + w\|^2 \\ &\geq \|u\|^2,\end{aligned}$$

hence $d(P, Q) \geq d(P_0, Q_0)$. By specializing the preceding reasoning to the case in which $P = P_1$ and $Q = Q_1$, we have

$$\|Q_1 - P_1\|^2 = \|u\|^2 + \|v + w\|^2 = \|Q_0 - P_0\|^2 + \|v + w\|^2.$$

Since, by hypothesis, we have $d(P_1, Q_1) = d(P_0, Q_0)$, we must have $\|Q_1 - P_1\|^2 = \|Q_0 - P_0\|^2$, hence $\|v + w\|^2 = 0$. It follows that $v + w = \mathbf{0}$, hence $w = -v$. This shows that $Q_1 - P_1 = u + v + w = u + v - v = u = Q_0 - P_0$, which is what we wanted to prove. □

Proposition 6.5.8. *Let $\mathbb{L} = (\mathscr{L}, L, +)$ and $\mathbb{M} = (\mathscr{M}, M, +)$ be two (non empty) affine subspaces of a Euclidean affine space $\mathbb{A}$. Then there exist points $P_0 \in \mathbb{L}$ and $Q_0 \in \mathbb{M}$ such that*

$$d(P_0, Q_0) \leq d(P, Q),$$

for every $P \in \mathbb{L}$ and every $Q \in \mathbb{M}$.

Proof. If $\mathbb{L}$ and $\mathbb{M}$ intersect, it suffices to take $P_0 = Q_0 \in \mathbb{L} \cap \mathbb{M}$. We can then assume that the two affine subspaces $\mathbb{L}$ and $\mathbb{M}$ have no common point.

Let us consider any two points $P_1 \in \mathbb{L}$, $Q_1 \in \mathbb{M}$ and let us denote by u_1 the vector $Q_1 - P_1$. Note that u_1 does not belong to the sum of the two subspaces L and M. In fact, if we had $u_1 = Q_1 - P_1 \in L + M$, then $Q_1 - P_1 = v + w$, for some $v \in L$, $w \in M$. From this it would follow that the point $R = P_1 + v = Q_1 - w$ lies in both $\mathbb{L}$ and $\mathbb{M}$, which is false because we have assumed that $\mathbb{L}$ and $\mathbb{M}$ do not intersect.

Let us denote by u_1' the orthogonal projection of u_1 on the subspace $L + M$. We have

$$u_1 = u_1' + u_1'',$$

with $u_1' \in L + M$ and $u_1'' \in (L + M)^\perp = L^\perp \cap M^\perp$ (see the Proposition 5.4.10 in Chapter 5). Note that from what we have seen it follows that $u_1'' \neq \mathbf{0}$. Since $u_1' \in L + M$ we can write $u_1' = v + w$, for some $v \in L$ and $w \in M$ (note that this decomposition is not unique, in general; it is unique only if $L \cap M = \{\mathbf{0}\}$, i.e., when the two affine subspaces $\mathbb{L}$ and $\mathbb{M}$ are skew). Now we set $P_0 = P_1 + v$ and $Q_0 = Q_1 - w$; we obviously have $P_0 \in \mathbb{L}$ and $Q_0 \in \mathbb{M}$. We claim that P_0 and Q_0 are the points we are looking for. In fact we have

$$\begin{aligned} Q_0 - P_0 &= (Q_1 - w) - (P_1 + v) \\ &= (Q_1 - P_1) - (v + w) \\ &= u_1 - u_1' \\ &= u_1'' \in L^\perp \cap M^\perp, \end{aligned}$$

hence the vector $u = u_1'' = Q_0 - P_0$ is orthogonal to both affine subspaces $\mathbb{L}$ and $\mathbb{M}$. Now the conclusion follows by recalling the Lemma 6.5.7. □

The following result states that the distance between two non-intersecting affine subspaces $\mathbb{L}$ and $\mathbb{M}$ of a Euclidean affine space $\mathbb{A}$ is measured along a direction perpendicular to both subspaces. According to Lemma 6.5.7, such a direction turns out to be unique.

Proposition 6.5.9. *Let $\mathbb{L} = (\mathscr{L}, L, +)$ and $\mathbb{M} = (\mathscr{M}, M, +)$ be two (non-empty) affine subspaces of a Euclidean affine space $\mathbb{A}$ and let $P_0 \in \mathscr{L}$ and $Q_0 \in \mathscr{M}$ be two points such that $d(P_0, Q_0) = d(\mathbb{L}, \mathbb{M})$. Then, if we denote by u the vector $Q_0 - P_0$, we have $u \in L^\perp \cap M^\perp$, i.e., for every vector $v \in L$ and every $w \in M$, we have $u \cdot v = u \cdot w = 0$.*

Proof. If $\mathbb{L}$ and $\mathbb{M}$ intersect we have $P_0 = Q_0$, hence $u = Q_0 - P_0 = \mathbf{0}$. In this case the claim holds trivially.

Let us assume that $\mathbb{L} \cap \mathbb{M} = \varnothing$. Let $P_0 \in \mathscr{L}$ and $Q_0 \in \mathscr{M}$ be such that $d(P_0, Q_0) = d(\mathbb{L}, \mathbb{M})$: we have $d(P_0, Q_0) \le d(A, B)$, for every $A \in \mathscr{L}$ and every $B \in \mathscr{M}$. Let us set $u = Q_0 - P_0$ and suppose that $u \notin M^{\perp}$. This means that there exists a vector $w \in M$ such that $u \cdot w \neq 0$. This situation is represented in the following figure:

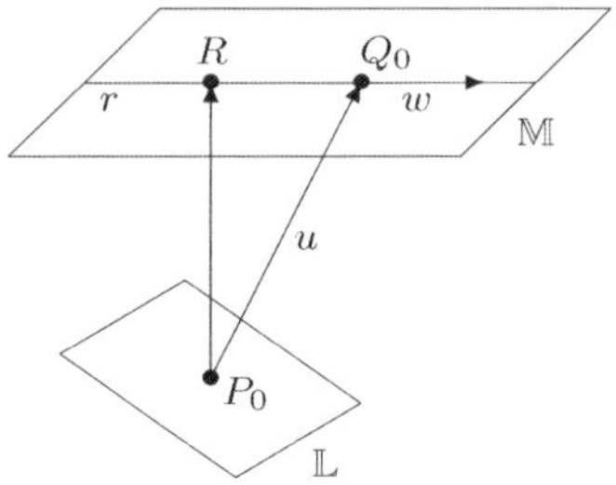

Let us denote by r the line passing through Q_0 and parallel to the vector w: this line is contained in $\mathbb{M}$, because $Q_0 \in \mathscr{M}$ and $w \in M$. Let R be the point of the line r such that the vector $\overrightarrow{P_0R}$ is orthogonal to the vector w; R is given by

$$R = Q_0 - \left(\frac{u \cdot w}{w \cdot w}\right) w.$$

The square of the distance between P_0 and R is:

$$\begin{aligned} d(P_0, R)^2 &= \|R - P_0\|^2 \\ &= \left\| u - \left(\frac{u \cdot w}{w \cdot w}\right) w \right\|^2 \\ &= \|u\|^2 - \left(\frac{u \cdot w}{\|w\|}\right)^2. \end{aligned}$$

Since we have assumed that $u \cdot w \neq 0$, we have

$$d(P_0, R) < \|u\| = d(P_0, Q_0),$$

but this contradicts the assumption that P_0 and Q_0 are the points of minimum distance of $\mathbb{L}$ and $\mathbb{M}$. This contradiction follows from the assumption that the vector u is not orthogonal to the subspace M; hence we must have $u \in M^{\perp}$. By exchanging the roles of $\mathbb{L}$ and $\mathbb{M}$ we can prove, in a similar way, that we also have $u \in L^{\perp}$. □

Remark 6.5.10. If the affine subspace $\mathbb{L}$ is reduced to a point P, the previous results allow us to conclude that there is a single point $Q \in \mathbb{M}$ such that

$d(P, Q) = d(P, \mathbb{M})$. In this case, moreover, the vector $\overrightarrow{PQ}$ turns out to be orthogonal to $\mathbb{M}$.

We can therefore conclude that, for every point P and every linear subspace $\mathbb{M}$ of a Euclidean affine space $\mathbb{A}$, there exists a single point $Q \in \mathbb{M}$ such that the vector $\overrightarrow{PQ}$ is orthogonal to $\mathbb{M}$; such point Q is called the *orthogonal projection* of P on $\mathbb{M}$. This defines a function

$$\pi_{\mathbb{M}} \colon \mathbb{A} \to \mathbb{M}$$

that associates to every $P \in \mathbb{A}$ its orthogonal projection $\pi_{\mathbb{M}}(P) = Q$ on the affine subspace $\mathbb{M}$. Note that, for every point $P \in \mathbb{M}$, we have $\pi_{\mathbb{M}}(P) = P$. It follows that $\pi_{\mathbb{M}}\big(\pi_{\mathbb{M}}(P)\big) = \pi_{\mathbb{M}}(P)$, for every $P \in \mathbb{A}$. This property is usually expressed with the following notation: $\pi_{\mathbb{M}}^2 = \pi_{\mathbb{M}} \circ \pi_{\mathbb{M}} = \pi_{\mathbb{M}}$.

The above results suggest a method for determining the distance between two linear subspaces $\mathbb{L}$ and $\mathbb{M}$, of dimensions r and s, respectively, of a Euclidean affine space $\mathbb{A}$. Consider a generic point $P \in \mathbb{L}$, whose coordinates will therefore depend on r parameters, and a generic point $Q \in \mathbb{M}$, whose coordinates will depend on s parameters. Compute the vector $u = Q - P$, whose components will therefore depend on $r + s$ parameters, and require that u be orthogonal to the subspaces $\mathbb{L}$ and $\mathbb{M}$. The orthogonality condition to $\mathbb{L}$ is expressed by imposing that the scalar product of u with the r vectors of a basis of the underlying vector subspace of $\mathbb{L}$ is zero (thus we obtain r linear equations). Similarly, the orthogonality condition to $\mathbb{M}$ is expressed by imposing that the scalar product of u with the s vectors of a basis of the underlying vector subspace of $\mathbb{M}$ is zero (hence we obtain s linear equations). We thus obtain a system of $r + s$ linear equations in $r + s$ unknowns, the solution of which allows us to determine two points $P_0 \in \mathbb{L}$ and $Q_0 \in \mathbb{M}$ such that the vector $Q_0 - P_0$ is orthogonal to $\mathbb{L}$ and $\mathbb{M}$. As seen above, the distance between P_0 and Q_0 coincides with the distance between $\mathbb{L}$ and $\mathbb{M}$.

Let us now illustrate the above with some examples.

Example 6.5.11 (Distance of a point from a line). Let $\mathbb{A}$ be a Euclidean affine space and let r be the line passing through a point A and parallel to a vector v. Given a point $P \in \mathbb{A}$, we want to determine the distance h of P from the line r.

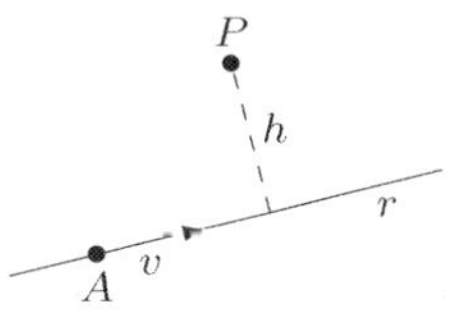

A generic point X of the line r is given by $X = A + \lambda v$, for $\lambda \in \mathbb{R}$. The vector $u = \overrightarrow{PX}$ is given by $u = X - P = (A - P) + \lambda v$. By requiring this vector to be orthogonal to the line r (i.e., to the vector v), we have

$$0 = u \cdot v = (A - P) \cdot v + \lambda\, v \cdot v,$$

from which it follows that

$$\lambda = -\frac{(A - P) \cdot v}{\|v\|^2}.$$

If we denote by Q the point of r corresponding to this value of λ, we have

$$Q = A - \frac{(A - P) \cdot v}{\|v\|^2}\, v.$$

Q is the point of r such that the vector $\overrightarrow{PQ}$ is orthogonal to the line r: hence Q is the point of intersection of r and the line orthogonal to r passing through P. It follows that we have

$$d(P, r) = d(P, Q) = \|Q - P\| = \left\|(A - P) - \frac{(A - P) \cdot v}{\|v\|^2}\, v\right\|.$$

From this formula we get

$$\begin{aligned} d(P, r)^2 &= \left((A - P) - \frac{(A - P) \cdot v}{\|v\|^2}\, v\right) \cdot \left((A - P) - \frac{(A - P) \cdot v}{\|v\|^2}\, v\right) \\ &= \|A - P\|^2 - 2\,\frac{\big((A - P) \cdot v\big)^2}{\|v\|^2} + \frac{\big((A - P) \cdot v\big)^2}{\|v\|^2} \\ &= \|A - P\|^2 - \frac{\big((A - P) \cdot v\big)^2}{\|v\|^2}, \end{aligned}$$

hence we obtain the following formula for computing the distance of a point from a given line:

$$d(P, r) = \sqrt{\|A - P\|^2 - \frac{\big((A - P) \cdot v\big)^2}{\|v\|^2}}.$$

The problem of determining the distance of a point from a straight line can also be approached in another way, as we will now explain.

Using the previous notations, we set $B = A + v$, $C = P + v$ and consider the parallelogram $ABCP$, having as base the segment AB and as height the distance h of the point P from the line r.

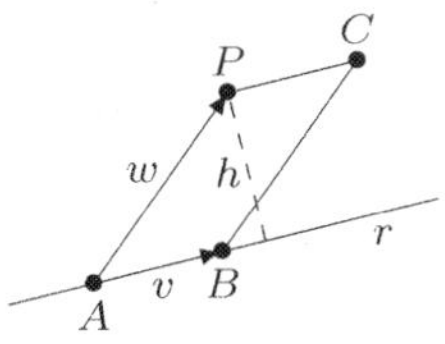

Since the area of a parallelogram is the product of its base and height, we have

$$h = \frac{\text{Area}(ABCP)}{AB}.$$

If we denote by w the vector $\overrightarrow{AP}$ and remember that $\overrightarrow{AB} = v$, the area of the parallelogram $ABCP$ is given by the following formula (see Chap. 5, Section 5.2)

$$\text{Area}(ABCP) = \sqrt{\det \begin{pmatrix} v \cdot v & v \cdot w \\ w \cdot v & w \cdot w \end{pmatrix}}.$$

It follows that:

$$h = \frac{\sqrt{\det \begin{pmatrix} v \cdot v & v \cdot w \\ w \cdot v & w \cdot w \end{pmatrix}}}{\|v\|} = \sqrt{\frac{1}{v \cdot v} \det \begin{pmatrix} v \cdot v & v \cdot w \\ w \cdot v & w \cdot w \end{pmatrix}}.$$

In the particular case where $\mathbb{A}$ is the Euclidean affine space $\mathbb{A}^3_{\mathbb{R}}$, the area of a parallelogram with sides v and w coincides with the norm of the vector product $v \times w$:

$$\text{Area}(ABCP) = \|v \times w\|.$$

In this case, the distance h of P from the line r can therefore be computed as follows:

$$h = \frac{\|v \times w\|}{\|v\|}.$$

In the next example we will obtain a useful formula for determining the distance of a point from a hyperplane in the Euclidean affine space $\mathbb{A}^n_{\mathbb{R}}$. But first we need the following result:

Proposition 6.5.12. *In the Euclidean affine space $\mathbb{A}^n_{\mathbb{R}}$ let us consider a hyperplane π of equation*

$$\pi : a_1x_1 + a_2x_2 + \cdots + a_nx_n + b = 0$$

and let us denote by u the vector whose components are the coefficients of the unknowns $x_1, x_2, \ldots, x_n$ in the equation of π,

$$u = (a_1, a_2, \ldots, a_n).$$

Then the vector u is orthogonal to the hyperplane π.

Proof. Let $P = (p_1, p_2, \ldots, p_n)$ be a point of π. We have

$$a_1p_1 + a_2p_2 + \cdots + a_np_n + b = 0.$$

For every vector $v = (\alpha_1, \alpha_2, \ldots, \alpha_n)$ belonging to the underlying vector subspace of π, the point $Q = P + v = (p_1 + \alpha_1, p_2 + \alpha_2, \ldots, p_n + \alpha_n)$ belongs to the hyperplane π, hence we have

$$a_1(p_1 + \alpha_1) + a_2(p_2 + \alpha_2) + \cdots + a_n(p_n + \alpha_n) + b = 0.$$

By subtracting the two previous equalities, we obtain

$$a_1\alpha_1 + a_2\alpha_2 + \cdots + a_n\alpha_n = 0,$$

i.e., $u \cdot v = 0$. Since this holds for every vector v belonging to the underlying vector subspace of π, we conclude that the vector u is orthogonal to the hyperplane π. □

Example 6.5.13 (DISTANCE OF A POINT FROM A HYPERPLANE). In the Euclidean affine space $\mathbb{A} = \mathbb{A}^n_{\mathbb{R}}$ let us denote by π the hyperplane of equation

$$\pi : a_1x_1 + a_2x_2 + \cdots + a_nx_n + b = 0$$

and let $P = (p_1, p_2, \ldots, p_n)$ be a point of $\mathbb{A}$. We want to determine the distance of P from the hyperplane π. By what we have previously seen, this distance is equal to the distance of P from the point $Q \in \pi$ such that the vector $\overrightarrow{PQ}$ is orthogonal to π.

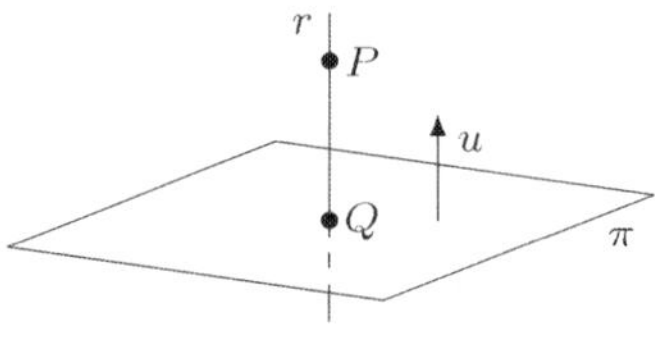

As we have seen in Proposition 6.5.12, the vector $u = (a_1, a_2, \dots, a_n)$ is orthogonal to the hyperplane π, hence the line r passing through the point P and orthogonal to π has equation

$$r : X = P + \lambda u.$$

It follows that the point Q we are looking for is the point of intersection of the line r and the hyperplane π:

$$Q = r \cap \pi.$$

Using coordinates, the parametric equations of the line r are

$$\begin{cases} x_1 = p_1 + \lambda a_1 \\ x_2 = p_2 + \lambda a_2 \\ \quad \cdot \quad \cdot \quad \cdot \\ x_n = p_n + \lambda a_n. \end{cases}$$

By inserting these expressions in the equation of π we get the equation

$$a_1(p_1 + \lambda a_1) + a_2(p_2 + \lambda a_2) + \cdots + a_n(p_n + \lambda a_n) + b = 0,$$

whose solution is

$$\bar{\lambda} = -\frac{a_1p_1 + a_2p_2 + \cdots + a_np_n + b}{a_1^2 + a_2^2 + \cdots + a_n^2}.$$

The point of r that corresponds to this value of λ is the point Q we are looking for:

$$Q = P + \bar{\lambda} u.$$

Hence we have

$$d(P, \pi) = d(P, Q) = \|Q - P\| = \|\bar{\lambda} u\| = |\bar{\lambda}|\|u\|.$$

From this formula we get

$$d(P, \pi) = \frac{|a_1p_1 + a_2p_2 + \cdots + a_np_n + b|}{\sqrt{a_1^2 + a_2^2 + \cdots + a_n^2}}.$$

By generalizing the previous examples, we now show how the distance of a point from any linear subspace of an affine Euclidean space can be determined.

Example 6.5.14 (DISTANCE OF A POINT FROM A LINEAR SUBSPACE). Let $\mathbb{A}$ be a Euclidean affine space and $\mathbb{L} = (\mathscr{L}, L, +)$ a linear subspace of dimension r of $\mathbb{A}$. Given a point $P \in \mathbb{A}$ we want to determine the distance of P from $\mathbb{L}$; this distance is equal to the distance of P from the point Q of $\mathbb{L}$ such that the vector $\overrightarrow{PQ}$ is orthogonal to $\mathbb{L}$.

Let A be a point of $\mathbb{L}$ and let us denote by $\{v_1, v_2, \dots, v_r\}$ an orthonormal basis[3] of L. A generic point X of $\mathbb{L}$ is given by

$$X = A + \lambda_1 v_1 + \cdots + \lambda_r v_r,$$

where $\lambda_1, \dots, \lambda_r \in \mathbb{R}$. If we denote by u the vector $X - P$ and by w the vector $A - P$, we have

$$u = w + \sum_{j=1}^{r} \lambda_j v_j.$$

The requirement that this vector be orthogonal to $\mathbb{L}$ is expressed by imposing that the dot product of u and the vectors of the basis of L is zero. Recalling that the vectors $v_1, \dots, v_r$ are an orthonormal basis of L, we have, for every $i = 1, \dots, r$,

$$0 = u \cdot v_i = w \cdot v_i + \sum_{j=1}^{r} \lambda_j v_j \cdot v_i = w \cdot v_i + \lambda_i,$$

hence

$$\lambda_i = -w \cdot v_i.$$

The point Q we are looking for is given by

$$Q = A - (w \cdot v_1)v_1 - \cdots - (w \cdot v_r)v_r.$$

Hence we have

$$d(P, \mathbb{L}) = d(P, Q) = \|Q - P\| = \|w - (w \cdot v_1)v_1 - \cdots - (w \cdot v_r)v_r\|.$$

From this we deduce that

$$d(P, \mathbb{L})^2 = \|w\|^2 - (w \cdot v_1)^2 - \cdots - (w \cdot v_r)^2,$$

and finally we obtain the following formula for the distance of P from $\mathbb{L}$:

$$d(P, \mathbb{L}) = \sqrt{\|w\|^2 - (w \cdot v_1)^2 - \cdots - (w \cdot v_r)^2}.$$

[3]The hypothesis that the basis $\{v_1, v_2, \dots, v_r\}$ is orthonormal is not indispensable, but it allows to simplify the computations.

Example 6.5.15 (DISTANCE BETWEEN TWO LINES). Let us consider two lines r and s in a Euclidean affine space $\mathbb{A}$. To determine the distance of r from s we try to find two points $P \in r$ and $Q \in s$ such that the vector $\overrightarrow{PQ}$ is orthogonal to both lines r and s.

Let v be a vector of r and let us choose a point $A \in r$: a generic point X of r is given by

$$X = A + \lambda v.$$

Similarly, if we choose a vector w of s and a point $B \in s$, a generic point Y of the line s is given by:

$$Y = B + \mu w.$$

Let u be the vector $\overrightarrow{XY}$. We have

$$u = (B - A) - \lambda v + \mu w.$$

The requirement that this vector be orthogonal to the lines r and s is expressed by imposing that the scalar products of u with v and w are zero:

$$\begin{cases} u \cdot v = (B - A) \cdot v - \lambda v \cdot v + \mu w \cdot v = 0 \\ u \cdot w = (B - A) \cdot w - \lambda v \cdot w + \mu w \cdot w = 0. \end{cases}$$

If the vectors v and w are linearly independent (i.e., if the lines r and s are not parallel) the determinant of the matrix of the coefficients of this system is different from zero, therefore the system admits one solution $\lambda = \bar{\lambda}$ and $\mu = \bar{\mu}$. Substituting these values into the expressions of X and Y yields the points P and Q we are looking for. If, on the other hand, the lines r and s are parallel, the determinant of the matrix of the coefficients of the previous system is zero. In this case the system has infinite solutions (that depends on one parameter). Any one of these solutions yields a pair of points P and Q such that $d(r, s) = d(P, Q)$.

6.5.1 Angles

Let's talk now about the concept of angle in a Euclidean affine space. As already observed in Chapter 5, the presence of a positive definite symmetric bilinear form on a real vector space V allows us to define the notion of an angle between two non-zero vectors of V. The same considerations therefore apply to the case of an affine Euclidean space. Using the notion of angle between two vectors it is then possible to define the angle between two intersecting lines.

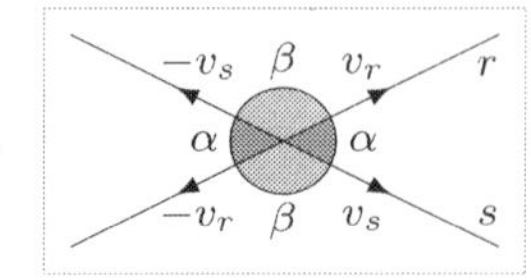

Let r and s be two intersecting lines in a Euclidean affine space $\mathbb{A}$ and let us denote by v_r and v_s, respectively, two generators of the underlying vector spaces of r and s.

Note that the vectors v_r and v_s are determined up to multiplication by a non-zero scalar. Recall now that the cosine of the angle α between v_r and v_s is given by the ratio

$$\frac{v_r \cdot v_s}{||v_r|| ||v_s||} \tag{6.5.1}$$

By replacing v_r and v_s with the vectors λv_r and μv_s, where $\lambda, \mu \in \mathbb{R}$, $\lambda, \mu \neq 0$, the previous ratio becomes

$$\frac{\lambda v_r \cdot \mu v_s}{|\lambda| ||v_r|| |\mu| ||v_s||} = \pm \frac{v_r \cdot v_s}{||v_r|| ||v_s||}$$

where the sign depends on the sign of the product $\lambda\mu$. In other words, this means that, given two lines r and s, the ratio (6.5.1) is well defined only up to the sign. This corresponds to the fact that, if we denote by α the angle between the vectors v_r and v_s, then the angle between v_r and $-v_s$, or between $-v_r$ and v_s, is $\beta = \pi - \alpha$, as shown in the previous figure.

Having said that, we can give the following definition:

Definition 6.5.16. With the previous notation, we will call *angle between the intersecting lines r and s* one of the angles α and $\beta = \pi - \alpha$, and precisely the one that lies in the interval $[0, \pi/2]$.

This definition can be generalized to the case of the angle between a straight line r and a linear subspace $\mathbb{L}$, of dimension ≥ 1, that intersects the line r.

Let $\mathbb{L}$ be a linear subspace of dimension ≥ 1 of a Euclidean affine space $\mathbb{A}$ and let us denote by $\pi_{\mathbb{L}}\colon \mathbb{A} \to \mathbb{L}$ the orthogonal projection on $\mathbb{L}$. Given a line r that intersects $\mathbb{L}$, we denote by $r' = \pi_{\mathbb{L}}(r)$ its orthogonal projection[4] on $\mathbb{L}$. If the line r is orthogonal to $\mathbb{L}$, then r' is a point, which is precisely the intersection point of r and $\mathbb{L}$; otherwise r' is a line.

[4] To determine the orthogonal projection of the line r on the linear subspace $\mathbb{L}$ it is sufficient to consider two points $P, Q \in r$ and determine their orthogonal projections $P' = \pi_{\mathbb{L}}(P)$ and $Q' = \pi_{\mathbb{L}}(Q)$ on $\mathbb{L}$. If $P' \neq Q'$, the line passing through P' and Q' is the line r' we are looking for, on the other hand, if $P' = Q'$ the line r is orthogonal to $\mathbb{L}$ hence its orthogonal projection on $\mathbb{L}$ is just a point.

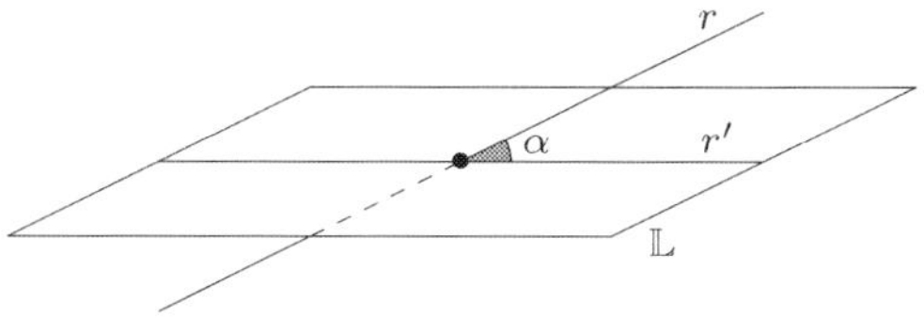

Definition 6.5.17. With the previous notation we will call *angle between the line* r *and the linear subspace* $\mathbb{L}$ the angle α between r and its orthogonal projection r' on $\mathbb{L}$, with the convention that, if r' is a point, α is a right angle.

If the linear subspace $\mathbb{L}$ is a hyperplane of $\mathbb{A}$, the orthogonal subspace $L^{\perp}$ of the underlying vector space of $\mathbb{L}$ has dimension 1. If we denote by P the point of intersection of the line r and the hyperplane $\mathbb{L}$, the line $s = P + L^{\perp}$ is the line orthogonal to $\mathbb{L}$ passing through the point P. If r is not orthogonal to $\mathbb{L}$, we denote by r' the orthogonal projection of r on the hyperplane $\mathbb{L}$. The three lines s, r and r' lie on the same plane and, moreover, s and r' are orthogonal. The situation is illustrated in the following figure:

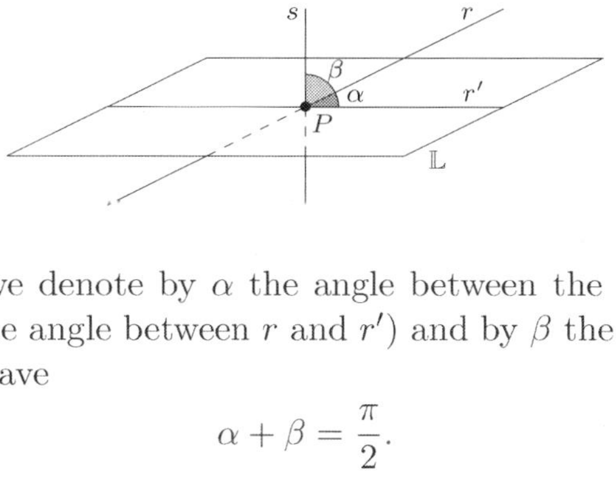

It follows that if we denote by α the angle between the line r and the hyperplane $\mathbb{L}$ (i.e., the angle between r and r') and by β the angle between the lines r and s, we have

$$\alpha + \beta = \frac{\pi}{2}.$$

This remark provides an alternative method to compute the angle α between the line r and the hyperplane $\mathbb{L}$: it is enough to determine the line s perpendicular to $\mathbb{L}$, compute the angle β between r and s and remember that $\alpha + \beta$ is a right angle.

Finally, we observe that the previous considerations allow us to define the notion of angle between two intersecting hyperplanes or, more generally, between a hyperplane and a linear subspace of dimension ≥ 1 that intersect it.

Definition 6.5.18. Let $\mathbb{L}$ and $\mathbb{M}$ be two intersecting hyperplanes in a Euclidean affine space $\mathbb{A}$. Let $P \in \mathbb{L} \cap \mathbb{M}$ and let us denote by r and s, respectively, the lines passing through the point P and orthogonal to the hyperplanes $\mathbb{L}$ and $\mathbb{M}$. We define the angle α between the hyperplanes $\mathbb{L}$ and $\mathbb{M}$, to be the angle between the lines r and s (if $\dim \mathbb{A} = 3$ the situation is illustrated in the following figure).

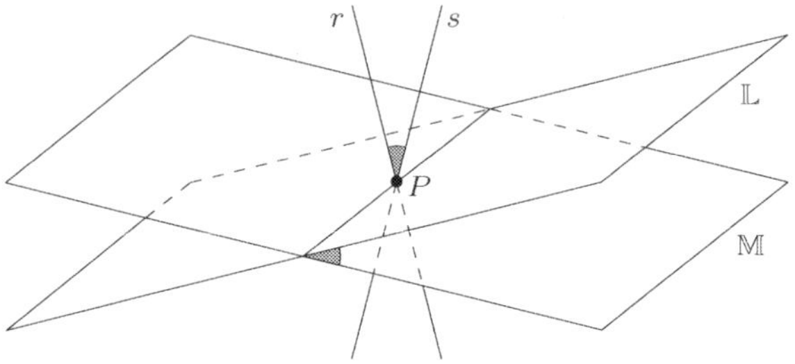

Definition 6.5.19. Let $\mathbb{L}$ be a hyperplane in a Euclidean affine space $\mathbb{A}$ and let $\mathbb{M}$ be a linear subspace of $\mathbb{A}$ of dimension ≥ 1 that intersects $\mathbb{L}$. Let $P \in \mathbb{L} \cap \mathbb{M}$ and let us denote by r the line orthogonal to $\mathbb{L}$ passing through the point P. The angle α between $\mathbb{L}$ and $\mathbb{M}$ is defined by setting $\alpha = \pi/2 - \beta$, where β is the angle between the line r and the linear subspace $\mathbb{M}$.

Exercises

Exercise 6.1. In the affine space $\mathbb{A}^3_{\mathbb{R}}$ let r be the line given by the equations

$$\begin{cases} 2x - 3y + 1 = 0 \\ 2x - 2y + z - 1 = 0. \end{cases}$$

Determine the line s passing through the origin and parallel to r, and the plane of the pencil with axis r passing through the origin.

Exercise 6.2. Let r_1 and r_2, respectively, be the lines passing through $P_1 = (0, 1, 2)$ and $P_2 = (2, -1, 0)$ and parallel to the vectors $v_1 = (2, 1, 1)$ and $v_2 = (-2, 3, 0)$. Let $\mathscr{F}_1$ and $\mathscr{F}_2$, respectively, be the pencils of planes with axis r_1 and r_2. Determine the planes π_1 of $\mathscr{F}_1$ and π_2 of $\mathscr{F}_2$ passing through the point $P = (1, 1, 1)$, and the intersections of the line $r = \pi_1 \cap \pi_2$ with the coordinate planes. Determine the triangle formed by the intersections of the lines r, r_1 and r_2 with the plane of equation $x = 0$.

Exercise 6.3. Determine the equation of the plane containing the lines r and s of equations

$$r : \begin{cases} y+1=0 \\ 2x+2z-3=0 \end{cases} \qquad s : \begin{cases} 2x-3y+2z-1=0 \\ x-2y+z=0. \end{cases}$$

Exercise 6.4. Let r be the line passing through the points $P=(1,0,-2)$ and $Q=(0,-1,3)$, and let s be the line passing through the point $R=(-1,1,0)$ and parallel to the vector $v=(1,-1,-1/2)$. Determine the plane π_1 passing through the points P, Q and R, the plane π_2 containing the line s and passing through the midpoint of the segment PQ, and finally a vector u parallel to the line $\pi_1 \cap \pi_2$.

Exercise 6.5. Determine the distance of the point $P=(2,0,-1)$ from the line r given by the parametric equations

$$\begin{cases} x=t+1 \\ y=2 \\ z=t-1. \end{cases}$$

Exercise 6.6. Determine the lines passing through the point $Q=(0,1,-1)$, having distance 1 from the point $P=(1,1,0)$ and contained in the plane π of equation $y+z=0$.

Exercise 6.7. Determine the equation of the locus of points of the Euclidean affine space $\mathbb{A}^3_{\mathbb{R}}$ that belong to the lines passing through the point $P=(2,1,1)$ that form an angle equal to $\frac{\pi}{6}$ with the plane π of equation $2x+y-z=0$.

Exercise 6.8. Determine the equation of the locus of points of the affine space $\mathbb{A}^3_{\mathbb{R}}$ that belong to the lines passing through the point $P=(3,3,3)$ that intersect the curve C of equations

$$\begin{cases} x^2+2y^2+3z^2=4 \\ x+z=0 \end{cases}$$

Exercise 6.9. Determine the lines of the plane π of equation $x-y+2z=0$ parallel to the vector $v=(1,3,1)$ having distance $\sqrt{6}$ from the point $P=(1,0,0)$.

Exercise 6.10. In the real 3-dimensional Euclidean space, determine the equation of the locus of points that belong to the lines having distance 1 from the point $P=(1,1,2)$ and that form an angle equal to $\frac{\pi}{3}$ with the planes $\pi_1 : x+y=0$ and $\pi_2 : x-z=0$.

Exercise 6.11. In the Euclidean affine space $\mathbb{A}^3_{\mathbb{R}}$ let π be the plane that contains the line r given by the equations $\frac{1}{2}(x-1)=y+1=\frac{1}{3}z$, and the point $P=(0,1,1)$. Determine the lines of the plane π passing through P and having distance $\sqrt{3}$ from the point $Q=(1,1,-1)$. Determine also the orthogonal projection Q_0 of Q on π and the lines of π passing through Q_0 and orthogonal to the lines previously found.

Exercise 6.12. In the Euclidean affine space $\mathbb{A}^3_{\mathbb{R}}$ determine the equations of the lines passing through $P = (1, 0, 0)$, intersecting the line r of equations

$$r : \begin{cases} y + z - 1 = 0 \\ x = 0 \end{cases}$$

and forming with r an angle equal to $\frac{\pi}{6}$.

Exercise 6.13. In the Euclidean affine plane $\mathbb{A}^2_{\mathbb{R}}$ let r_1 and r_2 be the lines of equations $x - 2y + 1 = 0$ and $3x - y + 2 = 0$, respectively. Determine the equations of the lines s_1 and s_2 that bisect the angles formed by r_1 and r_2.

A

Basic Notions of Algebra

In this chapter we will present, in a very concise way, some definitions and results concerning equivalence relations and sets endowed with operations. In particular, we will introduce the notions of *group*, *ring*, *field*, and we will describe the construction of the field of complex numbers.

A.1 Sets and relations

Sometimes some kind of *relation* can exist between the elements of a set. If we take, for example, all the books present in a library, there will be different books written by the same author, or different books published by the same publishing house, etc.; in a given set of people examples of relations are "being brothers," or "residing in the same city."

To describe a relation defined in a set A we can think of explicitly listing all the pairs (a, b) of elements of A such that a is in relation to b (if we denote by $\mathscr{R}$ the relation, to say that a is in relation to b we will write $a \mathscr{R} b$). We note that the order of the elements in a pair (a, b) does matter: for example, in the set of integers the relation $\mathscr{R} =$ "be less than" holds for the pair $(3, 5)$ (i.e., 3 is less than 5) but does not hold for the pair $(5, 3)$.

Since the relation $\mathscr{R}$ is described by the set of all pairs (a, b) of elements of A such that $a \mathscr{R} b$, we can identify the relation $\mathscr{R}$ with this set, hence we can say that a relation $\mathscr{R}$ in a set A *is* a set of pairs of elements of A, i.e., is a subset of the cartesian product $A \times A = \{(a, b) \mid a, b \in A\}$.

Definition A.1.1. A *relation* (or, more precisely, a *binary relation*) $\mathscr{R}$ in a set A is a subset $\mathscr{R} \subset A \times A$. The elements $a, b \in A$ are in relation, i.e., $a \mathscr{R} b$, if the pair (a, b) belongs to the subset $\mathscr{R}$.

Example A.1.2. Let $A = \{1, 2, 3\}$. The relation $\mathscr{R} =$ "be less then" is identified with the following subset of $A \times A$: $\mathscr{R} = \{(1, 2), (1, 3), (2, 3)\}$.

A.1.1 Equivalence relations

Definition A.1.3. A relation $\mathscr{R}$ in a set A is an *equivalence relation* if for every $a, b, c \in A$, we have:

- $a \mathscr{R} a$ (reflexive property)
- if $a \mathscr{R} b$ then $b \mathscr{R} a$ (symmetric property)
- if $a \mathscr{R} b$ and $b \mathscr{R} c$ then $a \mathscr{R} c$ (transitive property)

Example A.1.4. The *equality relation* (where $a \mathscr{R} b$ means $a = b$) is reflexive, symmetric and transitive, hence it is an equivalence relation.

Example A.1.5. We now describe a less trivial example of an equivalence relation. In the set $\mathbb{Z}$ of integers we define a relation $\mathscr{R}$, called *congruence modulo n*, where n is a fixed integer, as follows: for $a, b \in \mathbb{Z}$ we say that $a \mathscr{R} b$ (i.e., that a is congruent to b modulo n) if $a = b + nk$, for some integer k. Equivalently, we can say that a is congruent to b modulo n if $a - b$ is an integer multiple of n. It is easy to verify that this is an equivalence relation. Usually, to denote that a is congruent to b modulo n we write $a \equiv b \mod n$.

Definition A.1.6. Let $\mathscr{R}$ be an equivalence relation in a set A. The *equivalence class* of an element $a \in A$, denoted by $[a]$, is the set of all elements of A that are in relation to a:

$$[a] = \{\, b \in A \mid a \mathscr{R} b \,\}.$$

Any element b that belongs to the equivalence class $[a]$ is called a *representative* of this equivalence class (obviously a is a representative of $[a]$).

We note that two representatives of the same equivalence class are in relation to each other. Conversely, if two elements a and b are in relation to each other then they represent the same equivalence class. It follows that two different equivalence classes cannot have elements in common.

Definition A.1.7. Let $\mathscr{R}$ be an equivalence relation in a set A. The set of all equivalence classes is called the *quotient set* of A modulo the equivalence relation $\mathscr{R}$, and it is denoted by $A/\mathscr{R}$:

$$A/\mathscr{R} = \{\, [a] \mid a \in A \,\}.$$

The quotient set $A/\mathscr{R}$ is therefore a set whose elements are in turn sets which contain the elements of A which are in relation to each other.

Example A.1.8. In the set $\mathbb{Z}$ of integers we consider the relation of congruence modulo 4. The equivalence class of the number 0 is: $[0] = \{0 + 4k \mid k \in \mathbb{Z}\} = \{\dots, -8, -4, 0, 4, 8, \dots\}$. Similarly we have: $[1] = \{1 + 4k \mid k \in \mathbb{Z}\} = \{\dots, -7, -3, 1, 5, 9, \dots\}$, $[2] = \{2 + 4k \mid k \in \mathbb{Z}\} = \{\dots, -6, -2, 2, 6, 10, \dots\}$, $[3] = \{3 + 4k \mid k \in \mathbb{Z}\} = \{\dots, -5, -1, 3, 7, 11, \dots\}$. Since $4 \equiv 0 \mod 4$, the equivalence class of the number 4 is the same of the number 0. Similarly the equivalence class of 5 is equal to the equivalence class of 1, the equivalence class of 6 coincides with that of 2, etc.

Therefore the quotient set, denoted by $\mathbb{Z}/4\mathbb{Z}$, i.e., the set of all equivalence classes, contains only four elements:

$$\mathbb{Z}/4\mathbb{Z} = \{[0], [1], [2], [3]\}.$$

A.2 Sets with operations

Given two sets A and B we denote by $A \times B$ the set of ordered pairs (a, b), with $a \in A$ and $b \in B$; the set $A \times B$ is the *cartesian product* of A and B.

Let S be a non-empty set. An *internal composition law* (or simply an *operation*) on S is a function

$$*\colon S \times S \to S$$

that associates to any pair (a, b) of elements of S an element $c \in S$, that will be denoted by $c = a * b$.

Such an operation is:

- *associative* if, for every $a, b, c \in S$, we have

$$(a * b) * c = a * (b * c);$$

- *commutative* if, for every $a, b \in S$, we have

$$a * b = b * a.$$

An element $e_1 \in S$ such that

$$e_1 * a = a, \qquad \forall a \in S,$$

is called a *left neutral element*. Similarly an element $e_2 \in S$ such that

$$a * e_2 = a, \qquad \forall a \in S,$$

is called a *right neutral element.* If the operation $*$ is commutative every left neutral element is also right neutral, and vice versa.

In general, if the operation is not commutative, it is possible that there exists a left neutral element but no right neutral element (or vice versa). However, if there is both a left neutral element and a right neutral element, they must necessarily be equal. In fact, if e_1 is a left neutral element and e_2 is a right neutral element, we have $e_1 * a = a$, for every $a \in S$; in particular if we take $a = e_2$ we get $e_1 * e_2 = e_2$. On the other hand, being e_2 a right neutral element, we have $a * e_2 = a$, for every $a \in S$, hence if we take $a = e_1$ we obtain $e_1 * e_2 = e_1$. From this it follows that $e_1 = e_2$.

Example A.2.1. In the set $\mathbb{N} = \{0, 1, 2, 3, \dots\}$ of natural numbers, we consider the addition operation:

$$+ \colon \mathbb{N} \times \mathbb{N} \to \mathbb{N}, \quad (a, b) \mapsto a + b.$$

This operation is associative and commutative:

$$(a + b) + c = a + (b + c), \quad a + b = b + a, \quad \forall\, a, b, c \in \mathbb{N}.$$

The neutral element for the addition is 0, in fact we have

$$a + 0 = 0 + a = a, \quad \forall\, a \in \mathbb{N}.$$

Example A.2.2. In the set $\mathbb{N}^* = \mathbb{N} \setminus \{0\} = \{1, 2, 3, \dots\}$ of positive natural numbers, we consider the multiplication operation:

$$\cdot \colon \mathbb{N} \times \mathbb{N} \to \mathbb{N}, \quad (a, b) \mapsto ab.$$

This operation is associative and commutative:

$$(ab)c = a(bc), \quad ab = ba, \quad \forall\, a, b, c \in \mathbb{N}^*.$$

The neutral element for the multiplication is 1, in fact we have

$$a \cdot 1 = 1 \cdot a = a, \quad \forall\, a \in \mathbb{N}^*.$$

Definition A.2.3. A non-empty set S with an operation $*$ is called an *algebraic structure.* The algebraic structure $(S, *)$ is a *semigroup* if the operation is associative. If there is also a neutral element e, $(S, *)$ is called a *monoid.* If the operation is commutative, the semigroup (or monoid) is *commutative* or *abelian.*

Let $(S, *)$ be a monoid and denote by e the neutral element. An element $a \in S$ is *invertible* if there exists $b \in S$ such that

$$a * b = b * a = e.$$

As we will now see, an element b with the above property, if it exists, is necessarily unique; it is called the *inverse* of a and it is denoted by a^{-1}.

Suppose that, given $a \in S$, there are two elements $b_1, b_2 \in S$ such that $a * b_1 = b_1 * a = e$ and $a * b_2 = b_2 * a = e$. Using the associative property we can write

$$b_1 = b_1 * e = b_1 * (a * b_2) = (b_1 * a) * b_2 = e * b_2 = b_2,$$

which proves that $b_1 = b_2$.

Remark A.2.4. When the operation is denoted with the symbol $+$ (additive notation) the neutral element is denoted with the symbol 0. In this case the inverse of an element $a \in S$ is denoted by $-a$ and called the *opposite* of a.

We now give the following definition:

Definition A.2.5. A *group* is a monoid in which every element is invertible. In other words, a group is a non-empty set endowed with an associative operation, which admits a neutral element and in which each element has an inverse. If the operation is commutative the group is called *commutative* or *abelian*.

Example A.2.6. The set $\mathbb{Z} = \{\dots, -2, -1, 0, 1, 2, \dots\}$ of integers, with the addition operation, is an abelian group.

The set $\mathbb{Q}^* = \mathbb{Q} \setminus \{0\}$ of nonzero rational numbers is an abelian group for the multiplication.

We now consider sets with two operations.

Definition A.2.7. A *ring* is a non-empy set S endowed with two operations, that we denote by $\oplus$ and $\odot$, such that:

(*i*) S with the operation $\oplus$ is an abelian group (we denote by 0 the corresponding neutral element);

(*ii*) S with the operation $\odot$ is a semigroup;

(*iii*) the following *distributive properties* hold:

$$(a \oplus b) \odot c = (a \odot c) \oplus (b \odot c)$$
$$c \odot (a \oplus b) = (c \odot a) \oplus (c \odot b),$$

for every $a, b, c \in S$.

If the operation $\odot$ is commutative, the ring is *commutative*; if there exists a neutral element for the operation $\odot$ the ring is *unitary*.

Definition A.2.8. A *field* is a unitary commutative ring[1] in which every element different from 0 is invertible.

The sets $\mathbb{Z}$ of integers and $\mathbb{Q}$ of rational numbers are two unitary commutative rings (with the usual operations of addition and multiplication). Note that $\mathbb{Q}$ is a field while $\mathbb{Z}$ is only a ring.

Another example of field is the set $\mathbb{R}$ of *real numbers*. We will not describe here the construction of the real numbers starting from the set of rational numbers; we only recall that every real number $r \in \mathbb{R}$ admits a decimal representation of the form

$$r = (-)a_n a_{n-1} \dots a_1 a_0 . b_1 b_2 \dots b_i \dots$$

where the digits to the left of the decimal point are finite, while those to the right of the decimal point can also be infinite.[2]

Remark A.2.9. The sets $\mathbb{Q}$ and $\mathbb{R}$ are examples of infinite fields. Note that there are also fields with a finite number of elements. The most trivial example is given by the set $\{0, 1\}$ with the addition and multiplication defined as follows:

$$0 + 0 = 0, \quad 0 + 1 = 1 + 0 = 1, \quad 1 + 1 = 0,$$
$$0 \cdot 0 = 0 \cdot 1 = 1 \cdot 0 = 0, \quad 1 \cdot 1 = 1.$$

A.3 The field of complex numbers

In the field of real numbers every number ≥ 0 is the square of a real number, i.e., every non-negative real number admits a *square root*. However, this is not the case for negative numbers: for example, the equation $x^2 = -1$ has no solutions. To overcome this drawback it is necessary to appropriately enlarge the set of real numbers by "adding" new elements.

We therefore introduce a new symbol, which we will denote by the letter i, with the property that $i^2 = -1$; we will also write $i = \sqrt{-1}$. Obviously i is

[1] The unity is denoted by 1 and we assume that $1 \neq 0$. Therefore a field has at least two elements.

[2] The only restriction that is placed on the decimal representation of a real number is that no index j must exist such that $b_i = 9$ for every $i \geq j$. This is done to ensure the uniqueness of the decimal representation.

not a real number: we will call it the *imaginary unit.* Then we denote by $\mathbb{C}$ the set of all expressions of the form $a + ib$, where a and b are real numbers:

$$\mathbb{C} = \{a + ib \,|\, a, b \in \mathbb{R}\}.$$

The elements of the set $\mathbb{C}$ are called *complex numbers.* Given a complex number $z = a + ib$, the two real numbers a and b are called the *real part* and the *imaginary part* of z: $a = \Re(z)$, $b = \Im(z)$.

Obviously every real number $a \in \mathbb{R}$ can also be thought as a complex number, in fact we have $a = a + i0$, therefore the set $\mathbb{R}$ is a subset of $\mathbb{C}$. Now we want to extend the addition and multiplication operations defined in $\mathbb{R}$ to the set of complex numbers.

As for the addition, given two complex numbers $z_1 = a+ib$ and $z_2 = c+id$, we set

$$z_1 + z_2 = (a + c) + i(b + d).$$

It is easy to verify that this operation is associative and commutative, that $0 = 0+i0$ is the neutral element and that the opposite of the complex number $a + ib$ is $-a - ib$.

To define the product of two complex numbers we proceed in a purely formal way, using the distributive property and remembering that, by definition, $i^2 = -1$.

Given $z_1 = a + ib$ and $z_2 = c + id$, we have

$$\begin{aligned} z_1 z_2 &= (a + ib)(c + id) \\ &= ac + iad + ibc + i^2 bd \\ &= ac + iad + ibc - bd \\ &= (ac - bd) + i(ad + bc). \end{aligned}$$

It is easy to verify that this multiplication is associative and commutative and that the neutral element is $1 = 1 + i0$. We now prove that every complex number $z \neq 0$ has a multiplicative inverse.

Let $z = a + ib$ be a nonzero complex number. Since $z \neq 0$, the real numbers a and b are not both zero, hence $a^2 + b^2 \neq 0$. Then we can consider the complex number

$$w = \frac{a}{a^2 + b^2} - i\frac{b}{a^2 + b^2}.$$

Multiplying these two numbers we obtain

$$zw = (a + ib)\left(\frac{a}{a^2 + b^2} - i\frac{b}{a^2 + b^2}\right) = \frac{a^2}{a^2 + b^2} + \frac{b^2}{a^2 + b^2} = 1,$$

and this means that w is the inverse of z. Therefore we have

$$z^{-1} = \frac{a}{a^2 + b^2} - i\frac{b}{a^2 + b^2}$$

From what we have seen so far we can deduce that the set $\mathbb{C}$ of complex numbers, equipped with the operations of addition and multiplication, is a field.

Remark A.3.1. A complex number $z = a + ib$ can also be thought of as an ordered pair of real numbers (a, b). One could therefore define the set of complex numbers as the set of pairs of real numbers

$$\mathbb{C} = \{(a, b) \mid a, b \in \mathbb{R}\}.$$

The operations of addition and multiplication are then defined as follows:

$$(a, b) + (c, d) = (a + c, b + d)$$
$$(a, b) \cdot (c, d) = (ac - bd, ad + bc).$$

It is easy to verify that $\mathbb{C}$ with these two operations is a field. The neutral element for the addition is the pair $(0, 0)$, the opposite of the pair (a, b) is $(-a, -b)$, the neutral element for the multiplication is the pair $(1, 0)$ and the inverse of the pair $(a, b) \neq (0, 0)$ is the pair $\big(a/(a^2 + b^2), -b/(a^2 + b^2)\big)$. In this identification the imaginary unit i corresponds to the pair $(0, 1)$.

Remark A.3.2. We note that, unlike the field $\mathbb{R}$ of real numbers, the field of complex numbers is not an *ordered field*, i.e., given two complex numbers $z_1 \neq z_2$, in general they are not comparable; in other words, it makes no sense to ask which of the two is bigger and which is smaller. Therefore expressions like $z_1 \geq z_2$, or $z_1 < z_2$, with $z_1, z_2 \in \mathbb{C}$, make no sense.

To understand the reason for this let's consider, as an example, $z_1 = i$ e $z_2 = 0$. If it were possible to compare the two complex numbers i and 0, since $i \neq 0$, it should necessarily be $i > 0$ or $i < 0$. Recalling that the product of two positive numbers is a positive number, in the first case we would have $i^2 > 0$, i.e., $-1 > 0$, which is false. Then there remains only the possibility that $i < 0$; but in that case we would have $-i > 0$ hence $(-i)(-i) = i^2 = -1 > 0$, which is false. This reasoning shows that it can be neither $i > 0$ nor $i < 0$ and therefore the two numbers i and 0 are not comparable.

For a complex number $z = a + ib$ we define its *conjugate* $\bar{z}$ by setting $\bar{z} = a - ib$. The function that to each $z \in \mathbb{C}$ associates its conjugate $\bar{z}$ is called the *conjugation*:

$$\mathbb{C} \to \mathbb{C}, \quad z \mapsto \bar{z}.$$

In terms of the conjugation, the real and imaginary parts of a complex number are expressed as follows:

$$a = \Re(z) = \frac{z + \bar{z}}{2}, \qquad b = \Im(z) = \frac{z - \bar{z}}{2i}.$$

The conjugation satisfies the following properties, which can be verified with simple computations:

- $\overline{(\bar{z})} = z$, for every $z \in \mathbb{C}$;
- $\bar{z} = z$ if and only if $z \in \mathbb{R}$;
- $\overline{z_1 + z_2} = \bar{z}_1 + \bar{z}_2$, for every $z_1, z_2 \in \mathbb{C}$;
- $\overline{z_1 z_2} = \bar{z}_1 \cdot \bar{z}_2$, for every $z_1, z_2 \in \mathbb{C}$.

A consequence of these properties is the following: if

$$f(z) = a_0 + a_1 z + a_2 z^2 + \cdots + a_n z^n$$

is a polynomial with real coefficients $a_0, a_1, \ldots, a_n \in \mathbb{R}$, we have

$$\begin{aligned}
\overline{f(z)} &= \overline{a_0 + a_1 z + a_2 z^2 + \cdots + a_n z^n} \\
&= \bar{a}_0 + \bar{a}_1 \bar{z} + \bar{a}_2 \bar{z}^2 + \cdots + \bar{a}_n \bar{z}^n \\
&= a_0 + a_1 \bar{z} + a_2 \bar{z}^2 + \cdots + a_n \bar{z}^n \\
&= f(\bar{z}).
\end{aligned}$$

A corollary of this fact is the following: if f is a polynomial with real coefficients and if $z \in \mathbb{C}$ is such that $f(z) = 0$, then also $f(\bar{z}) = 0$. We note that it can certainly happen that $z = \bar{z}$, but in this case z is a real number.

A very important property of the field of complex numbers is expressed by the following theorem, known as the *Fundamental Theorem of Algebra*:

Theorem A.3.3. *Every polynomial of degree n with complex coefficients*

$$f(z) = a_n z^n + a_{n-1} z^{n-1} + \cdots + a_1 z + a_0, \qquad a_j \in \mathbb{C},$$

factors as a product of (not necessarily distinct) linear terms

$$f(z) = a_n (z - z_1)(z - z_2) \cdots (z - z_n),$$

with $z_1, z_2, \ldots, z_n \in \mathbb{C}$.

This means that every equation of degree n, of the type $f(z) = 0$, in the field of complex numbers always has n solutions, $z = z_1$, $z = z_2$, $\ldots$, $z = z_n$ (not necessarily distinct).

A.3.1 Representation of complex numbers in the plane

The interpretation of complex numbers as pairs of real numbers allows them to be represented as points on the Cartesian plane: we associate the point with coordinates (a, b) to the complex number $z = a + ib$.

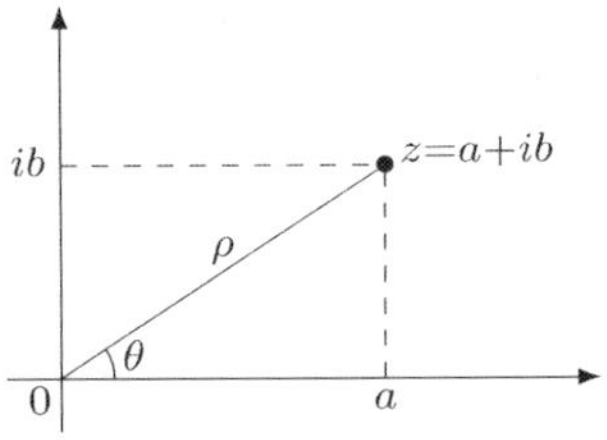

We denote by ρ the distance of this point from the origin of the axes and by θ the angle between the positive x-axis and the ray outgoing from the origin and passing through the point (a, b). The real numbers ρ and θ are the *modulus* and the *argument* of the complex number $z = a + ib$:

$$\rho = |z|, \qquad \theta = \arg(z).$$

Note that the argument of z is determined only up to multiples of 2π, furthermore, the argument of 0 is indeterminate. If $z = a + ib$ we have:

$$|z| = \rho = \sqrt{a^2 + b^2}, \qquad \tan(\theta) = b/a.$$

We observe that if z is a real number, the modulus of z coincides with its absolute value. We now list some properties of the modulus of a complex number, the verification of which is left to the reader:

(*i*) $|z| \geq 0$ for every $z \in \mathbb{C}$, and $|z| = 0$ if and only if $z = 0$;

(*ii*) $|\bar{z}| = |z|$, for every $z \in \mathbb{C}$;

(*iii*) $|z| = \sqrt{z\bar{z}}$;

(*iv*) $|z_1 z_2| = |z_1| \cdot |z_2|$, for every $z_1, z_2 \in \mathbb{C}$;

(*v*) $|z_1 + z_2| \leq |z_1| + |z_2|$, for every $z_1, z_2 \in \mathbb{C}$;

(*vi*) $z^{-1} = \frac{\bar{z}}{|z|^2}$, for every $z \neq 0$.

A.3.2 Trigonometric representation

Let $z = a + ib$ be a complex number and let $\rho = |z|$ and $\theta = \arg(z)$. From trigonometry we know that $a = \rho\cos\theta$ and $b = \rho\sin\theta$, hence we can write

$$z = \rho(\cos\theta + i\sin\theta).$$

This expression is called the *trigonometric representation* of the complex number z.

Using the trigonometric representation the formula for the product of two complex numbers becomes particularly simple.

Given two complex numbers

$$z_1 = \rho_1(\cos\theta_1 + i\sin\theta_1) \qquad \text{and} \qquad z_2 = \rho_2(\cos\theta_2 + i\sin\theta_2),$$

we have:

$$\begin{aligned} z_1 z_2 &= \rho_1\rho_2(\cos\theta_1 + i\sin\theta_1)(\cos\theta_2 + i\sin\theta_2) \\ &= \rho_1\rho_2\big((\cos\theta_1\cos\theta_2 - \sin\theta_1\sin\theta_2) + i(\cos\theta_1\sin\theta_2 + \sin\theta_1\cos\theta_2)\big) \\ &= \rho_1\rho_2\big(\cos(\theta_1+\theta_2) + i\sin(\theta_1+\theta_2)\big). \end{aligned}$$

Therefore, in the multiplication of two complex numbers, the respective moduli are multiplied together, while the arguments are added. For this reason, the trigonometric representation is particularly suitable for computing the powers of a complex number. In fact, if $z = \rho(\cos\theta + i\sin\theta)$, then for any integer n we have

$$z^n = \rho^n\big(\cos(n\theta) + i\sin(n\theta)\big).$$

Using this expression we can easily determine the so-called *n-th roots of unity*, i.e., the complex numbers z that are solutions of the equation $z^n = 1$. In fact, in the trigonometric representation the equation $z^n = 1$ can be written in the form

$$\rho^n\big(\cos(n\theta) + i\sin(n\theta)\big) = 1,$$

from which it follows that it must be $\rho = 1$ and $n\theta = 2k\pi$, hence $\theta = 2k\pi/n$, for $k = 0, 1, 2, \ldots$. Obviously the values $k = 0$ and $k = n$ determine the same complex number, therefore the equation $z^n = 1$ has exactly n distinct solutions, $z_0, z_1, \ldots, z_{n-1}$, corresponding to the values of k from 0 to $n-1$. The solutions are then given by

$$z_k = \cos\left(\frac{2k\pi}{n}\right) + i\sin\left(\frac{2k\pi}{n}\right), \qquad \text{for } k = 0, 1, 2, \ldots, n-1.$$

The n solutions $z_0, \ldots, z_{n-1}$ are the vertices of a regular polygon of n sides, inscribed in the circle of radius 1 centered at the origin, with $z_0 = 1$.

As an example, the third roots and fourth roots of unity are shown in the following figures.

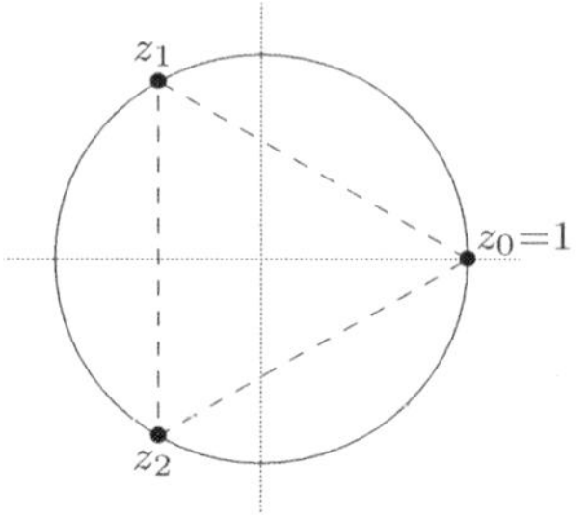

Third roots of 1

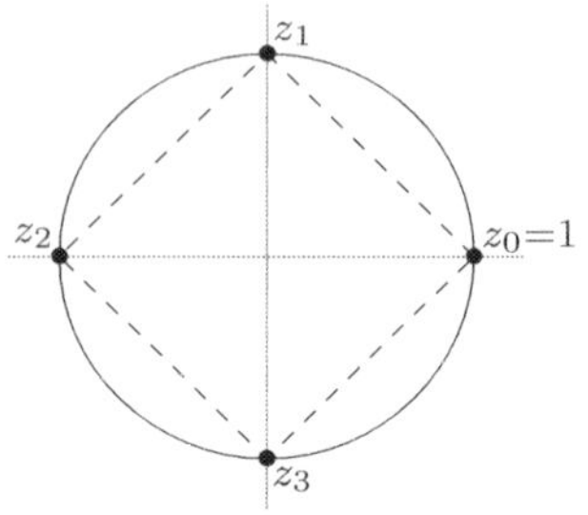

Fourth roots of 1

A.3.3 The complex exponential

In this section we will see how the exponential of a complex number can be defined. To this end we recall that the exponential function

$$\mathbb{R} \to \mathbb{R}, \quad x \mapsto e^x$$

admits a Taylor series expansion, given by

$$e^x = 1 + x + \frac{x^2}{2!} + \frac{x^3}{3!} + \cdots + \frac{x^n}{n!} + \cdots$$

This series can therefore be taken as the *definition* of the exponential of x, for every $x \in \mathbb{R}$.

Now we observe that only sums and products (powers) appear in this series, which make perfect sense even if, instead of a real number x, we consider a complex number z. It can also be verified that, for each $z \in \mathbb{C}$, the series

$$\sum_{n=0}^{\infty} \frac{z^n}{n!} = 1 + z + \frac{z^2}{2!} + \frac{z^3}{3!} + \cdots + \frac{z^n}{n!} + \cdots$$

converges.[3] We can therefore define the exponential of a complex number z by setting

$$e^z = 1 + z + \frac{z^2}{2!} + \frac{z^3}{3!} + \cdots + \frac{z^n}{n!} + \cdots$$

We are now able to prove an important identity, due to Euler.

[3] One can prove that the series $\sum_{n=0}^{\infty} \frac{z^n}{n!}$ converges absolutely and uniformly on any compact subset of $\mathbb{C}$.

Let x be a real number and consider the exponential of ix:

$$\begin{aligned} e^{ix} &= 1 + ix + \frac{(ix)^2}{2!} + \frac{(ix)^3}{3!} + \frac{(ix)^4}{4!} + \frac{(ix)^5}{5!} + \frac{(ix)^6}{6!} + \frac{(ix)^7}{7!} + \cdots \\ &= 1 + ix - \frac{x^2}{2!} - i\frac{x^3}{3!} + \frac{x^4}{4!} + i\frac{x^5}{5!} - \frac{x^6}{6!} - i\frac{x^7}{7!} + \cdots \\ &= \left(1 - \frac{x^2}{2!} + \frac{x^4}{4!} - \frac{x^6}{6!} + \cdots\right) + i\left(x - \frac{x^3}{3!} + \frac{x^5}{5!} - \frac{x^7}{7!} + \cdots\right) \end{aligned}$$

Recalling the Taylor series expansions of the sine and cosine

$$\begin{aligned} \sin x &= x - \frac{x^3}{3!} + \frac{x^5}{5!} - \frac{x^7}{7!} + \cdots \\ \cos x &= 1 - \frac{x^2}{2!} + \frac{x^4}{4!} - \frac{x^6}{6!} + \cdots \end{aligned}$$

we obtain the following identity:

$$e^{ix} = \cos x + i \sin x. \tag{A.3.1}$$

For example, if $x = 2\pi$, we have

$$e^{2\pi i} = 1.$$

Using the previous identity (and the properties of the exponential), it is possible to compute the exponential of any complex number $z = a + ib$, as follows:

$$e^z = e^{a+ib} = e^a \cdot e^{ib} = e^a(\cos b + i \sin b).$$

Remark A.3.4. Using the identity (A.3.1), the trigonometric expression of a complex number $z = \rho(\cos\theta + i \sin\theta)$ can be rewritten in the form

$$z = \rho e^{i\theta},$$

where $\rho = |z|$ and $\theta = \arg(z)$.

Exercises

Exercise A.1. Write the following complex numbers in the form $a + ib$:

$$(1-2i)^2, \qquad \frac{3-i}{1+2i}, \qquad \frac{1}{i}, \qquad \frac{1}{1-i}, \qquad \frac{3-2i}{2+5i},$$

$$i^4 - i^{13}, \qquad 1 - i + i^2 - i^3, \qquad \frac{(1-i)(1+i^{-7})}{2}.$$

Exercise A.2. Compute the modulus of the following complex numbers:

$$1-i, \qquad 3+4i, \qquad \frac{1-i}{1+i}, \qquad 1-i-i^2,$$
$$i^7-i^{10}, \qquad 3(2-i)+2(1+5i), \qquad \frac{3-2i+4i^2}{3-i}.$$

Exercise A.3. Compute the modulus and the argument of the following complex numbers:

$$3i, \qquad -4i, \qquad 2, \qquad -5, \qquad 3-i\sqrt{3}, \qquad \frac{1-i}{\sqrt{2}},$$
$$(1+i)^3, \qquad (1-i)^2, \qquad \frac{1}{1+i}, \qquad \frac{1}{(1+i)^2}.$$

Exercise A.4. Write the following complex numbers in trigonometric form:

$$2+2i, \qquad 3-3i, \qquad 1+i\sqrt{3}, \qquad -2\sqrt{3}+2i, \qquad 1+i^3,$$
$$(1+i)(1-i\sqrt{3}), \qquad (1-i)^2, \qquad \frac{-1+i}{1+i\sqrt{3}}, \qquad \frac{2-2i}{(1+i)^3}.$$

Exercise A.5. Solve the following equations in the unknowns $x, y \in \mathbb{R}$:

$$x+iy = x-iy, \qquad x+iy = |x+iy|, \qquad |2x+iy| = |x-2iy|,$$
$$(x+iy)^2 = (x-iy)^2, \qquad \frac{x+iy}{x-iy} = x-iy, \qquad \sum_{n=0}^{100} i^n = x+iy.$$

Exercise A.6. Give a graphical representation in the complex plane of the set of all $z \in \mathbb{C}$ that satisfy the following conditions:

$$|z|<1, \qquad z+\bar{z}=1, \qquad z-\bar{z}=i, \qquad |z+1|=|z-1|,$$
$$|z+i|=|z-i|, \qquad z+\bar{z}=|z|^2, \qquad |2z+3|<1,$$
$$|z+1|<|z-1|, \qquad |z| \leq |2z+1|.$$

Exercise A.7. Write the following complex numbers in the form $a+ib$:

$$e^{i\pi/2}, \qquad 2e^{-i\pi/2}, \qquad 3e^{i\pi}, \qquad -e^{-i\pi}, \qquad i+e^{2\pi i},$$
$$e^{i\pi/4}, \qquad e^{i\pi/4}-e^{-i\pi/4}, \qquad \frac{1-e^{i\pi/2}}{1+e^{i\pi/2}}.$$

Exercise A.8. Solve the following equations in the unknowns $x, y \in \mathbb{R}$:

$$x+iy = xe^{iy}, \qquad x+iy = ye^{iy},$$
$$e^{x+iy} = -1, \qquad \frac{1+i}{1-i} = xe^{iy}.$$

Exercise A.9. Find the 3 cubic roots of i.

Exercise A.10. Find the 4 fourth roots of $-i$.

Index

R

S

T

U

V

Bibliography

[1] M. Audin, *Geometry*, Springer-Verlag, 2002.

[2] N. Bourbaki, *Algèbre*, Masson, Paris, 1981.

[3] P.R. Halmos, *Finite-dimensional Vector Spaces*, Springer-Verlag, 1987.

[4] R. Hartshorne, *Geometry: Euclid and Beyond*, Springer-Verlag, 2000.

[5] K. Hoffman, R. Kunze, *Linear Algebra*, Prentice-Hall, 1971.

[6] L. Hogben, *Handbook of Linear Algebra*, Chapman & Hall/CRC, 2007.

[7] R.A. Horn, C.R. Johnson, *Matrix Analysis*, Cambridge University Press, 1985.

[8] N. Jacobson, *Lectures in Abstract Algebra*, Springer-Verlag, 1953.

[9] S. Lang, *Algebra*, Springer-Verlag, 2002.

[10] H. Lütkepohl, *Handbook of Matrices*, John Wiley & Sons, 1996.

[11] R.C. Lyndon, *Groups and Geometry*, Cambridge University Press, Cambridge, 1985.

[12] C.D. Meyer, *Matrix Analysis and Applied Linear Algebra*, SIAM, 2000.

[13] G.E. Shilov, *Linear Algebra*, Dover Publications, 1977.

[14] G. Strang, *Linear Algebra and its Applications*, Thomson Learning, 1988.

Made in United States
Orlando, FL
05 November 2026

90229833R00160